an introduction to
Industrial
Chemistry

Also edited by C. A. Heaton

The Chemical Industry
(a complementary volume)

Contents: Editorial introduction *C. A. Heaton*. Polymers *J. P. Candlin*. Dyestuffs *E. N. Abrahart*. The chlor-alkali, sulphur, nitrogen and phosphorus industries *D. R. Browning*. The pharmaceutical industry *C. W. Thornber*. Agrochemicals *C. A. Heaton*. Biological catalysis and biotechnology *M. K. Turner*. The future *C. A. Heaton*. References. Index.

an introduction to
Industrial Chemistry

Second Edition

Edited by
C R Heaton

Senior Lecturer and Industrial Chemistry Subject Tutor
School of Natural Sciences
Liverpool Polytechnic

Blackie
Glasgow and London

Blackie and Son Ltd.
Bishopbriggs Glasgow G64 2NZ
and
7 Leicester Place, London WC2H 7BP

Essex County Library

British Library Cataloguing in Publication Data

An introduction to industrial chemistry. – 2nd. ed.
1. Chemical engineering
I. Heaton, C. A. (C. Alan)
660

ISBN 0-216-92919-9

Filmset by Thomson Press (India) Limited, New Delhi
Printed in Great Britain by Bell and Bain Ltd., Glasgow

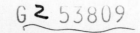

Preface to First Edition

The chemical industry is a major, growing influence on all our lives, encompassing household commodities and utensils, industrial materials and components, medicines and drugs, and the production of chemicals has become an essential factor in the economy of any industrialized nation. The scientists and engineers responsible for the efficient operation of the industry must have a sound knowledge not only of the physical and chemical principles, but also of the economic and environmental aspects and the cost-effective use of energy.

This book provides an introduction to these topics and includes detailed discussion of catalysis and petrochemicals. It is written as a basis from which students of chemistry and chemical engineering will be able to build an understanding and appreciation of the industry.

Acknowledgements
An undertaking of this nature requires teamwork and it is a pleasure to acknowledge the efforts and cooperation of the contributors. Thanks are also due to the publishers for their help and advice at all times. Finally, I wish to thank my wife Joy for typing part of the MS and for the support which she and our children, Susan and Simon, have given.

C.A.H.

Preface to Second Edition

The first edition of this book has been very well received and the few minor criticisms made by reviewers were largely answered by the publication of the complementary book—*The Chemical Industry* by C. A. Heaton (referred to as Volume 2)—which was in preparation at the time. This covers each of the major sectors of the chemical industry. They are designed to be used as a two volume set and the contents of Volume 2 are listed on page ii of this volume. We have, however, taken the opportunity in this second edition to add two new chapters: Chapter 1, Introduction to the chemical industry which gives both an overview of the industry and a lead into other chapters, and Chapter 9, Chlor-alkali products which provides a balance on the inorganic side to the Petrochemicals chapter on the organic side, plus leading into Chapter 3 of Volume 2 (The chlor-alkali, sulphur, nitrogen and phosphorus industries). Almost all statistics and tables have been updated as have references and bibliographies, where appropriate, and it is a pleasure to record that this has been done by the original team of authors. The new edition also reflects the changed situation of the industry which is currently riding high, in contrast to the recession when the first edition was written. Issues which have become more topical during the last few years, mostly environmental concerns, are also given increased coverage.

We hope you enjoy reading this new edition and find it both informative and interesting.

C.A.H.

Note

Where reference is made to West Germany this is because no figures were available for the newly combined Germany at the time of printing.

Contributors

D. G. Bew   Formerly of ICI Petrochemicals and Plastics Division, Wilton, Middlesbrough

C. A. Heaton   School of Natural Sciences, Liverpool Polytechnic

S. F. Kelham   ICI Chemicals and Polymers Ltd., Runcorn, Cheshire

J. McIntyre   Department of Chemistry, University of York

J. Pennington   Formerly of BP Chemicals Ltd, Hull

K. V. Scott   Consultant, Worthing, West Sussex

R. Szczepanski   Infochem Computer Services Ltd., London

Conversion factors

Mass

1 tonne (metric ton) = 1000 kilograms = 2205 pounds
= 0·984 tons
1 ton = 1016 kilograms = 2240 pounds
= 1·016 tonnes

Volume

1 litre = 0·220 gallons (U.K. or Imperial) = 1 cubic metre
1 gallon = 4·546 litres
1 gallon = 1·200 U.S. gallons = 0·00455 cubic metres
1 barrel = 42 U.S. gallons = 35 gallons = 0·159 cubic metres

(Densities of crude oil vary, but 7·5 barrels per tonne is an accepted average figure.)

1 cubic metre = 35·31 cubic feet
1 cubic foot = 0·02832 cubic metres

Pressure

1 atmosphere = 1·013 bar = 14·696 pounds per square inch
= $1·013 \times 10^5$ newtons per square metre
= $1·013 \times 10^5$ pascal

Temperature

Degrees Centigrade = 0·556 (degrees Fahrenheit − 32)
Degrees Fahrenheit = 1·80 (degrees Centigrade) + 32
Degrees Kelvin = degrees Centigrade + 273

Energy

1 therm = 100 000 British thermal units
1 British thermal unit = 0·252 kilocalories = 1·055 kilojoules
1 kilocalorie = 4·184 kilojoules
1 kilowatt hour = 3600 kilojoules = 859·8 kilocalories
= 3412 British thermal units.

Power

1 horsepower = 0·746 kilowatts
1 kilowatt   = 1·34 horsepower

Nomenclature of organic compounds

Common or trivial name	Systematic (or IUPAC) name	Structure
(a) *Classes of compounds*		
Paraffin	Alkane	—
Cycloparaffins or Naphthenes	Cycloalkanes	—
Olefins	Alkenes	—
Acetylenes	Alkynes	—
Methacrylates	2-Methylpropenoates	$CH_2{=}\underset{\underset{\displaystyle CH_3}{\displaystyle \mid}}{C}{-}CO_2R$

(b) *Individual compounds*		
Ethylene	Ethene	$CH_2{=}CH_2$
Propylene	Propene	$CH_3CH{=}CH_2$
Styrene	Phenylethene	
Acetylene	Ethyne	$H{-}C{\equiv}C{-}H$
Isoprene	2-Methylbuta-1,3-diene	$CH_2{=}\underset{\underset{\displaystyle CH_3}{\displaystyle \mid}}{C}{-}CH{=}CH_2$
Ethylene oxide	Oxirane	
Propylene oxide	1-Methyloxirane	
Methyl iodide	Iodomethane	CH_3I
Methyl chloride	Chloromethane	CH_3Cl
Methylene dichloride	Dichloromethane	CH_2Cl_2
Chloroform	Trichloromethane	$CHCl_3$
Carbon tetrachloride	Tetrachloromethane	CCl_4
Vinyl chloride	Chloroethene	$CH_2{=}CH{-}Cl$
Ethylene dichloride	1,2-Dichloroethane	$ClCH_2CH_2Cl$
Allyl chloride	3-Chloropropene	$CH_2{=}CH{-}CH_2{-}Cl$

Chloroprene	2-Chlorobuta-1, 3-diene	$CH_2{=}C{-}CH{=}CH_2$ $\quad\quad\ \ |$ $\quad\quad\ \ Cl$
Epichlorohydrin	1-Chloromethyloxirane	$CICH_2\overset{\overset{O}{\frown}}{CH{-}}CH_2$
Ethylene glycol	Ethane-1, 2-diol	$HOCH_2CH_2OH$
Propargyl alcohol	Prop-2-yn-1-ol	$H{-}C{\equiv}C{-}CH_2OH$
Allyl alcohol	Prop-2-en-1-ol	$CH_2{=}CH{-}CH_2OH$
iso-Propanol	2-Propanol	CH_3CHCH_3 $\quad\quad |$ $\quad\quad OH$
Glycerol	Propane-1, 2, 3-triol	$HOCH_2{-}CH{-}CH_2OH$ $\quad\quad\quad\ \ |$ $\quad\quad\quad\ \ OH$
sec-Butanol	2-Butanol	$CH_3CHCH_2CH_3$ $\quad\quad |$ $\quad\quad OH$
Pentaerythritol	2, 2-Di (hydroxymethyl) propane-1, 3-diol	$\quad\quad\quad CH_2OH$ $\quad\quad\quad\ |$ $HOCH_2{-}C{-}CH_2OH$ $\quad\quad\quad\ |$ $\quad\quad\quad CH_2OH$
Lauryl alcohol	Dodecanol	$CH_3(CH_2)_{10}CH_2OH$
Acetone	Propanone	CH_3COCH_3
Methylisobutyl ketone	4-Methylpentan-2-one	$CH_3COCH_2CHCH_3$ $\quad\quad\quad\quad\ \ |$ $\quad\quad\quad\quad\ \ CH_3$
Formaldehyde	Methanal	$HCHO$
Acetaldehyde	Ethanal	CH_3CHO
Chloral	2, 2, 2-Trichloroethanal	Cl_3CCHO
Propionaldehyde	Propanal	CH_3CH_2CHO
Acrolein	Propenal	$CH_2{=}CHCHO$
Butyraldehyde	Butanal	$CH_3CH_2CH_2CHO$
Formic acid	Methanoic acid	HCO_2H
Methyl formate	Methyl methanoate	HCO_2CH_3
Acetic acid	Ethanoic acid	CH_3CO_2H
Acetic anhydride	Ethanoic anhydride	$(CH_3CO)_2O$
Peracetic acid	Perethanoic acid	CH_3CO_3H
Vinyl acetate	Ethenyl ethanoate	$CH_2{=}CHO_2CCH_3$
Acrylic acid	Propenoic acid	$CH_2{=}CH{-}CO_2H$
Dimethyl oxalate	Dimethyl ethanedioate	CO_2CH_3 $\ |$ CO_2CH_3
Propionic acid	Propanoic acid	$CH_3CH_2CO_2H$
Methyl methacrylate	Methyl 2-methylpropenoate	$CH_2{=}C{-}CO_2CH_3$ $\quad\quad |$ $\quad\quad CH_3$

Maleic acid   cis-Butenedioic acid

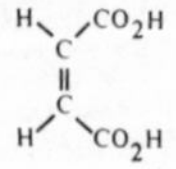

Maleic anhydride   cis-Butenedioic anhydride

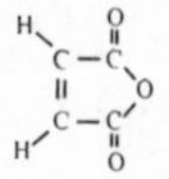

Citric acid   2-Hydroxypropane-1,2,3-tricarboxylic acid

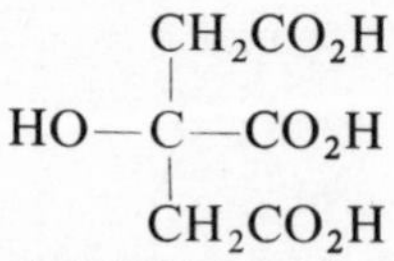

$$HO-\underset{\underset{CH_2CO_2H}{|}}{\overset{\overset{CH_2CO_2H}{|}}{C}}-CO_2H$$

Methyl laurate   Methyl dodecanoate $CH_3(CH_2)_{10}CO_2CH_3$

Stearic acid   Octadecanoic acid $CH_3(CH_2)_{16}CO_2H$

Acrylonitrile   Propenonitrile $CH_2{=}CH{-}CN$

Adiponitrile   Hexane-1,6-dinitrile $NC{-}(CH_2)_6{-}CN$

Urea   Carbamide H_2NCONH_2

Ketene   Ethenone $CH_2{=}C{=}O$

Toluene   Methylbenzene

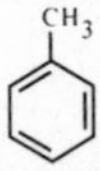

Aniline   Phenylamine

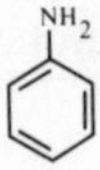

Cumene   iso-Propylbenzene

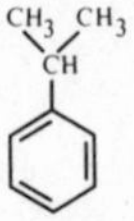

Benzyl alcohol   Phenylmethanol

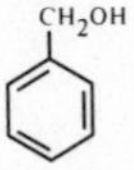

o-Xylene   1,2-Dimethylbenzene

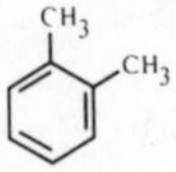

m-Xylene   1,3-Dimethylbenzene

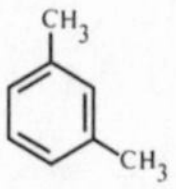

p-Xylene   1,4-Dimethylbenzene

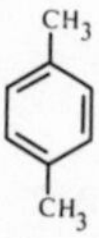

Phthalic acid   Benzene-1,2-dicarboxylic acid

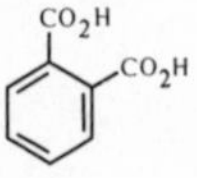

Isophthalic acid	Benzene-1, 3-dicarboxylic acid	
Terephthalic acid	Benzene-1, 4-dicarboxylic acid	
o-Toluic acid	2-Methylbenzoic acid	
p-Toluic acid	4-Methylbenzoic acid	
p-Tolualdehyde	4-Methylbenzaldehyde	
Benzidine	4, 4'-Biphenyldiamine	
Furfural	2-Formylfuran	

Contents

4 Organization and finance

D. G. Bew

7 Energy   233

J. McIntyre

EDITORIAL INTRODUCTION

The importance of industrial chemistry

Chemistry is a challenging and interesting subject for academic study. Its principles and ideas are used to produce the chemicals from which all manner of materials and eventually consumer products are manufactured. The diversity of examples is enormous, ranging from cement to iron and steel, and on to modern plastics which are so widely used in the packaging of consumer goods and in the manufacture of household items. Indeed life as we know it today could not exist without the chemical industry. Its contribution to the saving of lives and relief of suffering is immeasurable; synthetic drugs such as those which lower blood pressure (e.g. β-blockers), attack bacterial and viral infections (e.g. antibiotics such as the penicillins and cephalosporins) and replace vital natural chemicals which the body is not producing due to some malfunction (e.g. insulin, some vitamins), are particularly noteworthy in this respect. Effect chemicals also clearly make an impact on our everyday lives. Two examples are the use of polytetrafluoroethylene (polytetrafluoroethene Teflon or Fluon) to provide a non-stick surface coating for cooking utensils, and silicones which are used to ease the discharge of bread from baking tins. It should also be noted that the chemical industry's activities have an influence on all other industries, either in terms of providing raw materials or chemicals for quality control analyses and to improve operation, and to treat boiler water, cooling water and effluents. The general public is increasingly interested in the operations of the chemical industry, in its concern both about the safety of chemicals and the operation of chemical plant.

Industrial chemistry is a topic of growing interest and importance for all chemistry students. Indeed a survey[1] of all U.K. departments which offer a degree course in chemistry showed that almost two-thirds included some industrial chemistry in their courses and several offered a full degree in this subject.

Industrial chemistry is characterized by the very broad nature of the subject, spanning as it does several different disciplines. Apart from chemistry it includes topics such as organization and management of a company, technical economics, chemical engineering and environmental pollution control, and it would not be complete without an in-depth study of several particular sectors of the chemical industry. The latter would be selected as a representative cross-section of the entire industry.

1

Clearly a comprehensive treatment of all these topics would hardly be possible even in a full degree in industrial chemistry, let alone as an option or part of a chemistry degree course. Nor would this be appropriate. An understanding of the basic aspects of, and an appreciation of, the language of some of the above topics, and their linking with physico-chemical principles in the manufacture of chemicals is required.

The growing interest in the study of industrial chemistry in undergraduate courses clearly requires the availability of suitable accompanying student textbooks. In some areas of the subject excellent monographs are available (these are detailed in the 'further reading' section at the end of each chapter). However, for a broad introductory treatment to the whole of industrial chemistry they are too detailed and therefore inappropriate. Books presenting an overall introduction to industrial chemistry are few and far between. The few valiant attempts indicate the difficulties, since they either (a) attempt to be too comprehensive and are therefore somewhat superficial in the treatment of certain topics[2], (b) tend to be rather a catalogue of factual material[3], or (c) adopt an entirely different approach—that of process development—leading to the coverage of rather different topics[4]. A more recent two-volume publication[5] has much merit but its very high cost puts it beyond the reach of students.

The aim of writing this textbook is to provide a readable introduction to the very broad subject of industrial chemistry (or chemical technology) in a single volume of a reasonable length. Although the text is aimed primarily at chemists, much of its material will be of value to first- and second-year students studying for degrees in chemical engineering. Finally those graduates about to enter industry after taking a 'pure' chemistry degree course should find that it is a useful introduction to industry, and therefore provides a bridge between their academic studies and their first employment. Our approach has been to use a team of specialist authors comprising both practising industrialists and teachers. We hope to convey the challenge, excitement, and also the difficulties which are involved in chemicals manufacture. Emphasis will be given to factors vital in the production process. Examples are the economics, engineering and pollution-control aspects. However, it should be made clear that our aim is just to introduce these subjects and not provide an extensive course in them. This should equip the reader—assuming he or she is a chemist—not only with a broader appreciation of their subject but also with the understanding to enable them to converse with chemical engineers and technical economists. This is vital in the very large projects undertaken by the major chemical companies where co-operation within a team comprising scientific and commercial personnel of several different disciplines is essential. Those readers requiring a more detailed study are directed to the bibliography at the end of the appropriate chapter. Physico-chemical principles will be integrated with the above aspects, where appropriate. The interplay, and often the compromise, between these various factors will be discussed and emphasized. Even within the chemistry itself there are often substantial differences between college and industrial chemistry. The general area of oxidation reactions is a good example. In the academic situation sophisticated, expensive, or even toxic reagents are often used, e.g. osmium tetroxide, whereas industrially the same reaction will almost certainly be carried out catalytically and using the cheapest reagent of all—air. Another important point of difference is that in college decisions are usually made in a situation where the required background information is virtually complete. For example, assigning a particular mechanism to a reaction is usually carried out on the basis of a substantial amount of experimental supporting evidence. Although one can never prove a chosen mechanism correct, the more supporting evidence that is collected the more confidence one has in it. The manager in industry, in contrast, is invariably working in a situation of only limited information. Thus he may have to decide which of several new projects to back, on the basis of financial evaluations projected up to ten years ahead. The uncertainty in the figures is considerable since assumptions have to be made on the interest rates, inflation, taxation, etc. Nevertheless he has to use his managerial ability to come to a decision in a situation of limited availability of reliable information.

Statistics

Production statistics and prices are extensively used in certain chapters in this book in order to illustrate points such as scale of operation, or comparison of companies and national chemical industries. National and international statistics, due to their extensive nature, take some considerable time to collect and collate, and may refer to a period which ended a few years before the date of their publication. We therefore recommend that the reader overcomes these difficulties by regarding all figures not as absolute or currently correct but more as being indicative of orders of magnitude, trends or relative positions. In other words it is the conclusions which we can draw from the figures which are important rather than the individual statistics themselves. If current figures are of interest, however, they can often be gleaned from journals such as *Chemical and Engineering News*, *Chemical Marketing Reporter* or *European Chemical News*. More specific sources are detailed at the end of each chapter.

There are a number of points which should be borne in mind when considering statistics—particularly those relating to economic comparisons. Firstly, in terms of national and international statistics items included under the term, say 'chemicals' may vary from country to country. In other words there is not a single standard classification system. The information may also be incomplete since it may cover only the larger companies, and in some countries there may be a legal obligation to provide it whereas in others it may be purely voluntary. Similarly, sales of chemicals compared either internationally or by individual multinational corporations (a term which covers all large chemical companies) can be significantly affected by the currency exchange rate chosen in order to obtain all figures in U.S. dollars or

U.K. pounds. Because of its fluctuating nature, depending on the time chosen for the interconversion, it could favour some companies or countries and adversely affect others. This reinforces the suggestion made above to treat figures on a relative rather than an absolute basis.

Although figures relating to companies and national chemical industries are available for most of the world, those relating to the communist bloc of countries (USSR, Eastern Europe and China) are either difficult to obtain or are of limited reliability. For this reason these countries have not been considered in the text.

Costing details relating to specific processes are also difficult to obtain. Whilst the reluctance of companies to make these known, for commercial reasons, can to a large extent be appreciated, this does leave an important gap when putting together case studies for use in teaching. It must be acknowledged that U.S. companies are rather more forthcoming in this respect than their European counterparts. Fortunately journals like *European Chemical News* and particularly *Chemical and Process Engineering* do publish such information from time to time.

Units and nomenclature

There is a general trend in science towards a more systematic approach to both units and nomenclature, i.e. naming specific chemical compounds. For units the S.I. (Système International) system has been widely introduced, and science students in Europe are brought up on this. However, in industry a range of non-S.I. units are used. For example, weights may be expressed as short tons (2000 lbs), metric tons or tonnes (1000 kg or 2205 lbs) or long tons or tons (2240 lbs) or even (particularly in the U.S.A.) millions of pounds. It is therefore necessary to be bi- or even multilingual and to assist in this conversion factors are given at the beginning of this book.

There are arguments for and against whichever units are used but we have chosen to standardize on tonnes for weight (tonnes), degrees centigrade for temperature (°C), and atmospheres for pressure (atm). Both pounds sterling (£) and U.S. dollars ($) are used for monetary values because of the volatility of their exchange rates over the last two decades. Billions are U.S. billions, i.e. one thousand millions.

In naming chemical compounds the systematic IUPAC system is increasingly used in educational establishments. However in many areas of chemistry, e.g. natural products, trivial names are still far more important, as indeed they are in the chemical industry. Again it is desirable to be bilingual. To assist in this trivial names are used in this book, but the IUPAC name is usually given in brackets afterwards. A reference table for the two systems of naming compounds is also provided at the front of the book. Since only trivial names are used in the index in this book, this conversion table should be used to obtain the trivial name from its systematic counterpart.

General bibliography

A selection of some of the important sources of information on the chemical industry and its major processes is given below. These should be used in conjunction with the more specific references given at the end of each chapter.

(a) Reference works

(i) *Encyclopedia of Chemical Technology*, R. E. Kirk and D. F. Othmer, Interscience, New York. This multi-volume series is very comprehensive and is usually the first source to consult for information. Publication of the third edition has been completed.

(ii) *Riegel's Handbook of Industrial Chemistry*, 8th edn., J. A. Kent, Van Nostrand Reinhold, New York, 1982. A multiauthor survey of the chemical industry.

(iii) *The Chemical Process Industries*, 3rd edn., N. Shreve, McGraw-Hill, New York, 1967. A little dated now, strong on heavy inorganics and weak on organics.

(iv) *Chemical Technology*, 1st English edn., F. A. Henglein, Pergamon Press, London, 1969. Very strong on the technology of the German chemical industry.

(v) *Industrial Organic Chemicals in Perspective*, Vols. I and II, Harold A. Witcoff and Bryan G. Reuben, Wiley–Interscience, New York, 1980. An excellent account of the production and use of organic chemicals.

(b) Textbooks

These are detailed under references 2, 3, 4, and 5 below.

(c) Journals

A selection are given below, the first four giving a fairly general coverage and the remainder more specific coverage of the chemical industry.

(i) *European Chemical News*

(ii) *Chemistry and Industry*

(iii) *Chemical and Engineering News*

(iv) *Chemical Age*

(v) *Chemical Marketing Reporter*

(vi) *Chemical and Process Engineering*

(vii) *Hydrocarbon Processing*

(d) Patents

These are covered in *Chemical Abstracts* plus specialist publications such as those issued by Derwent Publications in the U.K.

References

1. Alan Heaton, *Chem. Brit.*, 1982, **18**, 162.
2. *The Chemical Economy*, B. G. Reuben and M. L. Burstall, Longman, 1973.
3. *Basic Organic Chemistry*, Part V: *Industrial Products*, J. M. Tedder, A. Nechvatel and A. H. Jubb, Wiley, 1975.
4. *Principles of Industrial Chemistry*, Chris. A. Clausen III and Guy Mattson, Wiley–Interscience, 1978.
5. *Industrial Organic Chemicals in Perspective*, Part I. *Raw Materials and Manufacture*, Part II, *Technology, Formulation and Use* Harold A. Witcoff and Bryan G. Reuben, Wiley–Interscience, 1980.

CHAPTER ONE

INTRODUCTION

C. A. HEATON

The aim of this first chapter is to give something of an overview of that diverse part of manufacturing industry which is called the chemical industry. In doing so, a number of topics are introduced fairly briefly, but are discussed in more detail later in the book. As well as being a lead in to the other chapters, it should give the reader an idea of what the industry does, which are the major chemical producing countries, the scale of operations, the major products, and briefly discuss environmental issues.

The chemical industry exists to increase wealth, or add value (primarily of the shareholders in the companies which make up the industry), by taking raw materials such as salt, limestone and oil, and turning them into a whole range of chemicals which are then either directly, or indirectly, converted into consumer products. These products arguably improve our lives and lifestyles, and we could not live the way we do without them. Examples such as synthetic fibres being made into garments which drip dry and do not need ironing, and the amazing range of their colours, testify to the achievements of the research chemists, engineers and technologists. Also modern fresh fruit and vegetables are of better quality and remain fresh longer thanks to the products of the agrochemicals sector. However, the latter applications of chemicals plus a number of other areas are not without controversy and these items are addressed in more detail in a number of sections in this book (see sections 1.6 and 8.6 particularly). Although some people may question the addition of chemicals to food, for example to hasten ripening of fruit or to extend the life of fresh vegetables, ther is no doubt that without the products of the chemical industry less food would be available to us and it would be of inferior quality and have a shorter life time.

We should also draw attention to the many life saving and therapeutic drugs and medicines produced by the pharmaceutical sector of the industry. These have made a major contribution to the dramatically increasing life expectancy rates during the 20th century. These topics are discussed in detail in Volume 2 (throughout this book references to Volume 2 refer to *The Chemical Industry* by C. A. Heaton, Blackie and Son Ltd.), Chapters 4 and 5. The few unfortunate tragedies and disasters—thalidomide, Seveso, Bhopal (which are discussed later in this chapter)—should not detract from the remarkable contribution which the chemical industry makes to our lives. Sadly in the past it has, by default, contributed to its own negative image by only speaking out in response to some pollution incident which has occurred. It is surely time that the industry took the initiative to put across to the public all the positive and beneficial things which it does and the remarkable products which it continues to develop. There are welcome signs that this change is starting to take place.

Clearly the chemical industry is part of manufacturing industry and within this it plays a central part even though it is by no means the largest part of the manufacturing sector. Its key position arises from the fact that almost all the other parts of the sector utilize its products. For example, the food industry relies on the chemical industry for its packaging materials; modern automobiles depend heavily on synthetic polymers and plastics, which also play an increasing role in the building industry. Nowadays all manufacturing industry must keep a careful check on the quality of the waste materials and effluents which are produced. This necessitates chemicals for analysis and probably also for treating the waste before discharge or for recovering by-products.

Having emphasized the central role which the chemical industry plays both in our lives and within manufacturing industry, and having been given examples to illustrate this, the reader should now look at section 3.2.1 to appreciate how to define the chemical industry. As you will see it's not as easy as it seems!

Several references to the importance of the chemical industry to our society have already been made and this is further emphasized in a number of places in this book, but particularly in sections 3.2.2 and 3.3.1.

1.1 Characteristics of the industry

In the developed world, Europe, U.S.A. Japan, etc., the chemical industry has now become a mature manufacturing industry, following its explosive growth in the 1960s and early 1970s. However, its rate of growth still exceeds that of most manufacturing industries. In most developed countries the ratio is 1.5–2 to 1 (see p. 64). As expected, growth rates are higher in countries which might be classed as developing countries in chemical terms. Examples are Korea, Mexico and Saudi Arabia. In the latter two cases readily and cheaply available supplies of crude oil and natural gas have created the stimulus for this growth.

One of the main factors responsible for the growth in the developed world is a major characteristic of the chemical industry—its great emphasis on research and development (R&D). This led in the 1950s and 1960s to the introduction of many new products which now command world markets of over 1 million tonnes annually. Examples are synthetic polymers such as polythene, PVC and nylon. Although the number of new products coming forward has declined since those halcyon days, the very high commitment to

R&D remains and expenditure as a percentage of sales income is double that of all manufacturing industry.

The chemical industry is very much a high technology industry with full advantage being taken of advances in electronics and engineering. Thus the use of computers is extremely widespread: from automatic control of chemical plant to automating and/or extending the abilities of analytical instruments. This also partly explains why it is capital rather than labour intensive.

The scale of operations within the industry ranges from quite small plants (a few tonnes per year) in the fine chemical area to the giants (100–500 thousand tonnes per year) of the petrochemical sector. Although the latter take full advantage of the economy of scale effect (section 5.7), if the balance between production capacity and market demand is disturbed the losses due to running at well under design capacity can be extremely high. This is particularly evident when the economy is depressed, and the chemical industry's business tends to follow the cyclical pattern of the economy with periods of full activity followed by those of very low activity.

The major chemical companies are truly multinational and operate their sales and marketing activities in most of the countries of the world, and they also have manufacturing units in a number of countries. For example ICI, the U.K.'s largest chemical company, has sites in 40 countries and sells to over 150 countries. This international outlook for operations, or globalization, is a growing trend within the chemical industry, with companies expanding their activities either by erecting manufacturing units in other countries or by taking over companies which are already operating there. Further discussion of these characteristics can be found in section 3.7.

1.2 Scale of operations

It is important to appreciate that although most of the discussion about the chemical industry tends to revolve around the multinational giants who are household names—Bayer, Ciba-Geigy, DuPont, ICI—the industry is very diverse and includes very many small-sized companies as well. There is a similar diversity in the sizes of chemical plants. By and large these divide according to whether the plant operates in a batch mode or a continuous one. Generally speaking, the batch type are used for the manufacture of relatively small amounts of a chemical, say up to 100 tonnes per annum. They are therefore not dedicated to producing just a single product but are multi-purpose and may be used to produce a number of different chemicals each year. In contrast continuous plants are designed to produce a single product (or a related group of products) and as the name suggests they operate 24 hours a day all the year round. Nowadays they are invariably controlled by computers. They have capacities in the 20 000 to 600 000 tonnes per annum range and are generally used to make key intermediates which are turned into a very wide range of products by downstream processing. Most examples (ethylene, benzene, phenol, vinyl chloride etc.) are to be found in the petrochemicals sector but another well known example is ammonia. These operations are discussed in detail in Chapter 11. Clearly such large and sophisticated plants require a very high capital investment, and this is illustrated by SHOP (Shell Higher Olefines Plant) which cost Shell £100 million to build at Stanlow on Merseyside in the early 1980s. In the late 1980s further investment was made to increase its capacity. In contrast small-scale batch plants are used to manufacture fine chemicals. These are chemicals which are needed in relatively small quantities and high purity. Examples are pharmaceuticals, dyestuffs, pesticides and speciality chemicals such as optical fibre coatings and aerospace advanced materials (speciality polymers).

1.3 Major chemical producing countries

Comparisons between the U.K. chemical industry and those in other countries are made in section 3.3.2. The U.S. chemical industry and other countries are also discussed, in sections 3.4, 3.5, and 3.6. However, it is useful at this stage to indicate which are the important chemical producing countries. The most important by a considerable margin is the United States, whose total production equates roughly with that of Western Europe. Within the latter area West Germany is the largest producer followed in turn by France, U.K., Italy and the Netherlands. Note, however, that the second most important chemicals producer (based on value of sales) is Japan, which has double the output of the third country, West Germany. The U.S.A.'s output is some 50% higher than that of Japan. Although reliable statistics are harder to obtain, the U.S.S.R. and the Eastern Bloc are also important chemical manufacturers.

1.4 Major sectors and their products

The major sectors of the chemical industry are those forming most of the chapter headings in *The Chemical Industry* (Volume 2). A similar categorization is shown in Table 1.1, even though this is based more on end uses of the chemicals. Note that here the comparison between the sectors is based on the value added, i.e. roughly the difference between the selling price and the raw material plus processing costs. This means that pharmaceutical products, which sell for very high prices per unit of weight (up to tens of thousands of pounds per tonne or more), stand out much more than petrochemicals (organics plus synthetics/plastics) which typically sell for several hundreds of pounds per tonne. Even the vastly greater tonnage of the latter does not reverse the positions. Some of the major sectors of the chemical industry are listed below.

Petrochemicals — Chlor-alkali products
Polymers — Sulphuric acid (sulphur industry)
Dyestuffs — Ammonia and fertilizers (nitrogen industry)
Agrochemicals — Phosphoric acid and phosphates (phosphorus
Pharmaceuticals — industry)

Table 1.1 Sectors of the U.K. chemical industry (1987) (gross value added)

	% share
Pharmaceuticals	27
Specialized chemical products—industrial/agricultural use	18
Organics	16
Soaps and toilet preparations	10
Synthetic resins, plastics and rubber	7
Paints, varnishes and printing inks	6
Dyestuffs and pigments	5
Inorganics	5
Specialized chemical products—household/office use	4
Fertilizers	2

The petrochemicals sector provides key intermediates, derived from oil and natural gas, such as ethylene, propylene and benzene. These are then used as raw materials for downstream processing in some of the other sectors listed. It is clearly one of the most important sectors of the industry and forms the subject of Chapter 11.

The polymers sector is the major user of petrochemical intermediates and consumes almost half the total output of organic chemicals which are produced. It covers plastics, synthetic fibres, rubbers, elastomers and adhesives, and it was the tremendous demand for these new materials with their special, and often novel, properties which brought about the explosive growth of the organic chemicals industry between 1950 and 1970.

Although the dyestuffs sector is much smaller than the previous two, it has strong links with them. This arose because the traditional dyestuffs, which were fine for natural fibres like cotton and wool, were totally unsuitable for the new synthetic fibres like nylon, polyesters and acrylics. A great deal of research and technological effort within the sector has resulted in the amazingly wide range of colours in which modern clothing is now available.

Agrochemicals (or pesticides) is an area of immense research effort with demonstrable success in aiding the production of more and better food. Along with pharmaceuticals it is a bluechip sector, i.e. very profitable for those companies which can continue to operate in it.

Pharmaceuticals has been the glamorous sector of the industry for some years now. This arises from an excellent innovative record in producing new products which has led, for many companies, to high levels of profitability. In addition, demand for its products is unaffected by the world's economy and therefore remains high even during recessions. This contrasts with the situation for most other sectors of the chemical industry. Indeed criticism seems to regularly surface that profits are too high, but this must be set against the R&D costs, which exceed £100 million to get one new drug to market launch. The chlor-alkali products sector produces mainly sodium hydroxide and chlorine, both of which are key basic chemicals and are discussed in Chapter 9. This chapter demonstrates nicely the influence of new technology and energy costs on chemicals production.

Sulphuric acid is the most important chemical of all in tonnage terms. Its production can be regarded as having reached maturity some years ago, but even now work is being done to remove the last traces of unreacted SO_2 (for environmental reasons).

Ammonia and fertilizers is a sector in which it has been difficult to achieve a balance between capacity and demand, and this has often led to major cost cutting and losses for many companies. In tonnage terms it is one of the most important sectors and it is based on the Haber process for ammonia. This is very energy demanding (moderately high temperatures and very high pressures) and a fortune is awaiting anyone who can find a viable alternative route. It will not be easy since no one has yet succeeded despite 70 years of intensive research effort!

Various phosphates are produced from phosphoric acid which is made either by adding sulphuric acid to phosphate rock (wet process) or by burning phosphorus in air to give phosphorus pentoxide, which is then hydrated. Major uses of phosphoric acid are the production of phosphate and compound fertilizers, formation of sodium tripolyphosphate (which is used as a builder in detergents where it forms stable water-soluble complexes with calcium and magnesium ions) and the production of organic derivatives like triphenyl and tricresyl phosphate. These are used as plasticizers for synthetic polymers and plastics.

Soaps and detergents represent an interesting and rather different sector. Interesting in that early production of soap, with its demand for alkali, can be viewed as the beginnings of the modern chemical industry. Different from the other sectors in that its products are sold directly to the public and market share probably has more to do with packaging and marketing than the technical properties of the product. Many of its products can be derived from both petrochemical intermediates and from animal and vegetable oils and fats, e.g. alkyl and aryl sulphonates. Chemicals from oils and fats are discussed in section 2.2.4.

1.5 Turning chemicals into useful end products

Although some chemicals, such as organic solvents, are used directly, most require further processing and formulating before they can be put to their end uses. In some cases, where novel materials have been discovered, major technological advances were required before they could be processed and their unique properties utilized. Such a material is polytetrafluoroethylene, which is better known as PTFE or under its trade names Fluon (ICI) and Teflon (DuPont). When this was first made its special properties of great chemical stability, excellent electrical insulation, very low coefficient of friction (hence its non-stick applications) and very wide working temperature range were quickly recognized. However, its use was delayed for several years because it could not be processed by conventional techniques and it had to await the development of powder metallurgy techniques.

In order to appreciate the downstream processing and technology, let us take as an example polyvinyl acetate and one of its applications as a binder in emulsion paint. Here its function is to bind the pigment, e.g. titanium dioxide, such that a homogeneous film is produced on evaporation of the water base. What processing steps are involved in making the polyvinyl acetate and in finally formulating the paint?

The story starts with crude oil or natural gas fractions, e.g. naphtha, which are cracked to give principally ethylene. The ethylene is then reacted with acetic acid and oxygen over a supported palladium catalyst to produce the vinyl acetate (see section 11.7.4). Finally this is polymerized to polyvinyl acetate which is then mixed with the other ingredients to produce the emulsion paint.

The above examples teach us an important lesson; although it is the chemists who make and discover the new chemicals which may have special properties, a considerable input from engineers and technologists may be required before the chemical can be processed and converted into a suitable form in which it can be used. This emphasizes an important aspect of research and technology in the chemical industry, namely the importance of inter-disciplinary teamwork.

1.6 Environmental issues

Ever since Rachel Carson drew attention to the adverse environmental effects of some pesticides in her book *Silent Spring* in 1962, there has been a growing concern and awareness of environmental issues. The environmental move-ment has grown very rapidly in recent years and this is evident from the establishment and rapid growth of the Green political parties in countries like West Germany and the U.K. It is therefore appropriate to examine the position of the chemical industry with regard to the environment since a number of the problems have been laid at the former's doorstep.

Some of the major problems are much wider than the chemical industry; acid rain (see section 8.6.1) and the greenhouse effect (section 8.6.3) are clearly problems created by the energy industry, although in the latter case burning of the rain forests is a significant and worrying contributor. Nuclear waste is also a result of the activities of part of the energy industry. It is interesting to speculate that if all our energy requirements were met by nuclear fission power both the acid rain and greenhouse effect (caused largely by the combustion of fossil fuels) would be considerably reduced, but would this merely replace one set of problems by another? The answer to all this would seem to be the generation of energy by nuclear fusion, but as we all know there are immense technical problems to be overcome before this is a viable process.

Let us now briefly look into several major environmental problems/disasters which clearly are associated with the chemical industry.

1.6.1 *Flixborough*

This major disaster occurred in 1974 at the Flixborough works of Nypro (U.K.) Ltd. The plant involved was part of the process for producing Nylon 6, and was used for the stage in which cyclohexane is oxidized to cyclohexanol plus cyclohexanone. One of the reactors had been removed for repair and temporarily replaced by a large diameter pipe which was inadequately supported. Cyclohexane began to leak and a very large cloud of it eventually ignited, causing a massive explosion. This resulted in 28 dead, almost 100 injured and damage to nearly 2000 factories, houses and shops in the neighbourhood.

It was the U.K. chemical industry's blackest day but note that it appeared to be caused by human error.

1.6.2 *Minamata Bay (Japan)*

This incident, in 1965, led to almost 50 deaths and to 100 seriously ill people. They displayed symptoms of mercury poisoning and the problem was traced to mercury which had been discharged into the bay by a chemical company. There it was converted into the very toxic dimethyl mercury by micro-organisms at the bottom of the sea. This substance concentrated in fish which were subsequently eaten by the victims.

Incidents such as this led to a major tightening up of pollution laws in Japan and nowadays up to 50% of the capital cost of a new plant can be earmarked for pollution control equipment.

1.6.3 *Thalidomide and drugs*

The thalidomide tragedy in 1961, in which some pregnant women who were prescribed the drug gave birth to grossly malformed babies, is one in which the companies involved (Chemie Grünenthal and Distillers) were criticized for not detecting this problem during testing of the drug before it was marketed. This was unjustified because at that time no one had ever envisaged that a drug could pass from the mother to the foetus and cause such dreadful results. Nowadays of course testing for this—known as tetratogenicity—is routinely carried out with several different mammals for all potential drugs and pesticides. Where the companies were quite rightly criticized was in not withdrawing the product quickly enough and in not compensating the victims (until after litigation taking many years), once the link between the drug and these side effects had been established.

Drug abuse involving e.g. barbiturates and tranquilizers, is an area where the industry sometimes comes in for criticism. This is quite wrong because it is a social problem of our whole society. It is also very important to place these problems in context. These few tragedies must be viewed against the hundreds of millions of lives which have been saved and prolonged by the thousands of new drugs and medicines which the pharmaceutical industry has produced.

1.6.4 *Seveso, Bhopal and pesticides*

Over the years pesticides have probably attracted more adverse comment than any other chemical products. Whilst one should quite rightly discuss the problems associated with products such as Dieldrin, DDT and Agent Orange, and have these replaced by less toxic, more environmentally acceptable products, these difficulties must again be put into context. These few problem pesticides must be viewed against the many thousands which are in regular use and have not caused any difficulties but have helped crop yields increase dramatically. It is an accepted projection that if all pesticides were banned world food production would fall by at least 50%. Those products which caused problems did so because they were not selective enough. However, the selectivity of some recent pesticides is quite remarkable; for example, herbicides now exist which kill wild oats (a weed) growing amongst the oat or barley crop, leaving the crop totally unaffected.

In Seveso in Northern Italy in 1976, a plant used for manufacturing the herbicide 2,4,5-T (2,4,5-trichlorophenoxyacetic acid), which was a component of Agent Orange (used as a defoliant in Vietnam), blew up and released about 2 kg of dioxin. This is one of the most stable and toxic chemicals known and is also teratogenic, like thalidomide. Hundreds of people living nearby were evacuated and the contaminated soil removed and destroyed. Although no one died or appeared to suffer as a result of this accident, it was a potential disaster. Like Flixborough, the cause again appears to be human error. Most of the above points are discussed in more detail in Chapter 5 of Volume 2.

More recently an accident at a plant in Bhopal in India accidently released methyl isocyanate onto the surrounding population with terrible consequences—some reports have put the final death toll as high as 3000 and many thousands more were seriously injured. After 4 years of legal wrangling the company owning the plant, Union Carbide, in 1989, agreed to pay $470 million to the victims. Yet again human error appears to have been the cause of the tragedy.

1.6.5 *CFCs (chlorofluorocarbons)*

CFCs are the remarkably inert and non-toxic chlorofluorocarbons such as CFC11, ($CFCl_3$) and CFC12, (CF_2Cl_2). Their trade names are Freons (DuPont) and Arctons (ICI). They have been widely used for many years as refrigerants, aerosol propellants and polyurethane foam-blowing agents. Ironically it is their very stability which has proved to be their undoing because they rise up into the stratosphere unchanged. There they are decomposed on exposure to the sun's short wave U.V. radiation into Cl^- and these radicals cause the breakdown of the ozone layer which screens us on earth from this dangerous U.V. radiation. More detail is given on this topic in section 8.6.2.

Most of the world's developed nations, the major users of CFCs, plus some third world countries, signed an agreement known as the Montreal Protocol which is a timetable for reducing and phasing out the use of CFCs. Pressure is being increased to speed up the phase out so that it is complete by the end of this century.

All the major producers, about 10 worldwide, are working flat out to produce a family of alternatives which are known as HFCs (hydrofluorocarbons). These generally do not contain chlorine and should therefore have a much smaller effect on the ozone layer. The front runner is HFC 134a (CF_3CH_2F) and several companies have collaborated on joint toxicity tests for this compound. The race is on to get the first manufacturing unit up and running to produce this product in the most economical way and the investment is huge. ICI, for example, has a team of 60 scientists and engineers working on this project and by the time the first plant comes on stream in 1991 they will have invested a total of more than £100 million. Companies like DuPont and Hoechst will have made similar commitments. There are also some major technical difficulties to be overcome to ensure that the new products are compatible with, for example, the mineral oils and gaskets used in compressors, etc.

In concluding this section, although attention has been drawn to some serious pollution problems these must be seen against the background of an industry operating very many potentially very hazardous processes every day. Clearly the vast majority operate safely and efficiently. There does also seem to be a greater concern for the environment by the companies and although one can see that they have to a degree been pushed towards this by organizations like Greenpeace, they can take a good deal of credit themselves for their changing attitudes to safety and the environment. After all, the people who work for the companies live in the same environment as the rest of us.

CHAPTER TWO

SOURCES OF CHEMICALS

C. A. HEATON

2.1 Introduction

The number and diversity of chemical compounds is remarkable: over ten million are now known. Even this vast number pales into insignificance when compared to the number of carbon compounds which is theoretically possible. This is a consequence of catenation, i.e. formation of very long chains of carbon atoms due to the relatively strong carbon–carbon covalent bonds, and isomerism. Most of these compounds are merely laboratory curiosities or are only of academic interest. However, of the remainder there are probably several thousand which are of commercial and practical interest and this text will demonstrate the very wide range of chemical structures which they encompass. It might therefore be expected that there would be a large number of sources of these chemicals. Although this is true for inorganic chemicals, surprisingly most organic chemicals can originate from a single source such as crude oil (petroleum).

Since the term 'inorganic chemicals' covers compounds of all the elements other than carbon, the diversity of origins is not surprising. Some of the more important sources are metallic ores (for important metals like iron and aluminium), and salt or brine (for chlorine, sodium, sodium hydroxide and sodium carbonate). In all these cases at least two different elements are combined together chemically in the form of a stable compound. If therefore the individual element or elements, say the metal, are required then the extraction process must involve chemical treatment in addition to any separative methods of a purely physical nature. Metal ores, or minerals, rarely occur on their own in a pure form and therefore a first step in their processing is usually the separation from unwanted solids, such as clay or sand. Crushing and grinding of the solids followed by sieving may achieve some physical separation because of differing particle size. The next stage depends on the nature and properties of the required ore. For example, iron-bearing ores can often be separated by utilizing their magnetic properties in a magnetic separator. Froth flotation is another widely used technique in which the desired ore, in a fine particulate form, is separated from other solids by a difference in their ability to be wetted by an aqueous solution. Surface active (anti-wetting) agents are added to the solution, and these are typically molecules having a non-polar part, e.g. a long hydrocarbon chain, with a polar part such as an amino group at one end. This polar grouping attracts the ore, forming a loose bond. The hydrocarbon grouping now repels the water, thus preventing the ore being wetted, and it therefore floats. Other solids, in contrast, are readily wetted and therefore sink in the aqueous solution. Stirring or bubbling the liquid to give a froth considerably aids the 'floating' of the agent-coated ore which then overflows from this tank into a collecting vessel, where it can be recovered. The key to success is clearly in the choice of a highly specific surface-active agent for the ore in question.

Chemical treatment depends on the nature of the ore, but for metal oxides and sulphides, reaction with a reducing agent like coke may be employed, as in the blast-furnace where iron oxide is converted into iron. More recent developments have centred on extracting metals from waste-heaps of old mine workings. When these materials (or tailings) were first co-mined they were discarded because the desired metal ore content was too low to make the extraction economically viable. However over recent years the prices of metals have increased and this, coupled with new processing methods, means that the economics have now become favourable. An example is extraction of copper from an aqueous solution of its nitrate using a selective complexing agent or even using a specific micro-organism which concentrates a particular metal.

Atypically, there are a few materials which occur in an elemental form. Perhaps the most notable example is sulphur, which occurs in underground deposits in areas such as Louisiana, Southern Italy and Poland. It can be brought to the surface using the Frasch process in which it is first melted by superheated steam and then forced to the surface by compressed air. This produces sulphur of high purity. Substantial quantities of sulphur are also removed and recovered from natural gas and crude oil (petroleum). This amounted to 21 million tonnes out of a total world sulphur production of 35 million tonnes in 1980, and clearly demonstrates the vast scale on which the oil and petrochemical industries operate since crude oil normally contains between 0.1 and 2.5% of sulphur, depending on its source. Desulphurization of flue gases from some U.K. power stations will be another source of sulphur in the future (see section 8.1.1.1). Over 80% of all sulphur is converted into sulphuric acid, and approximately half of this is then used in fertilizer manufacture.

A second example of the occurrence in nature of materials in elemental form is air, which may be physically separated into its component gases by liquefaction and fractional distillation. In this way substantial amounts of nitrogen and oxygen, plus small amounts of the inert gases argon, neon, krypton, and xenon are produced. A recent development has been the use of zeolites (p. 320) for carrying out this separation.

In contrast to inorganic chemicals which, as we have already seen, are derived from many different sources, the multitude of commercially important organic compounds are essentially derived from a single source. Nowadays in excess of 90% (by tonnage) of all organic chemicals is obtained from crude oil (petroleum) and natural gas via petrochemical processes. This is a very interesting situation—one which has changed over the years and will change again in the future—because technically these same chemicals could be obtained from other raw materials or sources. Thus aliphatic compounds, in particular, may be produced via ethanol, which is obtained by fermentation of carbohydrates. Aromatic compounds on the other hand are isolated from coal-tar, which is a by-product in the carbonization of coal. Animal and vegetable oils and fats are a more specialized source of a limited number of aliphatic compounds, including long-chain fatty acids such as stearic (octa-decanoic) acid, $CH_3(CH_2)_{16}CO_2H$, and long-chain alcohols such as lauryl alcohol (dodecanol), $CH_3(CH_2)_{11}OH$.

The relative importance of these sources of chemicals, or chemical feed-stocks, has changed markedly over the past thirty years. In 1950, in the U.K., coal was the source of 60% of all organic chemicals, oil accounted for 9% and carbohydrates the remainder. Since 1970, oil and natural gas have dominated the scene, providing the source for over 90% of chemicals. Coal and carbohydrates complete the total, the latter contributing < 1% of total production. The relative positions in the next century could be quite different because supplies of oil are limited and at the present rates of usage, even allowing for the current discoveries of new oilfields, it is forecast that they will be exhausted some time during the next century. Coal is in a similar situation, although because of its lower rate of use and its vast reserves, the time-scale is greater and is measured in hundreds of years before supplies run out.

Figure 2.1 shows the economically recoverable reserves of fossil fuel, i.e. oil, gas and coal reserves[1]. The formation of these fossil fuels takes millions of years and once used they cannot be replaced. They are therefore referred to as *non-renewable resources*. This contrasts with carbohydrates which, being derived from plants, can be replaced relatively quickly. A popular source is sugar-cane—once a crop has been harvested and the ground cleared, new material may be planted and harvested, certainly in less than one year. Carbohydrates are therefore described as *renewable resources*. The total *annual* production of dry plant material has been estimated[2] as 2×10^{11} tonnes.

Fossil fuels—natural gas, crude oil and coal—are used primarily as energy sources and not as sources of organic chemicals. For instance various petroleum fractions are used as gas for domestic cooking and heating, petrol or gasoline for automobiles, and heavy fuel oil for heating buildings or generating steam for industrial processes. Typically only around 8% of a barrel of crude oil is used in chemicals manufacture. Thus the price of crude oil is affected by the world supply/demand for energy. The lower-boiling fractions of value as feedstock for the chemical industry have alternative uses as premium

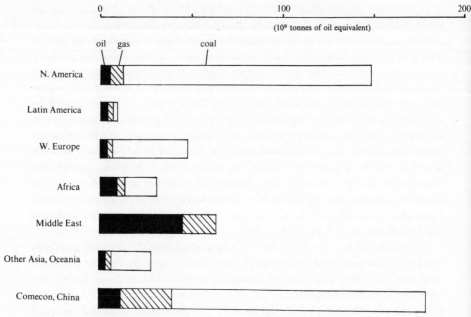

Figure 2.1   Economically recoverable fossil fuel reserves.

fuels. Prices have risen faster than energy prices in general. Thus the increasing popularity of petroleum as a world energy source has had a double impact on the organic-chemical industry, increasing both the price of feedstock and the energy required for chemical conversion and separation processes. The following figures demonstrate why the chemical industry can compete with the fuel- or energy-using industries for the crude oil:

Form of oil	Relative value of oil
Crude oil	1
Fuel	2
Typical petrochemical	10
Typical consumer product	50

This competition for the precious, limited, resource of crude oil will have to be resolved in favour of its use for chemicals manufacture. Alternative energy sources are available such as nuclear and, in certain locations, hydroelectric. Despite varying degrees of opposition from some sections of the public because of the possible hazards associated with the use of radioactive materials, nuclear energy makes a significant contribution to the total energy requirements of most nations in the developed world. A typical figure for Western Europe is around 10%, with France leading the way by producing 40% of its needs by nuclear fission. If the immense practical problems of bringing about, and controlling, nuclear fusion can be solved then almost unlimited supplies of energy will become available. The process of nuclear fusion is not only a very important natural process but an essential one for life on earth, since the fusion of two hydrogen nuclei together to produce a helium nucleus occurs in the sun and is accompanied by the release of vast amounts of solar radiation energy, which is required to warm our planet and enable plants to carry out photosynthesis and hence grow to provide food. Tremendous worldwide efforts are being made to satisfactorily harness many natural forms of energy, notably solar, wind, tidal and wave. Again, despite the practical problems, the long-term rewards will make a significant diminution in the world use of fossil fuels and nuclear energy. Another energy medium which is attractive from an environmental point of view, since it is clean, is hydrogen. Combustion of this yields only water. There seems, however, to be a psychological barrier to its use since it is thought of as a dangerous material due to its flammability. Solid metallic tanks are required for its storage, but the principal drawback is its currently expensive method of production by electrolysis of water. Development of an efficient process for the photodecomposition of water is the key, and work towards this is progressing slowly. However if some of these barriers can be overcome its attractions are obvious.

To demonstrate and assess the feasibility of hydrogen, a village has been constructed in the United States which uses hydrogen as its only energy medium—for cooking, heating and even powering automobiles and buses. A West German company has also developed, and is currently evaluating, a hydrogen-powered bus. Although it is a little early to pronounce judgements on these experiments, nevertheless it would appear that most of the technical problems have been solved and the only difficulties are economic and psychological.

A further reason to discontinue the use of fossil fuels for energy generation is that they produce mainly carbon dioxide and water by complete combustion. Because this has taken place on such an enormous scale over many years, the quantity of carbon dioxide released into the atmosphere has been very large indeed. Considerable concern has been expressed for the consequences of this. It has been suggested that as all this carbon dioxide diffuses into the earth's upper atmosphere it will reduce the amount of screening of the sun's rays, causing the so-called greenhouse effect (section 8.6.3).

Clearly alternative energy sources to fossil fuels are now available if we have the will to use them, and we can confidently expect other alternatives to become available in the not too distant future. It is therefore essential that we retain our precious oil supplies for chemicals production. The statement that 'the last thing you should do with oil is burn it' becomes more valid every year. It is interesting, and salutory, to note that as early as 1894 Mendeleyev (the Russian chemist who developed the Periodic Table) reported to his government that 'oil was too valuable a resource to be burned and should be preserved as a source of chemicals'.

A further benefit of replacing oil as an energy source is that most of the known reserves—around 370 thousand million barrels out of a free-world total (i.e. excluding the Communist bloc) of about 550 thousand million barrels—are located in the Middle East, which is an area noted for its political instability. Thus security of oil supply and gradual changes in its price cannot be guaranteed. Therefore reducing dependence on oil to a large extent and substituting for it, if possible, an indigenous material like coal not only secures supplies but helps the balance of payments. A country like West Germany, which does not have any oil of its own but has large coal reserves, could clearly derive great benefit from such change, provided it were economically viable. Even a change to nuclear energy would help, since the overall cost of the raw material—the uranium ore—would be smaller due to the much smaller quantities required, and the security of supply should be less of a problem since the ore occurs throughout the world, particularly in North America, Southern Africa, Australia and Sweden. The much higher capital costs for the nuclear power station are an important factor which has to be taken into consideration. Prototype fast-breeder reactors have been operated in the U.K. for several years now and when fully developed they could substantially improve the economics of nuclear energy. This is because they enable more energy to be extracted from 'waste' uranium and in addition utilize the plutonium produced in conventional reactors as fuel.

2.2 Sources of organic chemicals

As we have already seen, these are surprisingly few in number and it is technically possible for each to act as the raw material for the synthesis of the majority of all the organic chemicals of commercial importance. The choice between them is therefore largely a matter of economics, which has been greatly influenced by the scale of operation. The dominant position of oil and natural gas as the source of more than 90% of all organic chemicals is due in considerable measure to the very large scale on which the petrochemical industry operates. This feature is shared with the oil-refining industry from which it developed and which provides its feedstocks. For example, Shell's Stanlow refinery on Merseyside—a typical large modern refinery—has the capacity to process 50 000 tonnes per *day* of crude oil, i.e. 18 million tonnes per annum.

A different job is carried out at the very large on-shore establishments at the U.K. end of the pipelines from North Sea oilfields. They are primarily engaged in removing the dissolved hydrocarbon gases in order to stabilize the oil for export in tankers. Petrochemical plants having capacities between 100 and 500 thousand tonnes per year are commonplace. In times of high production levels this allows the full benefits of the economy-of-scale effect (see chapter 5) to be realized, resulting in a relatively low (but still profitable) selling price for the chemical. Unfortunately the recession (or period of low demand) at the start of the 1980s shows the other side of the coin, where once production levels on these very large continuous plants fall below something like 70% of capacity considerable losses are incurred, and these rise drastically the further this figure falls. At the present time (1989), production is buoyant and operations are quite profitable. However, some writers are already forecasting that there will be overcapacity again by the mid 1990s, since in western Europe seven new ethylene plants have been announced during the past eighteen months and virtually all existing plants are having their capacity increased to varying degrees. This contrasts completely with the early 1980s when only two new ethylene crackers were built, at Ludwigshafen (BASF) in 1981 and Mossmorran (Exxon/Shell) in 1986. During this period major rationalization of businesses occurred as a result of heavy losses being sustained.

Natural gas is also a major petrochemical feedstock. Because it has a similar chemical nature it will be included with oil (petroleum) in subsequent discussions in this chapter. However it is less versatile than petroleum because carbon–carbon bonds have to be built up. Carbonylation technology is available (see section 11.4) and natural gas is far superior to coal as a source of carbon monoxide. Nevertheless, although coal is much more difficult to extract from the ground and handle than fluid hydrocarbons, it will grow in importance as a source of organic chemicals as petroleum and natural gas stocks decrease because of overall world shortage or political starvation (coal is currently the major source of chemicals and hydrocarbon fuels in South Africa). Carbohydrates (now generally called biomass) also involve solids processing. They are mainly available in tropical countries where there is a much smaller concentration of chemical processing plants.

Each of these chemical feedstocks—oil (and natural gas), coal, carbohydrates, and also animal and vegetable oils and fats—will be considered in turn. Treatment of oil and natural gas will be more limited than its importance merits because they are considered in detail in Chapter 11 on petrochemicals.

2.2.1 *Organic chemicals from oil and natural gas*

The petroleum or crude-oil processing industry dates from the 1920s, and in these early days its operations were confined to separation of the oil into fractions by distillation. These various fractions were then used as energy sources, but the increasing sales of automobiles pushed up demand for the gasoline or petrol fraction. Development of processes such as cracking and reforming was stimulated and by this means higher-boiling petroleum fractions, for which demand was low, were converted into materials suitable for blending as gasoline. Additionally these processes produced olefins or alkenes, and at this particular time there was no outlet for these as petroleum products. Subsequent research and development showed that they were very useful chemical intermediates from which a wide range of organic chemicals could be synthesized. This was the start of the petrochemical industry.

Up to the early 1940s the production of chemicals from petroleum was confined to North America. This was due in no small measure to the policy of locating refineries adjacent to the oilfields. From 1950 this policy was reversed and the crude oil was transported from oilfields (the major ones then being located in the Middle East) to the refineries, now located in the areas occupied by the major users of the end products. This coincided with, or led to, the development of the European petrochemical industry, followed in the 1960s by Japan. Oil-producing countries, such as Saudi Arabia, have continued this trend into chemicals production and in the 1970s and 1980s petro-chemicals production has become a truly worldwide activity.

The meteoric rise of oil as a source of chemicals has already been attributed mainly to economic factors—its remarkably stable price during the 1950–1970 period, when prices generally were rising *each* year due to inflation, and the benefits of the economy-of-scale effect (section 5.7) coupled with process improvement over the years. However, over the past decade two factors have had a marked effect leading to rising production costs, despite increased competition in petrochemicals. The first of these was the policy of OPEC (Organization of Petroleum Exporting Countries) which led in 1973 to the price of crude oil quadrupling within a very short period. Its effect, not only on chemicals and energy production, was dramatic on the economies of the world, particularly in Europe. The ramifications of the resulting swing of purchasing power from the oil-importing countries to the Arab countries has had a profound influence on world financial dealings. Since 1973 crude oil prices have fluctuated, sometimes quite markedly, in contrast to the period of great stability up to 1970 (see section 11.1). The second factor influencing production costs of petrochemicals and chemicals production in general, has been the increasing concern of the public about the environment and also for the safety of industrial plant. The need, and desirability, of conforming to regulations in attempting to ensure a clean and safe environment have had a significant influence in increasing processing costs. One extreme example of this quotes 70% of the total capital cost of a plant constructed in Japan being attributed to pollution-control equipment. These aspects are discussed in detail in Chapter 8.

Crude oil or petroleum has been discovered in many locations throughout the world, important reserves occurring in the Middle East, North America (including Alaska), the North Sea (U.K. and Norwegian) and North Africa (Algeria and Libya). The precise composition of the oil varies from location to location, and this may be apparent even from its odour and its physical appearance, particularly with reference to viscosity, since oils vary from those with very high viscosities, almost treacle-like in consistency, to much more fluid liquids. Indeed analysis of oil slicks at sea, by gas-liquid chromatography, has been used to identify the country of origin, and this, together with knowledge of tanker movements, has been used in the prosecution of skippers for illegal washing-out of empty crude-oil tanks at sea.

Crude oil consists principally of a complex mixture of saturated hydro-carbons—mainly alkanes (paraffins) and cycloalkanes (naphthenes) with smaller amounts of alkenes and aromatics—plus small amounts (< 5% in total) of compounds containing nitrogen, oxygen or sulphur. The presence of the latter is undesirable since many sulphur-containing compounds, e.g. mercaptans, have rather unpleasant odours and also, more importantly, are catalyst poisons and can therefore have disastrous effects on some refinery operations and downstream chemical processes. In addition, their combustion may cause formation of the air pollutant sulphur dioxide. They are therefore removed at an early stage in the refining of the crude oil (or else they tend to concentrate in the heavy fuel oil fraction). One way of achieving this is by hydrodesulphurization in which the hot oil plus hydrogen is passed over a suitable catalyst. Sulphur is converted into hydrogen sulphide, which is separated off and recovered.

The complex mixture of hydrocarbons constituting crude oil must first be separated into a series of less complex mixtures or fractions. Since the components are chemically similar, being largely alkanes or cycloalkanes, and ranging from very volatile to fairly involatile materials, they are readily separated into these fractions by continuous distillation. The separation is based on the boiling point and therefore accords largely with the number of carbon atoms in the molecule. Figure 2.2 shows the typical fractions which are obtained, together with an indication of their boiling range, composition and proportion of the starting crude oil.

Figure 2.2 Distillation of crude oil

	Fraction	Boiling range (°C) (at atmospheric pressure)	Number of carbon atoms in molecule	Approximate % by volume
	→GASES	< 20	1–4	1–2
	→LIGHT GASOLINES OR LIGHT NAPHTHA	20–70	5–6	20–40
	→NAPHTHA (MID-RANGE)	70–170	6–10	
CRUDE OIL	→KEROSENE	170–250	10–14	10–15
	→GAS OIL	250–340	14–19	15–20
	→DISTILLATE FEEDSTOCKS for LUBRICATING OIL and WAXES, or HEAVY FUEL OILS	340–500	19–35	40–50
	→BITUMEN	> 500 i.e. Residue	> 35	

In terms of producing chemicals it is the lower-boiling fractions which are of importance—particularly the gases and the naphtha fractions. However consideration of the nature of the components of these fractions (they are largely alkanes and cycloakanes) suggests that they will not be suitable for chemical synthesis as they stand. Alkanes are well known for their lack of chemical reactivity—indeed their old name, paraffins, is derived from two Latin words, *parum* and *affinis*, meaning, 'little affinity'. They need to be converted into more reactive molecules and this is achieved by chemical reactions which produce unsaturated hydrocarbons such as alkenes and aromatics. As expected, because of the alkanes' lack of reactivity, the reaction conditions are very vigorous, and high temperatures are required. Alkenes are produced by a process known as cracking, which may be represented in very simple terms as

$$2C_6H_{14} \xrightarrow[\text{catalyst}]{800-1000°C} CH_4 + 3C_2H_4 + C_2H_6 + C_3H_6$$

In practice the feedstock, being a crude-oil fraction such as gases or naphtha, is a mixture and therefore the product consists of a number of unsaturated and saturated compounds (cf. p. 356). Cracking is used to break down the longer-chain alkanes, which are found in (say) the gas-oil fraction, producing a product akin to a naphtha (gasoline) fraction. This process was developed because of the great demand for naphtha and the relatively low demand for gas oil, particularly in the U.S.

Aromatics are made from alkanes and cycloalkanes by a process aptly named reforming, which may be represented, again in over-simplified terms, as

$$C_6H_{14} \xrightarrow[\text{catalyst, e.g. Pt metal}]{\text{heat}} \bigcirc + 4H_2$$

As in cracking, the feedstock is a mixture of compounds. A substantial conversion (*c.*50%) to aromatic compounds is achievable. The principal components, benzene, toluene and the xylenes, are separated for further processing.

Separation and purification of products is a major cost item in industrial chemical processes. It is important not only to isolate the desired product but also to recover the by-products. The economic success of many processes involves finding uses for co-produced materials. Greater selectivity in the reaction will minimize by-product formation and hence reduce the purification requirements. It is often economically desirable to run a reaction at a lower conversion level in order to increase selectivity even though this increases the amount of recycling of reactants.

Natural gas is found in the same sort of geological areas as crude oil, and may occur with it or separately. It consists mainly of methane plus some ethane, propane and small amounts of higher alkanes, and has long been a very important chemical feedstock for ethylene (ethene) production in the U.S.A., although its importance has started and will continue to decrease as supplies dwindle. In Europe natural gas was first discovered in the province of Groningen in Holland in 1959. This turned out to be the largest natural gas field in the world, containing an estimated 60 million million cubic feet of gas and extended out under the North Sea. The gas consists almost entirely of methane, with very few other alkanes being present. This discovery stimulated searches in the other areas of the southern North Sea and was rewarded in 1965 when British Petroleum found natural gas. Further discoveries soon followed and this gas has been used as a fuel for heating (both domestic and industrial) and cooking. Drilling in the more exposed northern areas of the North Sea has resulted, during the 1970s, in the discovery of several other important oil and gas fields in areas east of the Shetland Isles. This has made natural gas readily available in the United Kingdom as a chemical feedstock. Steam reforming of natural gas (discussed in detail in section 11.4) is a very large scale and important reaction for producing synthesis gas (syngas), which is a mixture of carbon monoxide and hydrogen, viz.

$$\underset{\text{Natural gas}}{CH_4 + H_2O} \xrightarrow[850°C]{\text{Ni catalyst}} \underset{\text{syngas}}{CO + 3H_2}$$

The importance of syngas as an intermediate for the production of a variety of organic chemicals is demonstrated later (section 11.4). Large quantities of hydrogen, produced as indicated above, are used in ammonia synthesis using the high-temperature and high-pressure Haber process, viz.

$$N_2 + 3H_2 \rightleftharpoons 2NH_3 \quad \text{(see section 11.4.1.2)}$$

2.2.2 Organic chemicals from coal

Coal, like crude oil (petroleum), is a fossil fuel which forms over a period of millions of years from the fossilized remains of plants. It is therefore also a non-renewable resource. However, reserves of coal are several times greater than those of petroleum[1], and, in contrast to petroleum, most European countries have deposits of coal varying from significant to very large quantities. The United States also has large reserves of coal. Extraction and handling of the coal is more difficult, and expensive, than for oil.

Although the precise nature of coal varies somewhat with its source (like crude oil), analysis of a representative sample shows it to be very different from oil. Firstly, its H:C ratio is on average about 0.85:1 (the corresponding figure for oil being around 1.70:1). Secondly, it consists of macro, or giant, molecules having molecular weights up to 1000. Thirdly, a significant proportion of heteroatoms—particularly oxygen and sulphur—are present[1]. The complexities of coal suggest that coal chemistry will develop along different lines to that of oil[3], although the wealth of knowledge and experience gained with the latter will provide considerable and valuable help in this development.

2.2.2.1 *Carbonization of coal.* Traditionally, and even today to some extent, chemicals have been obtained from coal via its carbonization. This is brought about by heating the coal in the absence of air at a temperature of between 800 and 1200°C, viz.

$$\text{coal} \xrightarrow{800-1200°C} \text{coke} + \text{town gas} + \text{crude benzole} + \text{coal-tar}$$

The major product by far is the coke, followed by the town gas (a mixture of largely carbon monoxide and hydrogen) with only small amounts of crude benzole and coal-tar (~ 50 kg per 1000 kg of coal carbonized) being formed, but it is from these that chemicals are obtained. It is therefore clear that for the carbonization process to be viable there must be a market for the coke produced. The steel industry is the main outlet for this, and therefore demand for coal carbonization is closely linked to the fortunes of this industry. Until some twenty years ago demand was high, certainly in Europe, not only for coke but also for the town gas which was widely used for cooking and heating. With its replacement by natural gas there is now no demand at all for town gas in the United Kingdom. Coupled with this is a much lower demand for coke by the steel industry due to a marked decrease in the level of output of steel plus increases in the efficiency of the process which has reduced the quantity of coke required per tonne of steel produced. Thus the demand for coal carbonization has fallen considerably over the past few decades and it is not economically feasible to carry out this operation merely to obtain the crude benzole, coal-tar, and chemicals derived from them.

Coal-tar is a complex mixture of compounds (over 350 have been identified) which are largely aromatic hydrocarbons plus smaller amounts of phenols. The most abundant individual compound is usually naphthalene, but this only comprises about 8% of the coal-tar. Typical weights of major chemicals obtainable from coal (isolated from coal-tar plus crude benzole) are

$$\begin{array}{cccc}
\text{coal} & \text{benzene} + & \text{naphthalene} + & \text{phenol} \\
\text{1000 kg} \xrightarrow{} & 5.3 \text{ kg} & 2.9 \text{ kg} & 0.4 \text{ kg}
\end{array}$$

The initial step in the isolation of individual chemicals from coal-tar is continuous fractional distillation (cf. oil) which yields the fractions shown in Fig. 2.3.

Each of these fractions still consists of a mixture of compounds, albeit a

Figure 2.3 Distillation of coal-tar

	Fraction	Boiling range (°C) (at atmospheric pressure)	Approximate % by volume	Main components
	→ AMMONIACAL LIQUOR			
	→ LIGHT OIL	up to 180		benzene, toluene, xylenes (dimethyl benzenes), pyridine, picolines (methyl-pyridines)
COAL-TAR	→ TAR ACIDS (carbolic oil)	180–230	8	phenols, cresols (methylphenols), naphthalene
	→ ABSORBING OIL (creosote oil)	230–270	17	methylnaphthalenes, quinolines, lutidines (dimethylpyridines)
	→ ANTHRACENE OIL	270–350	12	anthracene, phenanthrene, acenaphthene
	→ RESIDUE (pitch)		60	

much simpler and less complex mixture than the coal-tar itself. Some are used directly, e.g. absorbing oil is used for absorbing benzene produced during carbonization of coal and it is also used—under the name creosote—for preserving timber. More usually they are subjected to further processing to produce individual compounds. Thus the light-oil fraction is washed with mineral acid (to remove organic bases such as pyridine, plus thiophene), then with alkali (to remove tar acids, i.e. phenols). The remaining neutral fraction is subjected to fractional distillation which separates benzene, toluene and the xylenes.

Figure 2.4 shows some of the main uses of primary products which are obtained from crude oil and from coal.

2.2.2.2 *Gasification and liquefaction of coal.* In view of the difficulties of obtaining chemicals from coal via carbonization and coal-tar (discussed above), it is not surprising that alternative routes starting from this source have been under very active investigation for some years now. These routes have all

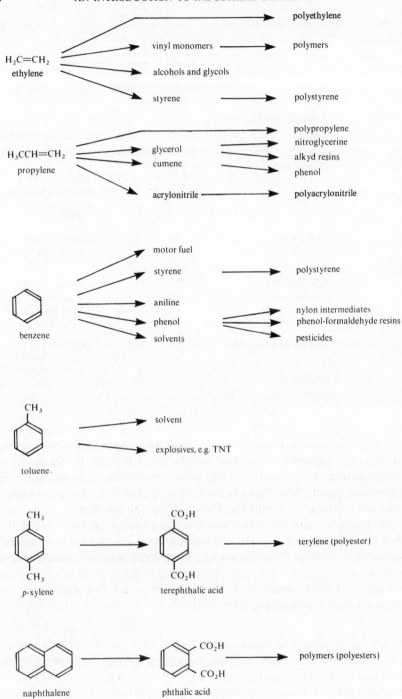

Figure 2.4 Some uses of primary products derived from coal and oil.

reached at least the pilot-plant stage of development and indeed some have achieved full commercialization. They may be grouped under two headings: (a) gasification and (b) liquefaction. Before considering representative examples of these methods it is important to realize that, as in the case of oil, the same products are also used as fuels, i.e. as energy sources. Although the technological problems in these coal conversion processes are being overcome fairly rapidly, the swing back from oil to coal as a source of organic chemicals will tend to be gradual. Even by the end of this century some estimates suggest that only between 10 and 30% of organic chemicals will be derived from coal. Reasons for this include the following.

(a) *Capital required.* Very large capital sums will be required to expand mining operations and transportation facilities. Due to the considerable differences in the nature of oil on the one hand, and coal and its conversion products on the other, much of the existing petrochemical plant may not be suitable, and in any case it may require replacing by the end of the century. This clearly requires a vast amount of capital expenditure.

(b) *Commercial availability.* As indicated, this varies from process to process, and although some are already available many are still only undergoing pilot-plant trials and it may be some years before they are suitable for full-scale manufacturing.

(c) *Incentives for change.* The principal advantage of oil is that (i) it contains molecules with linked carbon atoms, and (ii) it contains hydrogen already attached to carbon atoms. Other attractions of oil as a source of chemicals are its mobility due to its liquid nature and the scale and efficiency of the established processes for its conversion. Coal is less attractive because, as a solid, it is much more difficult to handle and transport. Recent events have reduced the momentum of the drive from oil to coal. Many of the oil-producing countries have maintained or even increased production levels (although Saudi Arabia is a notable exception) since the oil revenues are the cornerstone of their development and industrialization, but consumption of oil has been reduced due to energy conservation measures, and this has led to a current surplus. The present world recession is another major contributor to reduced oil usage. Stabilized, or reduced, prices resulted, maintaining the advantage of oil over coal. Certainly continued lower levels of consumption will prolong the time-scale for the availability of oil, postponing the swing back to coal. Any major switch away from oil to alternative sources of energy will have the same effect although the net effect would be much more dramatic.

There is no doubt about the switch from oil to coal as a source of organic chemicals; the only uncertainty concerns the time-scale and rate at which it happens.

Coal-conversion processes under development are directed towards producing either gaseous or liquid feedstocks which approximate in composition to petroleum-derived feedstocks. They can then be utilized directly in existing petrochemical plant and processes. To achieve this, however, two problems must be overcome, which are a consequence of the differing natures of coal and oil. Firstly, the H:C ratios are different for coal and for petroleum-derived liquid feedstocks. Secondly, significant amounts of heteroatoms are present in coal, particularly sulphur which may reach levels as high as 3%. The sulphur has to be removed for two reasons: (i) on combustion it will form the atmospheric pollutant SO_2, and (ii) it is a potent catalyst poison, and most of the downstream petrochemical processes are catalytic. However, its removal from coal is difficult and it is therefore removed from the conversion products instead.

Consider now the two types of coal conversion processes, (a) liquefaction and (b) gasification.

(a) *Liquefaction.* Liquefaction of coal, via hydrogenation, is quite an old process which was operated commercially in Germany during World War II when external fuel supplies were cut off. Tens of millions of tonnes of gasoline were produced in this way during this period. Interestingly, West Germany is today playing a leading part in the development of more efficient processes. For the resulting liquids to be suitable chemical feedstocks the H:C ratio must be improved in favour of more hydrogen. Clearly this can be achieved in two ways—either by adding hydrogen or by removing carbon. Although many of the new processes are only at the pilot-plant stage of development, their superiority over the old methods is due to increased sophistication of the chemical engineering employed plus improvements in the catalysts available. The basic problems, however, remain the same—poor selectivity in producing the desired fractions and a relatively high rate of consumption of hydrogen.

The fundamental processes involved are pyrolysis, solvent extraction and hydrogenation. Differences between the techniques being developed lie in how these fundamental processes are combined. Thus the *Solvent Refined Coal Process* (Gulf Oil) uses the minerals in the coal as the catalyst, and hydrogenation with hydrogen is effected in the liquefaction reactor at 450°C and 140 atm. pressure. In contrast the *Exxon Donor Solvent* process uses tetralin (1, 2, 3, 4-tetrahydronaphthalene) as the source of hydrogen in the liquefaction reactor, which also operates at 450°C and 140 atm. pressure. Further hydrogenation of the product liquids is carried out in a separate reactor, and the overall yield is about 0·4 tonnes of liquids per tonne of coal used. The National Coal Board in the U.K. is developing a supercritical direct solvent extraction of the coal process. This process uses toluene (the best solvent) at 350–450°C and 100–200 atm pressure, and it is claimed that separation of liquids from the solid material remaining is easier and the solvent can be recycled. Hydrogenation is effected in an additional step. Note that in the above processes the main product is a highly carbonaceous solid material known as char. This is either burnt to provide process heat or reacted with water and oxygen to produce hydrogen. Figure 2.5 summarizes some typical coal liquefaction processes.[1]

The liquids produced by coal liquefaction are similar to fractions obtained by distillation of crude oil (although they are much richer in aromatics), and therefore require further treatment, e.g. cracking, before being used for synthesis.

(b) *Gasification.* Coal gasification is commercially proven—for example in the SASOL plants in South Africa—and may be directed towards producing either high-energy fuel gas which is rich in methane, or synthesis gas (syngas) for chemicals production. Some well-known processes are summarized in Fig. 2.6[1]. In these processes the coal, in a suitable form, plus steam and oxygen enter the gasification reactor where they undergo a complex series of reactions, the balance between these depending on the temperature employed. Thus at high temperatures, i.e. 1000°C,

$$2C + 2H_2O \longrightarrow 2CO + 2H_2$$

dominates.

At lower temperatures competition with

$$CO + H_2O \longrightarrow CO_2 + H_2$$

and

$$CO + 3H_2 \longrightarrow CH_4 + H_2O$$

take place.

Hence higher temperatures are used to produce predominantly mixtures of carbon monoxide and hydrogen, or syngas. The Texaco process operated by Ruhrkohle in Germany produces a gas consisting mainly of carbon monoxide and hydrogen and only 1% methane. The production and value of syngas as a chemical and fuel feedstock is summarized in Fig. 2.7. Syngas, ammonia and methanol are considered in more detail in section 11.4. Note that small amounts of alcohols and esters are produced in the Fischer–Tropsch process. Synthesis of specific oxygenated products is described in chapter 11.

Although the Fischer–Tropsch process (based on coal) has been uneconomic for many years compared to oil-based routes, the peculiar political situation of South Africa allied with its large reserves of cheap coal led it to continue operating this process. Indeed, the experience gained led in the 1970s to the building of a second-generation plant, (the scheme for this SASOL II plant is shown in Fig. 2.8[4]) and a SASOL III plant is now on stream.

As indicated in Fig. 2.7, methanol is readily obtained from syngas, and its conversion to hydrocarbons over a metal-oxide catalyst is well known. However, this gives a wide range of hydrocarbons, and the catalyst's lifetime is relatively short. Mobil gained a breakthrough with their discovery in the 1970s

Figure 2.5 Typical coal liquefaction processes

Process	Typical catalyst	Conditions		Hydrogenation		Comments
		$T(°C)$	P (atm)	Prime source	Method	
Bergius (original process developed early 1900s)	Iron oxide	465	200	H_2	in liquefaction reactor	Most severe conditions; catalyst discarded
Solvent Refined Coal (Gulf Oil)	Minerals in coal (none added)	450	140	H_2	in liquefaction reactor	Recycle of portion of product liquid to reactor; lack of hydrogenation specificity
H-Coal® (Hydrocarbon Research Inc.)	$CoO—MoO_3/Al_2O_3$	450	200	H_2	in liquefaction reactor	Catalyst ages rapidly
Exxon Donor Solvent	Minerals in coal in liquefaction reactor $CoO—MoO_3/Al_2O_3$ in separate hydrogenation reactor	450	140	Tetralin in liquefaction reactor. Recycled after hydrogenation in separate reactor		Further hydrogenation of product liquids in separate reactor—catalyst deactivation slow. Typical product yields 0·3 to 0·4 te liquid/te coal feed
National Coal Board supercritical extraction	In separate hydrogenation reactor	350–450	100–200	H_2 in separate reactor		Supercritical gas extraction of portion of coal with PhMe as solvent

Figure 2.6 Gasification processes (commercially proven)*

Process	Conditions	Typical products (vol. %)				Comments
		CH_4	H_2	CO	CO_2	
Lurgi	Fixed bed reactor ~1000°C; 30 atm	12	37	18	32	Production of by-product heavy tar (~1%) restricts coal to 'non-caking' types. 'Slagging Gasifier' under development by British Gas Corporation to enable the difficult 'caking' coals to be handled
Koppers–Totzek	Entrained bed reactor ~1800°C; 1 atm	—	34	51	12	Can handle all coals; high temperatures destroy heavy organic tars. 'Shell–Koppers' pressurized version (15–30 atm) under development
Winkler	Fluidized bed reactor ~900°C; 1 atm	3	42	36	18	Higher pressure process (15 atm) under development
Texaco	Entrained bed reactor ~1200°C; 20–80 atm					Commercially successful process for partial oxidation of fuel oil to synthesis gas being developed to handle coal as coal/water or coal/oil slurries

*The field of coal gasification is in a very active state of development. Nearly 20 other processes at various stages of development have been described; see A. Verma, 1978, *Chemtech*, 372 and 626.

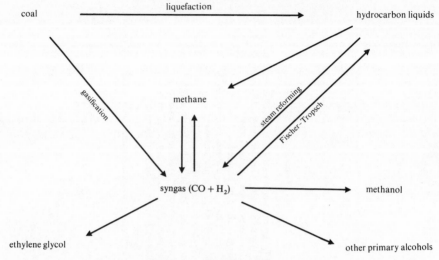

Figure 2.7   Uses of syngas.

of the ZSM-5 zeolite catalysts, which are much more selective, have longer lifetimes and can produce ethylene as well. Their selectivity is towards lower-molecular-weight hydrocarbons ($< C_{10}$). Also, the product is rich in aromatics ($\sim 25\%$) and the alkanes present show a high degree of branched chains (both of these contrast with the Fischer–Tropsch product). The product therefore has a high octane rating and is very suitable for use as gasoline. The sharp cut-off at C_{10} products is associated with the small pore size (5–6 Å) of the catalyst. A commercial plant utilizing this process came on stream in 1985 in New Zealand.

As indicated, the process can be directed towards ethylene production and hence chemical synthesis. The use of ZSM-5 catalysts for direct conversion of syngas into hydrocarbons (i.e., without the need to produce methanol first) and selective preparation of benzene, toluene, and xylene aromatics only are already being actively investigated.

2.2.3 *Organic chemicals from carbohydrates (biomass)*

The main constituents of plants are carbohydrates which comprise the structural parts of the plant. They are polysaccharides such as cellulose and starch. Starch occurs in the plant kingdom in large quantities in foods such as cereals, rice and potatoes; cellulose is the primary substance from which the walls of plant cells are constructed and therefore occurs very widely and may be obtained from wood, cotton, etc. Total dry biomass (i.e. plant material) production has been estimated at 2×10^{11} tonnes *annually*[2]. In some regions of the world—particularly in underdeveloped countries—biomass in the form of wood is the sole energy source. Also in all parts of the world much biomass is

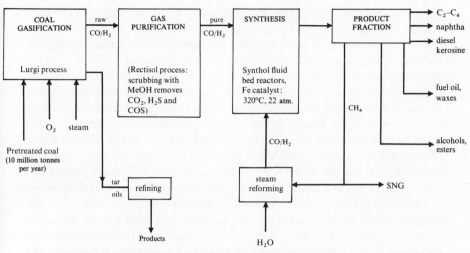

Figure 2.8 Fischer–Tropsch process—Sasol II plant.

grown and harvested for food, and the material remaining when the food has been extracted can be utilized for chemicals production. An example is molasses which is left after the sugar (sucrose) has been extracted from sugar cane. Thus, not only is the potential for chemicals production from biomass considerable, as the figure above demonstrates, but the feedstock is renewable. The major route from biomass to chemicals is via fermentation processes. However these processes cannot utilize polysaccharides like cellulose and starch, and so the latter must first be subjected to acidic or enzymic hydrolysis to form the simpler sugars (the mono- or disaccharides, e.g. sucrose) which are suitable starting materials.

Fermentation processes utilize single-cell micro-organisms—typically yeasts, fungi, bacteria or moulds—to produce particular chemicals. Some of these process have been used in the domestic situation for many thousands of years, the best-known example being fermentation of grains to produce alcoholic beverages. Indeed up until about 1950 this was the most popular route to aliphatic organic chemicals, since the ethanol produced could be dehydrated to give ethylene, which is the key intermediate for the synthesis of a whole range of aliphatic compounds. This is illustrated in Fig. 11.2. Although chemicals production in this way has been declining there is a lot of interest in producing automobile fuel in this way. This has been led by Brazil, which has immense resources of biomass, e.g. from the Amazon jungle, to utilize in alcohol production, and thus reducing its dependence on imported oil. Since 1930 it has been mandatory to use 5% of alcohol in gasoline or petrol in Brazil, and under the 'Proalcool' programme commenced in 1975 this has been increased to 15%. More recently 96%-alcohol gasoline has become available for engines which have been specially modified to run on this fuel. This seems likely to be followed by other developing nations, particularly if the price of crude oil continues to rise; Kenya's first gasohol plant, having a capacity of 18 million litres per year, has recently come on stream.

The current low level of production of chemicals from carbohydrates, or biomass, is largely a consequence of the unfavourable economics *vis à vis* oil. Disadvantages reflected in this can be divided into two parts (a) raw materials (b) the fermentation process. Raw-material costs are higher than that of crude oil, because biomass is an agricultural material and therefore in comparison its production and harvesting is very labour-intensive. Also, being a solid material transportation is more difficult and expensive. Major disadvantages of fermentation compared with petrochemical processes are, firstly, the time scale, which is usually of the order of days compared to literally seconds for some catalytic petrochemical reactions, and secondly, the fact that the product is usually obtained as a dilute aqueous solution ($< 10\%$ concentration). The separation and purification costs are therefore very high indeed. Since the micro-organism is a living system, little variation in process conditions is permitted. Even a relatively small increase in temperature to increase the reaction rate may result in death of the micro-organism and termination of the process.

On the other hand particular advantages of fermentation methods are that they are very selective and that some chemicals which are structurally very complex, and therefore extremely difficult to synthesize, and/or require a multi-stage synthesis, are easily made. Notable examples are various antibiotics, e.g. penicillins, cephalosporins and streptomycins. In addition to the antibiotics, citric acid (2-hydroxypropane-1, 2, 3-tricarboxylic acid) is a good example of this:

$$C_{12}H_{22}O_{41} \quad \xrightarrow[\text{6–12 days}]{\textit{Aspergillus niger}} \quad HO-\overset{\displaystyle CH_2CO_2H}{\underset{\displaystyle CH_2CO_2H}{\overset{|}{\underset{|}{C}}}}-CO_2H$$

sucrose   citric acid

The feedstock for these fermentations is usually a carbohydrate, but hydrocarbon fractions, or methanol derived from petroleum sources, have also been used in fermentation processes to produce single-cell protein. There is a shortage of protein for human consumption in many of the developing nations and fermentation protein could make a valuable contribution as a foodstuff supplement in these cases. However concern over safety, particularly with regard to even low concentrations of hydrocarbon residues, means that none of this protein has yet been approved for human consumption. There is an important psychological barrier to be overcome in these considerations and the difficulties have already forced BP to dismantle its 100 000 tonne per annum hydrocarbon-based plant, which used yeast as the micro-organism, in Sardinia. In contrast ICI have been operating a plant since 1980 at their Billingham works, using a methanol feedstock and the micro-organism

Methylophilus methylotropus to produce their protein 'Pruteen'[5], which is sold as an animal feed. Since methanol is readily soluble in water, any that is not converted does not contaminate the 'Pruteen'. This plant's capacity is 60 000 tonne per annum, but this plant has closed down because it could not compete with the cheaper protein obtained from soyabean. The USSR is reported to have several alkane-based plants operating with capacities up to 200 000 tonne per annum. In contrast there has been little interest in single-cell protein in the U.S.A because it is the world's leading soyabean-protein producer.

Provided that the immense practical problems associated with the rapidly developing field of genetic engineering, where micro-organisms such as bacteria are 'tailor-made' to produce the required chemical, can be overcome, then the interest in fermentation methods will be very considerable. Eli Lilly is one of the companies showing the way, and has started producing insulin at its Indianapolis and Liverpool plants via recombinant DNA techniques in which the genes to produce chain A of the insulin are inserted into one batch of the bacterium *E. coli*. A second batch of the bacterium receives the genes to produce chain B of the structure. After isolation and purification the two chains are then joined chemically to give the insulin. In tests this has proved identical to human insulin and its production will end total dependence on animal-derived insulin. This will be of considerable benefit to diabetes sufferers since it should reduce any allergic or side-effects, and eliminate, or at least minimize, long-term effects such as blindness. However it seems unlikely that bulk chemicals, i.e. those required in very large quantities such as ethylene and benzene, will be produced in this way in the foreseeable future because of the slow reaction rate and the very high product separation costs.

2.2.4 *Organic chemicals from animal and vegetable oils and fats*

Animal and vegetable oils and fats—commonly known as lipids—are composed of mixtures of glycerides, which are esters of the trihydric alcohol, glycerol (propane-1, 2, 3-triol). They have the general structure

$$
\begin{array}{l}
\quad\quad\quad\quad\; O \\
\quad\quad\quad\quad\; \| \\
CH_2-O-C-R \\
\quad\quad\quad\quad\; O \\
\quad\quad\quad\quad\; \| \\
CH-O-C-R^1 \\
\\
CH_2-O-C-R^2 \\
\quad\quad\quad\quad\; \| \\
\quad\quad\quad\quad\; O
\end{array}
$$

The groups R, R^1, R^2, which may be similar or different, are straight-chain aliphatic hydrocarbon groupings containing between 10 and 20 carbon atoms. They may be saturated or unsaturated, e.g.

$$CH_2-O-\overset{\overset{O}{\|}}{C}-(CH_2)_7-CH=CH-(CH_2)_7CH_3$$
$$CH-O-\overset{\overset{O}{\|}}{C}-(CH_2)_{16}-CH_3$$
$$CH_2-O-\underset{\underset{O}{\|}}{C}-(CH_2)_{16}-CH_3$$

There are many different sources of these oils and the nature and proportion of the R groups varies with the source. Some popular sources are soya, corn, palm-kernel, rapeseed and olive, animal fats and even sperm whales. Table 2.1 gives an indication of the distribution of R groups. The oils are isolated by solvent extraction and considerable quantities are used in the food industries as cooking oils and fats, and for production of butter, margarine and various other foodstuffs such as ice-cream. There is still controversy about the effect of the R-groups in these foodstuffs on human health, particularly on high cholesterol levels in blood which may lead to high blood pressure and heart disease. Opinion now seems to favour a high proportion of unsaturated groups as being beneficial in lowering cholesterol levels and reducing the risk of heart attacks. This has led to a trend away from cooking fats and ordinary butter or margarine (which are all rich in saturated R groups) to cooking oils and the use of margarines rich in polyunsaturates.

Being esters, the use of lipids for chemicals production starts with hydrolysis. Although this can be either acid- or alkali-catalysed, the latter is preferred since it is an irreversible reaction, and under these conditions the process is known as saponification, viz.:

$$CH_2-O-\overset{\overset{O}{\|}}{C}-(CH_2)_{16}CH_3$$
$$CH-O-\overset{\overset{O}{\|}}{C}-(CH_2)_{16}CH_3 \quad\xrightarrow[\text{heat}]{\text{NaOH}}\quad$$
$$CH_2-O-\underset{\underset{O}{\|}}{C}-(CH_2)_{16}CH_3$$

$$CH_2-OH$$
$$CH-OH \qquad +3CH_3(CH_2)_{16}CO_2^-Na^+$$
$$CH_2-OH$$

stearin   glycerol   sodium stearate

Table 2.1 Distribution of R groups in various oils

R group	(corresponding fatty acid)	Palm oil	Corn oil	Soybean oil	Cod-liver oil
Saturated groups	C_{14} (myristic)	1·0	1·0	0·5	4·0
	C_{16} (palmitic)	47·5	9·0	9·0	10·5
	C_{18} (stearic)	5·0	2·5	3·5	0·5
Unsaturated groups	C_{18} (oleic) with one C=C bond	38·0	40·0	28·0	28·0
	C_{18} (linoleic) with two C=C bonds	8·5	45·0	55·0	0[†]

[†]cod-liver oil contains a large proportion of C_{20} and $> C_{20}$ R groups

Salts of long-chain carboxylic, or fatty, acids such as sodium stearate, are the basic ingredients of soaps. The long hydrocarbon chain dissolves oily or greasy dirt, whereas the polar part of the molecule—the carboxylate anion—dissolves in the water enabling the dirt to be washed away.

The hydrolysis can be effected without the aid of a catalyst, but as expected requires more vigorous reaction conditions, i.e. heating strongly under pressure with steam. This procedure is known as 'fat splitting':

$$
\begin{array}{ccc}
\underset{\text{stearin}}{
\begin{array}{l}
CH_2{-}O{-}\overset{\overset{O}{\|}}{C}{-}(CH_2)_{16}CH_3 \\
| \qquad \overset{O}{\|} \\
CH{-}O{-}\overset{}{C}{-}(CH_2)_{16}CH_3 \\
| \\
CH_2{-}O{-}\underset{\underset{O}{\|}}{C}{-}(CH_2)_{16}CH_3
\end{array}}
& \xrightarrow{\;H_2O\;} &
\underset{\text{glycerol}}{
\begin{array}{l}
CH_2{-}OH \\
| \\
CH{-}OH \\
| \\
CH_2{-}OH
\end{array}}
\quad \underset{\text{stearic acid}}{+3CH_3(CH_2)_{16}CO_2H}
\end{array}
$$

A similar result can be obtained under milder conditions using acidic catalysts, e.g. sulphuric acid. Long-chain fatty acids, such as stearic acid, are difficult to synthesize and this is therefore the route for their manufacture.

Hydrogenolysis of either glycerides themselves or the corresponding methyl esters (easily prepared by an ester interchange reaction with methanol) produces long-chain or 'fatty' alcohols e.g.

$$CH_3(CH_2)_{10}CO_2CH_3 \xrightarrow{\;H_2\;} CH_3(CH_2)_{10}CH_2OH$$
$$\text{methyl laurate} \qquad\qquad \text{lauryl alcohol}$$

This reaction is brought about by reaction with hydrogen over a copper chromite catalyst at a pressure of about 250 atm. Under these conditions carbon–carbon double bonds in the glycerides or methyl esters are also reduced by the hydrogen. These alcohols can also be made from ethylene but the vegetable-oil-based route remains competitive at the present time.

It is important to note that for the sake of simplicity the examples given above of saponification, hydrolysis (fat splitting) and hydrogenolysis reactions have each used a single glyceride (or methyl ester). In practice, the vegetable oil which is used is a mixture of various glycerides and the product is therefore a mixture which requires separating.

2.3 Sources of inorganic chemicals

The diversity of sources of inorganic chemicals was touched on in the introduction to this chapter and the book entitled *The Modern Inorganic Chemicals Industry*[6] discusses, in detail, the production of a number of important inorganic chemicals. Comment here therefore will be restricted to a

Table 2.2 Major sources of inorganic chemicals[7]

Source	World chemical consumption 1975 (millions of tonnes)	Examples of uses
Phosphate rock	120·0	Fertilizers, detergents
Salt	120·0	Chlorine, alkali production
Limestone	60·0	Soda ash, lime, calcium carbide
Sulphur	50·0	Sulphuric acid production
Potassium compounds	25·0	Caustic potash, fertilizers
Bauxite	8·0	Aluminium salts
Sodium carbonate	8·0	Caustic soda, cleaning formulations
Titanium compounds	4·4	Titanium dioxide pigments, lightweight alloys
Magnesite	3·0	Magnesium salts
Borates	2·4	Borax, boric acid, glazes
Fluorite	1·5	Aluminium fluoride, organofluorine compounds

summary of world consumption and major uses of the more important inorganic raw materials, plus a brief study of the changing raw-material usage for sulphuric acid production and the reasons for this. Between 1960 and 1975 four alternatives have been utilized in sulphuric acid production. They are sulphur, anhydrite ($CaSO_4 \cdot \frac{1}{2}H_2O$), zinc concentrates (largely ZnS), and pyrites (largely FeS). Nowadays almost all sulphuric acid production is based on sulphur as the raw material.

The raw material for sulphuric acid production via the lead-chamber process or the contact process is sulphur dioxide. However, whereas the former could accept impure sulphur dioxide, in the contact process (which is the only process now operated) this is not possible since the impurities would poison the vanadium pentoxide catalyst used in the conversion to sulphur trioxide. Hence the disappearance of zinc concentrates and pyrites as sources today. In contrast, elemental sulphur, extracted from the earth or obtained from petroleum sources, is very pure and is ideal for burning to sulphur dioxide. Additionally it is favoured on economic grounds over anhydrite. Burning sulphur to form sulphur dioxide is a highly exothermic process and this makes the overall process for manufacturing sulphuric acid energy-producing. This excess energy is used for steam generation. Sulphuric acid plant managers are probably the only people (excepting the oil producers) who do not like to see the price of oil fall—this lowers energy costs and reduces their net credit for the sale of their energy! This results in an increase of their manufacturing costs for the sulphuric acid.

Due to its many uses sulphuric acid can be used as a measure of a country's industrial development, since consumption of the former increases with the latter. Also, because production responds rapidly to changes in consumption it is also a barometer of industrial activity, even in highly developed countries such as our own.

2.4 Recycling of materials

The finite nature of most of our sources of chemicals, e.g. oil, coal, metallic ores, suggests that every effort should be made to conserve these valuable resources.

Limiting demand for them is clearly one way of approaching this problem but it may not be practicable if competition, which stimulates demand, is intense. Since many materials can be recycled, this is an alternative, or indeed complementary, approach. Some materials have been recycled for many years, ferrous metals being perhaps the most notable examples. Thus at the end of their useful life automobiles are crushed and the (rusted) metal returned as part of the feed to the blast furnace where it is reconverted into iron and steel. Paper and cardboard are other well-known examples.

More recently many municipal authorities, in conjunction with glass-making companies, have started to provide 'bottle banks'. Benefits to the community accrue in two ways as a result. Firstly, the municipal authority receives a cash payment related to the amount of glass recovered. Secondly, despoilment of the environment due to unsatisfactory disposal of the bottles is eliminated, or at least considerably reduced.

Some municipal authorities recover organic waste from refuse, and although it is not actually recycled, it is combusted and the heat utilized in central heating systems for complete housing estates. McCauliffe of Manchester, U.K., has developed a method of converting organic refuse into high-quality crude oil, and in collaboration with the local authority is operating a pilot plant to demonstrate its commercial viability.

Smaller-scale operations are the recycling of automobile engine oil and also plastics. The latter scheme is made very difficult because the diversity of plastics makes their identification and separation troublesome.

References

1. R. Pearce and M. V. Twigg, in *Catalysis and Chemical Processes* (eds. Ronald Pearce and William R. Patterson), Leonard Hill, 1981, chapter 6.
2. August Vlitos, in *The Chemical Industry* (eds. D. Sharp and T. F. West), Ellis Horwood, 1982, p. 315.
3. Joseph Haggin, *C&EN*, 1982 (Aug. 9), pp. 17–22.
4. As ref 3, p. 167.
5. *C & I*, 1980 (24), 921.
6. *The Modern Inorganic Chemicals Industry*, ed. R. Thompson, The Chemical Society, 1977, pp. 21–25.
7. B. M. Coope, *Industrial Minerals for Inorganic Chemicals*, ECN Large Plants Supplement, 1976.
8. Annual Report of the National Sulphuric Acid Association Limited, 1982.

Bibliography

Industrial Organic Chemicals in Perspective, Part 1, *Raw Materials and Manufacture*, Harold A. Wittcoff and Bryan G. Reuben, Wiley–Interscience, 1980, chapters 2 and 3.

An Introduction to Industrial Organic Chemistry, Peter Wiseman, Applied Science, 1972, pp. 5–20.

The Chemical Industry (eds. D. Sharp and T. F. West), Ellis Horwood, 1982, chapters 20, 21, 25 and 26.

Catalysis and Chemical Processes (eds. Ronald Pearce and William R. Patterson), Leonard Hill, 1981, chapters 5 and 6.

The Modern Inorganic Chemicals Industry, R. Thompson, The Chemical Society, 1977.

Introduction to Industrial Chemistry, Howard L. White, Wiley, 1986, Chapters 2 and 4.

CHAPTER THREE

THE WORLD'S MAJOR CHEMICAL INDUSTRIES

C. A. HEATON

3.1 History and development of the chemical industry

3.1.1 Origins of the chemical industry

The use of chemicals dates back to the ancient civilizations. For example, many chemicals were known and used by the ancient Egyptians—they used soda (known to them as 'natron') mixed with animal fats as soap to wash corpses, and on its own in the mummifying process which followed. Glass objects and glazed pottery, which were buried with the mummies for their use in their assumed after-life, were made from soda and sand.

Evolution of an actual chemical industry is much more recent, and came about, as with many other industries, during the industrial revolution, which occurred in the U.K. around 1800 and rather later in other countries. Its initial development was stimulated by the demand of a few other industries for particular chemicals. Thus soapmaking required alkali for saponification of animal and vegetable oils and fats; the cotton industry required bleaching powder; and glassmaking required sand (silica) and soda (sodium carbonate). A key advance had been made by Roebuck and Gardner in Birmingham in 1746 when they substituted lead chambers for glass reaction vessels, which had been used until that time, because the construction in glass had previously limited the scale of vitriol (sulphuric acid) manufacture. The vitriol was required in the Leblanc process for making soda from salt, the chemical reactions being

$$2NaCl + H_2SO_4 \rightarrow Na_2SO_4 + 2HCl$$

$$Na_2SO_4 + CaCO_3 + 2C \rightarrow Na_2CO_3 + CaS + 2CO_2$$

In addition to the uses mentioned above, and as washing soda, sodium carbonate was also a ready source of sodium hydroxide, viz.

$$Na_2CO_3 + Ca(OH)_2 \rightarrow 2NaOH + CaCO_3$$

In the early days of the process, condensing the hydrochloric acid fumes proved difficult (or was ignored!) as this reference to the Liverpool Works of James Muspratt (one of the founding fathers of the chemical industry in the North West of England) shows[1]. It is from a letter, published in the *Liverpool*

45

Mercury dated 5th October 1827, which stated that Muspratt's works poured out 'such volumes of sulphureous smoke as to darken the whole atmosphere in the neighbourhood; so much so that the church of St. Martin-in-the-Fields now erecting cannot be seen from the houses at about one hundred yards distance, the stones of which are already turned a dark colour from the same cause. The scent is almost insufferable, as well as injurious to the health of persons residing in that neighbourhood.' Public outcry against this serious atmospheric pollution—which also rotted curtains and clothes—led to the passing of the Alkali Act in 1863, the first legislation in the world concerned with emission standards. Clearly environmental pollution is not merely a problem of the 20th century. Thus at the beginning of the 19th century the chemicals industry produced a small number of inorganic chemicals, with the Leblanc process for making soda from salt (summarized above) at its heart.

Many of these early processes had evolved by trial and error, and there was a considerable art in getting and keeping them working. Advances in scientific theories and knowledge during the 1800s—from Dalton's atomic theory onwards—provided a much more solid foundation from which the industry could develop a clearer understanding of the basis of a process and advances in technology were the result. Recognition of the importance and application of scientific principles needed men of great foresight, and one such person was John Hutchinson, who may be considered a pioneer of modern chemicals manufacture. In 1847 he established a chemical works on Merseyside, not far from Liverpool. However, he had the vision to recognize the importance of scale and scientific control to his manufacturing processes. He also proved to be an excellent manager, certainly in his judgement of men. Not only did he engage J. W. Towers to develop analytical methods, but also J. T. Brunner as office manager and Ludwig Mond from Germany to assist in the scientific development of the manufacturing processes. Friendship between the latter two led to the formation of the Brunner Mond Company in 1872. This in turn became a founder member of the well-known chemical giant ICI, Imperial Chemical Industries. Mond appreciated at a very early stage the importance of the challenge of the new Solvay ammonia–soda process to the Leblanc process. This not only overcame to a large degree the considerable effluent problems of the Leblanc process, but was cheaper in terms of raw materials and labour, although the capital cost of the plant was considerably higher. The first Solvay plant in Britain started production for Brunner Mond in 1874, and due to its economic advantages had captured 20% of the country's alkali production within 10 years. It went from strength to strength, and eventually in 1914 had 90% of the U.K.'s total production.

The process itself may be summarized as

$$NH_3 + H_2O + CO_2 \rightarrow NH_4HCO_3$$

$$NaCl + NH_4HCO_3 \rightarrow NaHCO_{3\downarrow} + NH_4Cl$$

and
$$2NaHCO_3 \xrightarrow{heat} Na_2CO_3 + H_2O + CO_2$$

Treatment of the ammonium chloride liquor with calcium hydroxide regenerates ammonia for the use in the first reaction:

$$2NH_4Cl + Ca(OH)_2 \rightarrow CaCl_2 + 2NH_3 + 2H_2O$$

The calcium hydroxide is obtained from limestone:

$$CaCO_3 \overset{roast}{\longrightarrow} CaO + CO_2$$
$$\text{limestone}$$

$$CaO + H_2O \rightarrow Ca(OH)_2$$

There was no organic chemicals side of the industry to speak of before 1856, since up to that time any organic materials were obtained from natural sources. Examples are animal and vegetable oils, fats and colouring matter, and natural fibres such as cotton and wool. In 1856, as has happened many times since (sometimes leading to major advances in the subject!), some planned research work did not give the results expected. William Henry Perkin was trying to synthesize the antimalarial drug quinine, a naturally-occurring compound found in the bark of *Cinchona* trees. Using sodium dichromate he was attempting to oxidize aniline (phenylamine) sulphate to quinine, but instead obtained a black precipitate. Rather than just rejecting this reaction which had obviously failed in its purpose, he extracted the precipitate and obtained a purple compound. It showed great promise as a dye and 'mauve', as it was named, became the first synthetic dyestuff. Rather ironically, the mauve was formed because Perkin's aniline was impure—he had obtained it in the standard way by nitrating benzene and then reducing the product, but his benzene had contained significant amounts of toluene. Although only 18, and a student in London, he had the confidence to terminate his studies in order to manufacture mauve, which he did very successfully. The synthetic dyestuffs industry grew rapidly from this beginning and was dominated by Britain into the 1870s. However, chemical research in Britain by this time tended to be very academic, whereas in Germany the emphasis was much more on the applied aspects of the subject. This enabled the Germans to forge ahead in the discovery of new dyes, early successes being alizarin and the azo-dyes. So successful were they that by the outbreak of World War I in 1914 they dominated world production, capturing over 75% of the market.

The very sound base which their success in dyestuffs provided for the large German companies—BASF, Bayer, Hoechst—particularly in terms of large financial resources and scientific research expertise and skills, enabled them to diversify into and develop new areas of the chemical industry. By the early years of the 20th century advances into synthetic pharmaceuticals had been made. Early successes were Salvarsan, an organoarsenical for treating syphilis, and aspirin.

BASF concentrated on inorganic chemistry and achieved notable successes in developing the contact process for sulphuric acid and the Haber process for ammonia production. The latter particularly was a major technological breakthrough, since it required novel, very specialized plant to handle gases at high temperatures and pressures. Thus by 1914 Germany dominated the world scene and was well ahead in both applied chemistry and technological achievement.

However, the outbreak of World War I changed this situation dramatically. Firstly, in both Britain and Germany stimulus was given to those parts of the industry producing chemicals required for explosives manufacture, e.g. nitric acid. Secondly, Germany was particularly isolated from its raw material supplies and could not of course export its products, such as dyestuffs, which led to shortages in Britain and the U.S.A. The result was a rapid expansion in the manufacture of dyestuffs in these countries. In contrast Germany had to rapidly develop production of nitrogen-based compounds—particularly nitric acid for explosives manufacture and ammonium salts for use in fertilizers—since its major source, sodium nitrate imported from Chile, had been removed. Commercialization of the Haber process (catalytic oxidation of ammonia leading to nitric acid) had come at just the right time to make this possible.

Overall, therefore, the effect of World War I on the chemical industries of the world was to stimulate home-based production.

3.1.2 *Inter-war years, 1918–1939*

The war had alerted governments to the importance of the chemical industry, and the immediate post-war years were boom years for the British and American industries. An important factor aiding this was the 'protectionist' policy operated by their governments at the time. Thus Britain brought in a heavy import duty on most synthetic organics, and the introduction of a tariff system—the American Selling Price—effectively closed the American market to most exporters. During the war and immediately afterwards most countries had, largely out of necessity, expanded their chemical industry. By the early 1920s considerable overcapacity was the result and international competition became correspondingly fierce. With its large and expanding home market and lack of competition from imports the American chemical industry grew rapidly during the 1920s. This was in complete contrast to the situation in most other countries at this time. Germany continued to dominate the international scene, and in 1925 an event occurred which was to not only support this position, but to set a trend which was to characterize the inter-war period, and remain with us up to the present time. This was the amalgamation of the major German dyestuff companies to form one giant company—I. G. Farben. This immediately became the largest chemical company in the world, with a very broad financial base, enormous applied scientific and technological expertise, and dynamic management. Certainly no European, and few American firms could hope to compete with it.

The consequences of this were soon appreciated in Britain and in the following year, 1926, amalgamation of Brunner Mond, United Alkali Company, British Dyestuffs Corporation and Nobel Industries led to the formation of another chemical giant—Imperial Chemical Industries (ICI). Although the integration and rationalization process was not without its difficulties, the result was a much more sound company. This was demonstrated in the early 1930s, during the very severe world recession, when although the chemical industry also suffered severely (from overcapacity caused by the big fall in demand) the effect on ICI was reduced because it had been able to participate in market-sharing agreements (or cartels) with other large companies. Within the continuing protection policies which it enjoyed the British industry grew steadily up to 1939, becoming virtually self-sufficient in the most important chemicals and utilizing and gaining expertise in new technology such as high-pressure gas reactions, e.g. the Haber process. It also became more innovative, introducing several important new dyestuffs.

As indicated previously, during this time the American industry was growing steadily, although largely in isolation. Up until 1940 it was also the only country to have petrochemical plants. The availability for some years of their own supplies of crude oil, and refineries, led to interest in the use of petroleum fractions for the synthesis of organic compounds and started in the 1920s with the manufacture of iso-propanol from the refinery off-gas propylene. However up until 1940 this synthetic scope was limited to producing oxygenated solvents.

Apart from the start of the trend towards very large companies, and bigger manufacturing units, the inter-war period will be remembered as marking the beginning of two areas which have had profound effects on the industry's subsequent development, growth, and profitability. The first, petrochemicals, has already been referred to above. Its all-pervading influence is apparent from the text and the devotion of the final chapter of this volume to it. The second, the synthetic polymer sector, has over the last few decades owed much to the development of petrochemicals. As we shall see later this has grown into the most important sector of the whole chemical industry.

The chemical industry's interest in polymers dates back to the 19th century. In those days it was a case of synthetically modifying natural polymers with chemical reagents to either improve their properties or produce new materials with desirable characteristics. Notable examples were nitration of cellulose giving the explosive nitrocellulose, production of regenerated cellulose (rayon or artificial silk) via its xanthate derivative, and vulcanization of rubber by heating with sulphur. Manufacture of acetylated cellulose (cellulose acetate or acetate rayon) developed rapidly from 1914 onwards with its use both as a semi-synthetic fibre and as a thermoplastic material for extrusion as a film.

The first fully synthetic polymer to be introduced (1909) was a phenol-formaldehyde resin known as bakelite. Although the reaction had been discovered some twenty years previously, it was only after a careful and systematic study of it that it was properly controlled to give a useful thermosetting resin. In the late 1920s two other types of thermosetting resins followed, namely urea-formaldehyde, and alkyd resins.

Assisted by advances in high-pressure technology, a more scientific approach to the study of synthetic polymers (macromolecules) led to the discovery and production of several important polymers in the 1930s and early 1940s. These plastics, elastomers and synthetic fibres have had a profound effect not only on the chemical industry but also on the quality of all our lives.

Germany, in the guise of I. G. Farben, had the foresight to realize the importance of this area, and they dominated the early research and development work. Their commercial successes ranged from rubber-like materials—the co-polymers poly(styrene-butadiene) and poly(acrylonitrile-butadiene)—to thermoplastics like poly(vinyl chloride) and polystyrene.

In the U.K., Gibson of ICI was studying high-pressure reactions of alkenes in autoclaves. His reaction with ethylene appeared to have failed since he obtained a small amount of a waxy white solid. Fortunately his curiosity led him to investigate this. It proved to be polyethylene, first produced commercially by ICI in 1938.

America's interest was led by the Dupont Company and its major success was the commercial introduction of the first nylon in 1941. This certainly was a reward for perseverance and effort since from Carother's first researches on the subject to commercial production took 12 years and reportedly cost $12 million.

3.1.3 Second World War period, 1939–1945

In general terms the impact on the chemical industry was similar to that of the First World War. Thus Germany especially was cut off from its raw-material supplies and therefore relied entirely on synthetic materials, e.g. poly(styrene-butadiene) rubber, and gasoline produced from coal. Britain and America were not affected to quite the same extent but demand for polymers like nylon and polyethylene for parachutes and electrical insulation was high. By the end of the war facilities for synthetic polymer production had expanded considerably in all three countries.

3.1.4 Post-1945 period

The Second World War had two major effects on the German chemical industry (apart from the consequences of the political separation into East and West Germany). Firstly, much of the chemical plant was either destroyed or damaged by bombing and process details were freely acquired by the British and Americans. Secondly, the Allies broke up I. G. Farben into a number of smaller companies such as Bayer, Hoechst and BASF, and therefore eliminated its competitiveness. As we shall see later, this turned out to be merely a temporary setback.

The major changes in the world-wide chemical industry since 1945 have been concerned with organic chemicals, in particular the raw materials used to produce key intermediates such as ethylene (ethene), propylene (propene), benzene and toluene. In the U.K. up to and including 1949 coal was the major source of raw materials, followed by carbohydrates, with oil accounting for under 10% of total production. However, the situation was starting to change due to events in the oil industry. Worldwide demand for petroleum products was increasing, this being most apparent for transport. Increasing ownership of automobiles and rapid development of air transport raised demand for gasoline. Since this demand was in the developed countries, oil refineries were now located in them, e.g. in Europe. Processes were developed in the refineries to convert low-demand fractions, e.g. gas oil, into high-demand lower-boiling fractions, e.g. gasoline. Alkenes such as ethylene and propylene were produced as by-products. Thus they were available for chemical synthesis—and at a very cheap price largely because of the vast scale of operation. Note that until this time ethylene had been obtained by dehydration of ethanol which in turn had been made by fermentation of carbohydrates. The other important source of aliphatics was acetylene (ethyne) which was obtained from calcium carbide. Although there were no technical difficulties in these routes, the ethylene and acetylene were both rather expensive to produce in this way. By 1959 oil was on a par with coal as the major source of chemicals in Europe, mirroring the position of the American industry some two decades previously. During this time the price of oil had hardly increased whereas that of coal was steadily increasing because its extraction is labour-intensive. Also, demand for the carbonization of coal was starting to decrease with falling demand for coke and town gas. This meant less benzole, and by-product coal-tar, from which the chemicals were obtained, was likely to be available. Although the processing costs of the oil tended to increase this was counterbalanced by the ever-increasing size of the operations (hence the benefits of the economy-of-scale effect) and the improved efficiency of the processes. Thus the economic advantages were swinging very much in oil's direction, and by 1969, oil dominated the scene, and this it has continued to do ever since. Petrochemicals had truly come of age.

Let us consider further the reasons for the explosive growth of the petro-chemicals industry. As indicated, oil-refinery processes such as cracking supplied key chemical intermediates—ethylene and propylene—at low prices compared with traditional methods for their preparation. In real terms these prices were more or less maintained for a considerable period. However this factor alone cannot account completely for the vast increase in the tonnage of organic chemicals produced from petroleum sources. Much of the credit may be placed with research chemists, process-development chemists and chemical engineers. Once it was realized that abundant quanties of ethylene and propylene were available, research chemists had the incentive to develop processes for the production of many other compounds. Success in the laboratory led to process development and eventually construction of manufacturing units. Chapter 11 demonstrates the versatility of the alkenes, and the achievements of the scientists. Consideration of the chemical reagents and reactants used in many of these processes shows the importance and influence on the development of the petrochemical industry of one particular area of chemistry, namely catalysis. This influence has been a two-way process: research into catalysis has been stimulated by the needs of the industry and, in turn, better, more efficient catalysts have been discovered which have improved process efficiency and economics, and may even have rendered an existing process obsolete by permitting introduction of an entirely new, more economic route. Introduction of these catalysts has made possible not only entirely new routes and interconversions of compounds, but also immensely shorter reaction times (sometimes now just a few seconds) compared with the corresponding non-catalytic route. Another major advantage has been that catalysts have introduced better control of the reaction, particularly improved selectivity towards the desired product. This is a vital aim in chemicals production, since very few processes yield only the desired product and by-product formation represents material and efficiency loss. Furthermore separation of product from by-products can be a major cost item in the overall process economics. A notable example is the introduction of catalytic cracking units in place of purely thermal cracking units for ethylene production. Evolution of manufacturing processes utilizing these new catalysts has often necessitated technological developments. Examples are design and construction of equipment to handle very high-pressure gas reactions, facilities for rapid quenching of products (as in cracking to cheat the kinetics and obtain predominantly alkenes rather than alkynes when the process has to be operated under conditions favouring the latter) and development of fluidized-bed reactors. We emphasize the importance of catalysts and catalysis by devoting Chapter 10 to this topic.

The major industrial developments in organic chemicals initiated in the 1930–1940 period have continued since that time. Most important of all is the introduction of purely synthetic polymers. Table 3.1 shows the growth in importance of these materials over the period 1950 to 1988. Although much of this growth was due to the increased demand for the polymers introduced in the 1930s and 1940s—urea-formaldehyde resins, nylon, polyethylene (low-density), poly(vinyl chloride), and butadiene co-polymers—new polymers

Table 3.1 World consumption of synthetic polymers (millions of tonnes)

	1950	1960	1970	1978	1986	1988
Plastics and resins	1·0	6·0	25·0	42·0	58·5	73·0^e
Fibres (excluding cellulosics)	1·5	1·5	5·0	12·0	11·4	15·3
Rubbers	1·0	3·0	5·0	4·0	5·4	10·1
Total	3·5	10·5	35·0	58·0	75·3	98·4

e = estimated

have also made a significant contribution. The most familiar examples (together with their year of commercial introduction and the company responsible) are polyacrylonitrile (1948, Du Pont); Terylene (1949, ICI); epoxy resins (1955, Du Pont); and polypropylene (1956, Montecatini). Their special properties and economic advantage have enabled them to displace traditional materials. Thus plastics such as poly(vinyl chloride) have replaced wood in window frames and metal drainpipes, since they are unaffected by weathering, lighter in weight, and maintenance-free. Synthetic fibres like nylon and terylene have enabled non-drip, machine-washable, and crease-resistant clothing to be introduced, and we are made aware many times each day of the widespread usage of synthetic plastics in all facets of the packaging industry. Indeed, so profound is the influence of plastics and polymers on our lives (something we tend to take for granted) that it is difficult to imagine our existence without them. Not only are all these polymers well established now, but so are the processes for making them. It must also be mentioned in passing that the actual use of these polymers required major technological advances in developing processes to get the material in the correct form for its particular applications. Techniques such as injection-moulding and blow-moulding are the result. Existing dyestuffs would not adhere to these synthetic materials and so new ones were in turn discovered and developed. Polymer science and technology has grown into a large field of study.

These 'standard' polymers are well established nowadays and are produced by many companies; this is therefore a very competitive area of the chemical industry. However, the overall market for these products is still expanding, as the figures for 1986 and 1988 clearly show.

A major advance in polymer chemistry was provided by the work of Karl Ziegler and Giulio Natta, which led in 1955 to the introduction of some revolutionary catalysts which bear their name. The great significance of this event was highlighted by them being awarded a Nobel Prize in 1963 for their work. Ziegler–Natta catalysts are mixtures of a trialkyl aluminium plus a titanium salt and they bring about the polymerization by a coordination mechanism in which the monomer is inserted between the catalyst and the growing polymer chain. Industrial interest in this centred on the fact that it gave more control over the polymerization process. Significantly, it allowed the polymerization to occur under milder conditions *and* produced a very stereoregular polymer. Taking polypropylene as an example, the Ziegler–Natta catalyst gives the isotactic product whereas a free-radical process leads to the atactic stereochemical form. These are shown below:

isotactic (stereoregular)

atactic (stereochemistry random)

These stereochemical forms show differences in their properties; for example, the isotactic polymer chains can pack together better and this form has a higher degree of crystallinity and hence a higher softening point than the atactic variety. Ziegler–Natta polymerization of ethylene gives high-density polyethylene (HDPE) which is different in some properties from the low-density form (LDPE) produced by free-radical polymerization, as is shown in Table 3.2.

During the late 1970s production of linear low-density polyethylene (LLDPE) was commercialized. Here the polymer chain is much more linear containing only short branching chains, in contrast to conventional LDPE. The polymer has greater strength and toughness, particularly in film applications, than ordinary low-density polyethylene.

Over recent years in addition to continuing production of the bulk polymers described above many companies have developed certain speciality polymers. These can be thought of as polymers whose structure has been tailor-made to yield certain specific properties, and therefore have limited, very specialized applications. For example, polymers which are exceedingly thermally and oxidatively stable are required for space capsules which will re-enter the earth's atmosphere. To develop these requires particular knowledge and skills. There is therefore little competition in producing these polymers; the market for them, although small, is assured and the product therefore commands a high price by comparison with 'standard' polymers. The polymer sector is the subject of Chapter 1 in Volume 2.

A very important trend during the post-Second World War period has been the increasing size of plants, and nowhere is this more apparent than in the petrochemicals sector, where capacities of 100 000 tonnes per annum are commonplace. Large integrated chemical complexes have evolved, due to this increase in scale, and the need to locate petrochemical plants adjacent to refineries, so as to minimize transportation of vast quantities of chemicals.

Table 3.2 Some properties of LDPE and HDPE

Property	LDPE	HDPE
Specific gravity/$g\,cm^{-3}$	0·920	0·955
% Crystallinity	55	80
Softening point/K	360	400
Tensile strength/atm.	85–136	204–313

Many of the downstream processes, which utilize petrochemical intermediates, are located on the same site, and this arrangement can lead to reduced utility costs. For example, a large complex of units can justify its own power station to produce electricity and steam. In the U.K., in times of low demand surplus electricity may be fed into the national grid and this generates a small payment to the company from the Central Electricity Generating Board. In contrast, when demand is high electricity is taken from the national grid—but at a much higher cost.

This trend to larger-sized individual chemical plants and large complexes of chemical plants has been matched by movement to bigger chemical companies by mergers, or takeovers, or both, as we have already seen happening in the 1920s with the formation of ICI and I. G. Farben. Nowadays the multinational giants dominate the international chemical scene.

The chemical industry has been to the fore during the last decade or two in utilizing the tremendous advances in electronics. Thus complete automation (even full computer control) of large continuous plants is commonplace. An additional advantage is the automatic data collection of throughput, temperatures, pressures, etc. This can be invaluable when subsequent analysis of the data may suggest small adjustments to the plant to improve its efficiency and hence reduce product costs and improve profitability. The rapid developments in microelectronics which have led to reduced selling prices for microprocessors is making their introduction for fairly small plants an economical and practical proposition.

Control of pollution, i.e. effluent control and treatment, is another important factor which has markedly influenced the industry over the past decade. Indeed there are legal requirements to be met in control of the emission or discharge of effluent. Although, as indicated previously, utilization of by-products is a feature of the industry, nevertheless some effluent is generated and due to the very large scale of operation the quantity can be substantial. Its treatment and disposal can be a considerable cost item of the process. Equipment for the prevention of loss of untreated wastes can also add significantly to the capital and running costs of a plant. Unfortunately the legislation controlling treatment varies quite a lot from country to country and those with strict requirements may place companies which manufacture there at a disadvantage, in terms of production costs, compared with companies in other countries. Chapter 8 takes up these aspects in detail.

Its widening international base has also been a recent feature of the chemical industry. As in so many other areas of manufacturing industry the shining star has been Japan. From a position of insignificance in 1945 it has grown to be the second most important chemical industry in the world today. Its growth is even more remarkable when one realizes that it has no oil of its own—it must all be imported. In more recent years countries having oil reserves, such as the Arab countries, have moved from merely exporting the oil to refining it, and now into processing it to produce petrochemicals (downstream operations).

Though small at present these industries are likely to grow steadily.

Since the rapid rise in the price of oil in 1973, energy costs have been a major concern of chemical producers. The chemical industry is a major energy consumer and therefore even small percentage savings in energy costs can mean tens or hundreds of thousands of dollars or pounds, and the difference between profit and loss. It is no surprise to find therefore that immense efforts are now being made to assess and analyse the energy usage of plants to see where savings can be made. We have emphasized this importance in the text: energy is a recurring theme and we have devoted a complete chapter (Chapter 7) to it.

The most exciting area of development at the start of the 1980s, arising from the immense amount of research and development work which the chemical and related industries carry out each year, is biotechnology. In biotechnology genes are manipulated to ensure that the micro-organism—often *E. coli*—will synthesize the desired specific chemical. Although the risks and practical problems are considerable, the rewards for their solution are immense. The technology is still in its infancy and its effect on the chemical industry as a whole is difficult to predict. Clearly it will make a major contribution to pharmaceuticals—as the successful development of insulin has already shown—and production of interferon and antibiotics are on the near horizon. In contrast it is difficult to see biotechnology being a competitive route to bulk chemical intermediates like ethylene and benzene, because of its slowness and the high product separation costs. However, it is an area whose development chemists are watching with interest and fascination. This topic is discussed in more detail in Chapter 6 of Volume 2.

3.2 The chemical industry today

3.2.1 *Definition of the chemical industry*

At the turn of the century there would have been little difficulty in defining what constituted the chemical industry since, as shown earlier in this chapter, only a very limited range of products was manufactured and these were clearly chemicals, e.g. alkali, sulphuric acid. At present, however, many thousands of chemicals are produced, from raw materials like crude oil through (in some cases) many intermediates to products which may be used directly as consumer goods, or readily converted into them. The difficulty comes in deciding at which point in this sequence the particular operation ceases to be part of the chemical industry's sphere of activities. To consider a specific example to illustrate this dilemma, emulsion paints may contain poly(vinyl chloride)/poly(vinyl acetate). Clearly, synthesis of vinyl chloride (or acetate) and its polymerization are chemical activities. However, if formulation and mixing of the paint, including the polymer, is carried out by a branch of the multinational chemical company which manufactured the ingredients, is this still part of the chemical industry or does it now belong in the decorating industry?

It is therefore apparent that, because of its diversity of operations and close links in many areas with other industries, there is no simple definition of the chemical industry. Instead each official body which collects and publishes statistics on manufacturing industry will have its definition as to which operations are classified as 'the chemical industry'. It is important to bear this in mind when comparing statistical information which is derived from several sources. Perhaps the best known international definition for chemicals is that contained in Section 5 of the United Nations Standard International Trade Classification. Individual countries' definitions will differ from this to varying degrees, as will the companies' trade organizations—such as the Chemical Industries Association (CIA) in the U.K.—in each country.

3.2.2 *The need for a chemical industry*

As indicated in Chapter 1, the chemical industry is concerned with converting raw materials, such as crude oil, firstly into chemical intermediates, and then into a tremendous variety of other chemicals. These are then used to produce consumer products, which make our lives more comfortable or, in some cases such as pharmaceutical products, help to maintain our wellbeing or even life itself. At each stage of these operations value is added to the product and provided this added value exceeds the raw material plus processing costs then a profit will be made on the operation. It is the aim of chemical industry to achieve this.

It may seem strange in a textbook like this one to pose the question 'do we need a chemical industry?' However, trying to answer this question will provide (i) an indication of the range of the chemical industry's activities, (ii) its influence on our lives in everyday terms, and (iii) how great is society's need for a chemical industry. Our approach in answering the question will be to consider the industry's contribution to meeting and satisfying our major needs. What are these? Clearly food (and drink) and health are paramount. Others which we shall consider in their turn are clothing, and (briefly) shelter, leisure and transport.

(*a*) *Food.* The chemical industry makes a major contribution to food production in at least three ways. Firstly, by making available large quantities of artificial fertilizers which are used to replace the elements (mainly nitrogen, phosphorus and potassium) which are removed as nutrients by the growing crops during modern intensive farming. Secondly, by manufacturing crop protection chemicals, i.e. pesticides, which markedly reduce the proportion of the crops consumed by pests. Thirdly, by producing veterinary products which protect livestock from disease or cure their infections.

These and related topics are discussed in Chapter 5 of Volume 2.

(b) *Health.*   We are all aware of the major contribution which the pharmaceutical sector of the industry has made to help keep us all healthy, e.g. by curing bacterial infections with antibiotics, and even extending life itself, e.g. β-blockers to lower blood pressure.

This topic is discussed more fully in Chapter 4 of Volume 2.

(c) *Clothing.*   The improvement in properties of modern synthetic fibres over the traditional clothing materials (e.g. cotton and wool) has been quite remarkable. Thus shirts, dresses and suits made from polyesters like Terylene and polyamides like Nylon are crease-resistant, machine-washable, and drip-dry or non-iron. They are also cheaper than natural materials.

Parallel developments in the discovery of modern synthetic dyes and the technology to 'bond' them to the fibre has resulted in a tremendous increase in the variety of colours available to the fashion designer. Indeed they now span almost every colour and hue of the visible spectrum. Indeed if a suitable shade is not available, structural modification of an existing dye to achieve this can readily be carried out, provided there is a satisfactory market for the product.

Other major advances in this sphere have been in colour-fastness, i.e. resistance to the dye being washed out when the garment is cleaned. For example, the Procion dyes, developed by ICI, actually chemically bond to cotton, rather than attaching by the more usual physical type of adherence to the fibre. This clearly leads to greater colour-fastness.

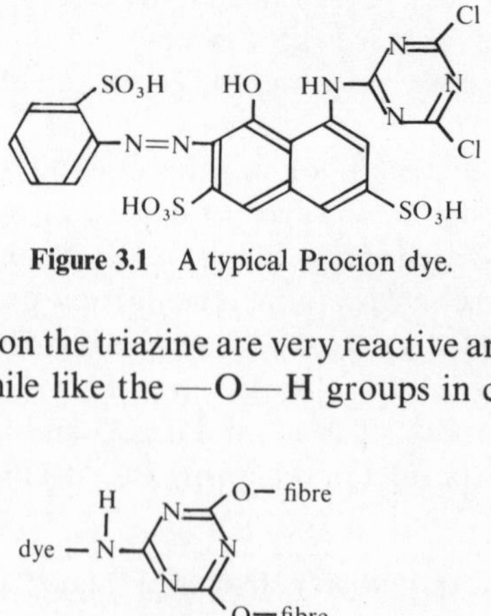

Figure 3.1   A typical Procion dye.

The chlorine atoms on the triazine are very reactive and can be displaced by even a weak nucleophile like the —O—H groups in cotton, giving

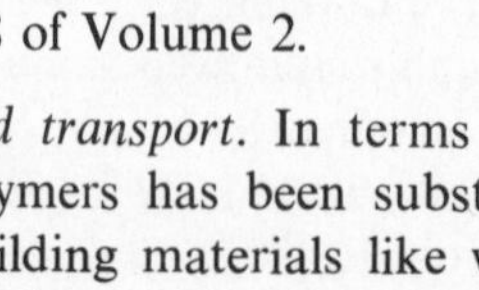

See also section 2.6.1.8 of Volume 2.

(d) *Shelter, leisure and transport.*   In terms of shelter the contribution of modern synthetic polymers has been substantial. Plastics are tending to replace traditional building materials like wood because they are lighter, maintenance-free (i.e., they are resistant to weathering and do not need painting). Other polymers, e.g. urea-formaldehyde and polyurethanes, are important insulating materials for reducing heat losses and hence reducing energy usage.

Plastics and polymers have made a considerable impact on leisure activities with applications ranging from all-weather artificial surfaces for athletic tracks, football pitches and tennis courts to nylon strings for racquets and items like golf balls and footballs made entirely from synthetic materials.

Likewise the chemical industry's contribution to transport over the years has led to major improvements. Thus development of improved additives like anti-oxidants and viscosity index improvers for engine oil has enabled routine servicing intervals to increase from 3000 to 6000 to 12 000 miles. Research and development work has also resulted in improved lubricating oils and greases, and better brake fluids. Yet again the contribution of polymers and plastics has been very striking with the proportion of the total automobile derived from these materials—dashboard, steering wheel, seat padding and covering etc.—now exceeding 40%.

So it is quite apparent even from a brief look at the chemical industry's contribution to meeting our major needs that life in the developed world would be very different without the products of the industry. Indeed the level of a country's development may be judged by the production level and sophistication of its chemical industry.

Table 3.3 Most important chemicals in the U.S.A. (thousands of tonnes)

Position (1988)	Chemical	1977	1988
1	Sulphuric acid	30 229	38 803
2	Nitrogen	10 562	23 628
3	Oxygen	13 998	16 821
4	Ethylene (ethene)	10 830	16 581
5	Ammonia	14 214	15 370
6	Lime	16 619	14 667
7	Sodium hydroxide	9 227	10 871
8	Phosphoric acid	6 854	10 626
9	Chlorine	9 359	10 277
10	Propylene (propene)	5 518	9 057
11	Sodium carbonate	7 016	8 662
12	Nitric acid	6 490	7 157
13	Urea (carbamide)	3 950	7 147
14	Ammonium nitrate	6 138	6 522
15	Ethylene dichloride (1,2-dichloroethane)	4 605	6 190
16	Benzene	4 943	5 370
17	Ethylbenzene	3 208	4 508
18	Terephthalic acid (benzene-1,4-dicarboxylic acid)	2 272	4 354
19	Carbon dioxide	2 018	4 254
20	Vinyl chloride (chloroethene)	2 635	4 109

3.2.3 *The major chemicals*

Table 3.3 shows the 20 most important chemicals produced in the U.S.A. in 1988, together with the corresponding figures for 1977. Ranking order will be similar for other major chemical-producing countries. It is noticeable that inorganic chemicals occupy 8 out of the top 10 places. With the exception of lime, all chemicals had higher production figures in 1988 than in 1977. Particularly large percentage increases are evident for ethylene, propylene, phosphoric acid, urea and ethylene dichloride but pride of place goes to nitrogen whose production increased by well over 100%.

Also notable is MTBE (methyl t-butyl ether). This does not appear in Table 3.3 but demand for it is increasing exponentially. This is because of the rapidly increasing demand for lead-free petrol—MTBE is added to the petrol in place of tetraethyl lead in order to increase its octane rating. Its preparation is described in section 11.9.1 and demand for it in 1989 could reach 14 million tonnes, in which case it could become the number one organic compound.

3.3 The United Kingdom chemical industry

The origins and development of the U.K. chemical industry have already been dealt with in the early parts of this chapter. This section will therefore be devoted to a comparision with other U.K. manufacturing industries and with other countries' chemical industries. Major manufacturing locations will be considered, as will each of the major companies. This will necessarily include quite a lot of statistical data and in view of the difficulties in obtaining figures which are firstly up-to-date and secondly are truly comparable (particularly international sales figures which require currency conversion), it is better to regard them as relative rather than absolute. In any case it is the analysis of the figures that is important to us rather than individual statistics.

Two publications are extremely valuable sources of information. Firstly, the leaflet *U.K. Chemical Industry Facts* which is published annually by (and is obtained from) the Chemical Industries Association. Although small it contains a wealth of statistical and graphical information which has been obtained from many sources. A number of tables and graphs from the September 1989 issue are published, with permission, on the following pages. Secondly, another annual publication, *Chemicals Information Handbook* from the Shell International Chemical Company Ltd., London, is more international in outlook and also deals with important individual chemicals which are derived from crude oil. This not surprising since the Royal Dutch Shell companies' major activities are exploration, extraction, and refining of crude oil plus chemical processing of suitable fractions.

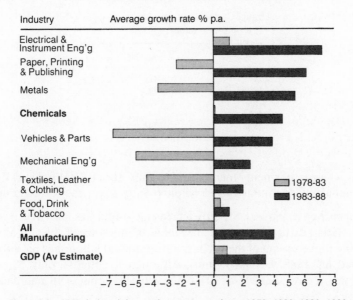

Figure 3.2 U.K. industrial growth rate comparisons 1978–1983, 1983–1988.

3.3.1 *Comparison with other U.K. manufacturing industries*

(i) *Growth rates.* Fig. 3.2 shows the growth rates for various industries over the periods 1978–83, and 1983–88. It is readily apparent that the chemical industry expanded much more rapidly than manufacturing industry generally—in fact almost twice as fast. As will become clear later, this is generally true throughout the world. Indeed chemicals growth rate exceeded most others in this list over the period 1982–83. For the period 1983–88, the annual growth rate of 4.6% was slightly better than for the manufacturing industry as a whole.

(ii) *Total sales.* In 1988 the value of exports of chemicals exceeded that of imports by £2020 million. Royalty payments received in 1988 totalled £210

Table 3.4 Total sales (gross output) of U.K. chemical industry

Year	Sales (£ millions, current prices
1970	3 448
1980	13 661
1986	20 547
1987	22 456
1988	24 791

Table 3.5 Number of employees in the
U.K. chemical industry

Year	Thousands
1972	426
1976	423
1980	402
1984	332
1988	346

million whereas payments made came to only about £100 million. For the U.K. manufacturing industry as a whole royalty payments exceeded income.

(iii) *Number of employees.* Steadily increasing output has been achieved with a fairly static, and in recent years a declining, labour force (Table 3.5) although following the recession of the 1980s some additional labour has recently been recruited (cf. 1988 v. 1984 figures). Although it employs only 6% of the manpower, it is responsible for 10% of the value of output of all manufacturing industry.

Linking employees and sales leads to

(iv) *Output per employee.* Figure 3.3 shows in value terms the much higher output by each employee in chemicals compared to all manufacturing. A significant part of this difference is attributed to the greater capital investment in the chemical industry—as Fig. 3.3 below shows.

(v) *Capital stock per employee.* See Fig. 3.3.

(vi) *Safety.* The chemical industry operates many hundreds of potentially very hazardous processes—particularly those which require extremely high pressures of several hundred atmospheres. Although a zero level is the only acceptable accident level, it is testimony to the good practices employed that

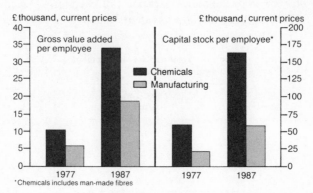

Figure 3.3 Gross value added and capital stock per employee comparisons, U.K.

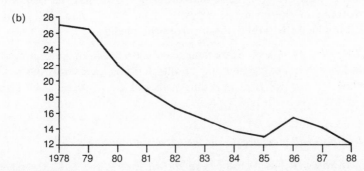

(a)	Year	Chemicals	Manufacturing
	1986/7	1.67	1.44

Figure 3.4 (a) Fatal and major injuries per thousand employees. (b) Accidents reported per thousand employees among CIA members (voluntary sample).

serious accidents are relatively rare. However, it is the occasional disaster which attracts the headlines in the newspapers and on television. The most serious such disaster of the recent past involved the caprolactam (the intermediate for Nylon-6 production) plant owned by Nypro (U.K.) Ltd., at Flixborough, in 1974 (see section 1.6.1).

However, let us place this in its proper context by looking at the industry's performance in two ways. Firstly, let us consider the incidence of reported accidents for chemicals versus all manufacturing industry (Figure 3.4).

The figures show a commendable trend towards a decrease in the number of accidents but the 1986–87 figure of 1.67 fatal and major injuries per thousand employees in the chemical industry still compares unfavourably to the equivalent figure of 1.44 in the manufacturing industry and must be improved.

Secondly, let us relate the chemical industry to some specific high-risk industries, and also everyday activities, by showing the risk of death. These

Table 3.6 Risk of death for various activities

Industry/activity	Deaths per thousand in a 40-year working life
chemical	3·5
all industry	4·0
fishing	35·0
coal-mining	40·0
staying at home	1·0
travelling (a) by train	5·0
(b) by automobile	57·0
(c) by motor-cycle	660·0

figures are calculated ones, and are not as reliable as those above, but are certainly acceptable in relative terms (Table 3.6). They were generated in 1975. Comparing these everyday activities certainly shows that the risk of death from working in the chemical industry is far lower than the qualitative impression which the public obtains from the media.

(vii) *Cost and price indices.* The chemical industry has been quite successful at keeping down the increases in selling or output prices—certainly this has been well below the general rises in selling prices of goods (details are given in Figure 3.5).

3.3.2 *International comparisons in the chemical industry*

(i) *Growth rates.* It has already been shown that in the U.K. the chemical industry has grown, since the early 1960s, at approximately twice the rate of all manufacturing industry. Table 3.7 gives the corresponding figures for all the major chemical-producing countries. The general pattern of chemicals growing at a much faster rate than industry in general is very clear—although Japan is an exception, due to the very rapid growth of a number of its manufacturing industries like electronics, iron and steel, and automobiles. Also apparent is the lower growth rate achieved by the U.K. in comparison with its international competitors.

(ii) *Chemical sales.* Figure 3.6 shows the value of chemical sales in both 1978 and 1988. Note firstly that there is a shortfall in certain figures because they do

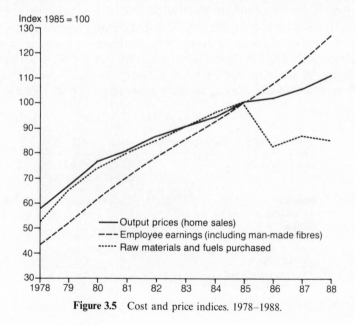

Figure 3.5 Cost and price indices. 1978–1988.

not include sales of man-made fibres. Secondly, all countries have achieved substantial increases in sales between 1978 and 1988. The U.S. industry dominates the scene and its sales, both in 1978 and 1988, were approximately the same as the combined efforts of the Western European or EEC (European Economic Community) countries.

Most countries increased sales by a factor of 1.8 to 2.7 times between 1978 and 1988. Although Japan's sales were approximately 50% higher than West Germany's in 1978, by 1988 the differential had expanded to almost 100%. As we will see later, when looking at the world's major chemical companies, this has been achieved by a chemical industry whose structure is very different to that of most other countries.

(iii) *U.K. external trade in chemicals.* Table 3.8 and Fig. 3.7 provide data on the U.K.'s international trading in chemicals.

Table 3.7 International growth rate comparisons 1978–88

Country	Average growth % p.a.		Ratio of Chemicals to All Industry
	Chemicals	All Industry*	
U.K.	2.4	1.4	1.7
W. Germany	1:9	1.5	1.2
France	3.3	1.1	3.0
Italy	3.0	1.9	1.6
Total EEC	2.7	1.8	1.5
U.S.A.	3.6	2.4	1.5
Japan	4.4	4.9	0.9

*Excluding construction

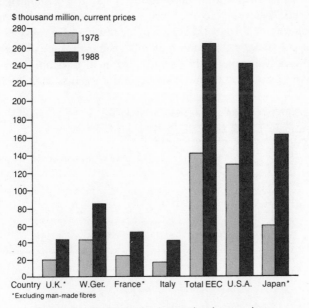

Figure 3.6 Chemical sales, international comparisons.

Table 3.8   U.K. external trade in chemicals

Annual growth rate in volume of chemical exports 1978 - 1988	4.3%
Annual growth rate in volume of chemical imports 1978 - 1988	7.2%
Chemical exports as a percentage of all UK visible exports, 1988	14.1%
Chemical imports as a percentage of all UK visible imports, 1988	9.2%

Trade with fellow EEC countries accounts for almost half the total but note the negative trade balance here. In contrast there is a positive trade balance with all the remaining areas and this leads to an overall favourable balance.

(iv) *Capital investment by EEC countries.* Attention has already been drawn to the highly capital-intensive nature of the chemical industry, particularly the petrochemicals sector, and this important point will again be emphasized later in this chapter. Table 3.9 supports this with some actual figures for the U.K. industry. It also expresses those as a percentage of all EEC capital investment and shows our percentage of total sales. This concludes our comparisons of the U.K. chemical industry with other U.K. manufacturing industries, and also with the chemical industry in other countries.

However, to complete our picture of the U.K. chemical industry the major manufacturing locations, plus a brief indication of the major U.K. chemical companies and their interests, will be considered.

3.3.3 *Major locations of the U.K. chemical industry*

The locations in which the chemical industry has grown and expanded have been governed by several factors. During the very early days the manufacturing location was dictated by the source of the raw materials plus the proximity of major users, since transportation of bulk chemicals was extremely difficult; at this time railways had only recently begun to develop. Thus the alkali industry grew up around the river Mersey in Lancashire and Cheshire, where

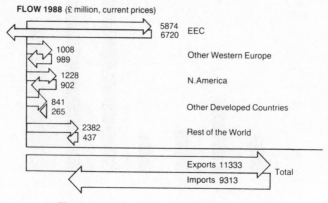

Figure 3.7   U.K. export and import flows, 1980.

Table 3.9 Total new capital invested by U.K. and EEC chemical industry 1978–1988 plus sales proportions

Year	U.K. capital investment (current prices, £m)	U.K. capital investment as % of total EEC	U.K. sales as % of total EEC
1978	967	20·2	15·0
1982	973	18·1	15·1
1986	1320	16·4	13·4
1988	1741	17·0	14·2

large deposits of salt occur. Close by were the major users of alkali, the textile- and oils and fats (particularly soap-making) industries. These factors are equally important now for chemicals which are produced in large quantities. Our best-known examples are again petrochemicals, and the large complexes for their manufacture have grown up next to the oil refineries which are the source of their raw materials. Also, processing plants to use their output, e.g. polymerization units, have been constructed either as part of the same complex or else nearby. This has also meant that transport facilities for products have been good. The complexes have invariably developed on river estuaries, since the large tankers bringing in crude oil need to get close to the refineries to discharge their very large cargo. Hence bulk chemicals transport by sea is readily available. Also, the road and rail network is good and well-developed.

There are several additional advantageous factors in this arrangement. Firstly the chemical industry, as we have already seen, is a very big energy user. Much of this is oil-derived, so that a location near to the refinery eases transportation of large volumes of fluids. Secondly, being located on rivers means that the vast quantities of cooling water which are needed by the chemical plants are readily available. Rivers, flowing a short distance into the sea, are useful for the discharge of treated effluent.

Almost all the areas in the U.K. where the chemical industry is concentrated are situated on river estuaries and have a very strong petrochemical link. Examples are Fawley (near Southampton), the Thames Estuary, Baglan Bay (south-west Wales), Merseyside, Humberside, Teesside and Grangemouth (near Edinburgh). Plans are also well advanced for petrochemical developments on the east coast of Scotland, well north of Aberdeen, where oil and gas from the North Sea oilfields are pumped ashore.

Avonmouth (near Bristol) is one of the few concentrations which does not have a strong petrochemical base. It is noted for fertilizer manufacture and other inorganic chemicals.

The above comments apply to the bulk of the chemical industry's operations. However, there are a few exceptions, and these are all concerned with products which are manufactured in small quantities, typically 10–100 tonnes per annum, but have a very high value per unit weight. For some products this might mean tens of thousands of pounds per tonne. Agrochemicals and pharmaceuticals are the best-known products of this type. Due to their high value and small bulk, transportation (and its costs) is no problem, and manufacturing units may therefore be sited anywhere. Quite often this means that they are located in attractive rural areas and many miles from the coast. Examples are Barnard Castle (county Durham), Loughborough, Macclesfield, and Ulverston (Cumbria).

3.3.4 Some major U.K. chemical companies

Although all the major multinational chemical giants have manufacturing facilities in the U.K., either directly or via subsidiary companies, we shall confine our attention here to those which are incorporated, or have their headquarters, in the U.K. They are Imperial Chemical Industries (ICI), Shell Chemicals, B.P. Chemicals, Croda International, Beechams, Glaxo, Fisons, Albright and Wilson, Laporte, Rio Tinto Zinc, and Unilever. Let us now briefly consider each of these in turn.

ICI. This was the U.K.'s largest private company for many years, until the privatization of British Gas and British Telecom. Its origins from the merger of companies engaged in alkali production were discussed in the early pages of this chapter. Like its multinational competitors it is engaged in the whole range of chemical activities. However, its performance has varied considerably from sector to sector, and major rationalizations and staff reductions were instituted in the early 1980s in order to improve productivity to at least the levels of its competitors. This is now paying off with the company being much more competitive and profitable. In 1988 pre-tax profits had reached £1470 m with all sectors of the business (except fertilizers) being profitable. In recent years this company has been very active in globalization of its activities with 151 acquisitions of companies, including several major ones, e.g. Stauffer in the U.S.A., in the last 5 years. There have also been 48 disposals. In the same period investment in R&D doubled to £565 m.

Although it has manufacturing facilities in very many locations, its major plants are located on Merseyside and Teesside.

Shell Chemicals. The Royal Dutch/Shell company is known best as one of the 'seven sisters' that dominate the oil industry. Developments into chemicals led to the setting up of the Shell Chemical Company in 1929. As one might expect, its major interests are petrochemicals and derived products. One of the latter, the liquid detergent Teepol, was the first petroleum-based chemical to be produced in Western Europe, in 1942. Their product range now covers several hundred chemicals, largely industrial organics, but also polymers (plastics, resins, synthetic rubbers) and agrochemicals, particularly herbicides. A small but growing area is speciality products, e.g. speciality rubbers and plastics. After the difficult period at the beginning of the 1980s, petrochemicals production is currently booming. Major locations are to be found at Carrington, Stanlow, Shell Haven and Mossmorran.

B.P. Chemicals. Like Shell, it is part of an oil giant—the British Petroleum Company. The move into chemicals manufacture in the 1950s and 60s was achieved by setting up several joint companies with established chemical manufacturers, and particularly with the Distillers Co. Ltd. This culminated with the takeover of almost all of Distillers chemical interests in 1967, including, logically, the Chemicals and Plastics Divisions. Incidentally, Distillers main interest, as their name suggests is production of whisky. Like Shell its major interests are petrochemicals and derived products e.g. plastics. Major manufacturing locations include Hull, Grangemouth and Baglan Bay.

Croda International. This group of companies started with the manufacture of lanolin from wool greases, and oleochemicals, i.e. glycerides, fatty acids and compounds derived from them, remain an area of major importance. Other areas of interest include coal-tar chemicals; heterocyclic compounds; polymers used in adhesives, inks, paints and resins; and foodstuffs. Manufacturing locations are centred on Humberside, Wolverhampton and Leek in Staffordshire.

Beecham. This group of companies produces proprietary and ethical pharmaceuticals. It also has extensive interests in related areas such as food, health beverages, soft drinks, confectionery, and toiletries. Many of its products are well known to the general public by their brand names, including Beecham's pills and powders (for counteracting the effects of colds, headaches, etc.) on which the company was founded in 1842. In terms of chemical achievement their development and introduction in 1963 of the first semi-synthetic penicillin antibiotics was notable. They have developed quite a number of these over the years, and they represent a substantial part of the company's earnings. Manufacturing tends to be concentrated in and around London.

Glaxo. This group is in many ways similar to the Beecham group, and in fact the latter made a takeover bid for Glaxo in 1971, but this was turned down by the Monopolies Commission. Since that time the company has gone from strength to strength with rapid growth and good profitability and it is now the fourth largest pharmaceutical company in the world. This is due in considerable measure to a big research effort leading to new products coming through and onto the market. Examples are the steroid Betnovate (used in treating skin conditions and some allergies), cephalosporin antibiotics and Zantac (an anti-ulcer drug) which has become a major sales success. Many vaccines and veterinary products are also produced, and baby and health foods are also important group products. Manufacturing plants include Annan, Ulverston, Barnard Castle, and Greenford.

Fisons. Ironically this company was founded on fertilizer manufacture and for many years this was a strength of the company. However, losses in recent years resulted in the whole fertilizer division being sold to Norsk-Hydro in 1982. This helped profits for the company to rise from £9·3 million in 1981 to £21·1 million in 1982. Its major interests are pharmaceuticals—where its most important success has been the anti-asthmatic drug Intal—and agrochemicals. Diversification into making scientific apparatus and laboratory chemicals, and more recently horticultural materials like composts and peat, appears successful to date.

The company's main location is at Loughborough.

Albright and Wilson. This company started by making phosphorus for matches in 1844. In the late 1960s a decision was taken to relocate phosphorus manufacture in a new plant in Newfoundland. However this was plagued with technical and pollution problems, which had such a serious effect on the company's finances that major help from the U.S. company Tenneco was needed and as a result it has become a subsidiary of the U.S. company. Its predominant interests are built on phosphorus and silicon chemistry but include other areas as well. Examples are detergents, shampoos, plasticizers, silicones, and flavours and essences.

Locations include the London area and Whitehaven.

Laporte. This company started by producing hydrogen peroxide in Yorkshire in 1888 for the textile industry. Hydrogen peroxide, other peroxides, and perborates continue to be a major interest today. They also produce a variety of predominantly inorganic chemicals, e.g. titanium dioxide, fuller's earth.

Major manufacturing plant is located at Widnes near Liverpool.

Rio Tinto Zinc. This is primarily a mining company and its chemical interests relate largely to the ores which it mines. It is therefore concerned with inorganic compounds of metals like copper, aluminium, iron, lead and zinc. The takeover of Borax Consolidated in 1968 took it into the area of boron compounds.

Avonmouth is the location of its lead and zinc smelting plant.

Unilever. Like Shell this is a joint U.K./Dutch company. It has major interests in the food industry (ice-cream, sausages, frozen foods, margarine), many of its subsiduaries being household names, and soaps and detergents. Its raw materials are largely animal and vegetable oils and fats and these are extracted and processed largely at its many works on Merseyside.

3.4 The U.S. chemical industry

The U.S. industry is by far the largest national chemical industry, being more than twice the size of its nearest rival Japan and approximately the same size as that of all the EEC countries combined. It has the advantage of a very large home market, and tariff barriers plus its physical remoteness from its major competitors combine to moderate external competition. In fact these restrictions virtually mean that any foreign company wishing to achieve a reasonable market penetration in the U.S. has to set up manufacturing facilities there. Another traditional advantage of U.S. chemical firms, certainly compared with their European counterparts, has been the ready availability of cheap raw materials, particularly natural gas and oil. This has been beneficial twice over. Firstly it has meant energy has been relatively cheap, and as we have already noted the chemical industry is a big energy user. Secondly, raw materials for bulk chemicals and intermediates such as petrochemicals and ammonia are cheap. However these advantages are gradually being eroded as natural gas supplies dwindle in the U.S.A. at the time that countries like the U.K. are making large discoveries of oil and natural gas around their shores. At the present time the U.S.A. still retains a significant advantage.

With their very solid home and financial base, U.S. companies have not found exporting to be too difficult despite the geographical remoteness of the major markets. They have tended to set up manufacturing units and subsidiary companies in Europe, and particularly in the U.K. In 1988 the U.S. exported chemicals to the value of $32.5 billion (compared to $21.8 billion in 1985) whereas imports were valued at $20.1 billion (cf. $14.5 billion in 1985), leading to a record balance of payments surplus in chemicals.

Several aspects of the U.S. chemical scene have been commented on in the earlier section of this chapter concerned with international comparisons with the U.K. chemical industry (section 3.3.2). These were items such as sales, growth rates and trade with the U.K. The remainder of this chapter will therefore be devoted to a brief consideration of some of the larger U.S. chemical companies. Bear in mind that, as with any very large chemical company, they are all multinationals, and their activities cover practically all the important sectors of chemicals manufacture. Comments on their interests are therefore confined to just a few areas that the company is particularly strong in, or noted for. Another point to remember is that, partly due to the physical size of the country, important manufacturing locations are numerous and, unlike the U.K., it is not reasonable to try and list them.

Du Pont. This has been the largest American chemical company for a number of years, and is noted for its high level of expenditure on research and development. It is the world's largest producer of man-made fibres. Its Teflon (polytetrafluoroethylene) is well known to the general public as the coating for non-stick cooking utensils. Since its takeover of the giant Conoco oil company in 1981, which has doubled total sales, the whole nature of the company has altered.

Dow Chemical. Another giant of the industry which produces a very wide range of both inorganic and organic chemicals. Basic inorganic and organic chemicals plus plastics account for 70% of sales, and pesticides and pharmaceuticals account for a further 10%. It is well known for its chlorinated products, such as herbicides like 2, 4-D and the much discussed 2, 4, 5-T, which was marketed under the trade name Agent Orange. (2, 4-D and 2, 4, 5-T are 2, 4-dichloro- and 2, 4, 5-trichlorophenoxyacetic acid.) Approximately half of total sales are made outside the U.S.A. It has many associated and subsidiary companies, and major manufacturing sites in Europe at Stade in West Germany and Terneuzen in the Netherlands. Rapid expansion in recent years has more than doubled its sales since 1976.

Union Carbide. As its name suggests, this company started by making calcium carbide. It ranks third in the U.S.A. for chemical sales, but is a diverse company with only about 40% of its revenue coming from chemical sales. Areas of particular interest are agricultural chemicals, industrial gases, and plastics. Although it has many overseas subsidiaries, it sold most of its European interests to BP in 1978.

Exxon. This is part of the giant Exxon Corporation, which is perhaps still better known by its former name of Standard Oil of New Jersey in the U.S.A. In the U.K. the company is well known for its petrol sales under the Esso name. As one might expect, therefore, its interests are primarily in petrochemicals and derived products. The majority of its sales are outside the U.S.A.

Monsanto. Another large company with a diversity of interests. It is known to the general public through its acrilan (polyacrylic) wear-dated fabrics. The company has experienced severe difficulties with its activities in the synthetic polymer field, and as a result it has abandoned nylon, polyester fibres and polystyrene in Europe during the last few years. Nevertheless one-third of its income is generated overseas. It also withdrew from the polyester fibre market in the U.S.A. in 1980. It has shown its interest in biotechnology by putting almost £5 million into the £9·5 million Advent Eurofund. This capital will be used to back new and established companies involved particularly in the biotechnology and genetics areas by acquiring a minority holding for between £100 000 and £500 000. Its pesticide 'Roundup' is a very important revenue earner.

Allied Chemicals. In order to reflect better its wide range of interests the company has dropped the word 'chemical' from its title, becoming just Allied Corporation. It was originally (in the 1920s) a major alkali and dyestuffs manufacturer. Currently its interests include basic inorganic and organic chemicals, plastics, fibres, fertilizers and pesticides. A very active and growing involvement in the energy field has been a feature of recent years and this area now accounts for one quarter of total sales. Diversification continues with the acquisition of a scientific instruments company.

American Cyanamid. Although its main post-war growth has been in pharmaceuticals it has interests also in pesticides and fertilizers, speciality chemicals (e.g. acrylic fibres and pigments) and consumer products (e.g.

laminates). It is noted for its part (with Pfizer) in the introduction of the tetracycline antibiotics. Overseas sales account for one-third of total sales.

Hoechst Celanese. This is one of the world's largest producers of cellulosic and synthetic fibres, which represent more than half of total company sales. Other interests are in a variety of organic chemicals.

Pfizer Inc. This is a more specialized company which concentrates on pharmaceuticals. Its specialities are fine chemicals and fermentation products, particularly penicillin and streptomycin antibiotics.

The U.S. chemical industry and the country in general tend to show a greater entrepreneurial flair, and are more willing to back risky projects, than their European counterparts. A recent example is the field of biotechnology. Several new companies, such as Biotech and Celltech, have been set up and recent share issues in them were oversubscribed many, many times—despite the fact that success, although potentially extremely rewarding, lies many years ahead owing to the formidable practical problems to be overcome.

3.5 Other chemical industries

From the international comparisons of the chemical industry made earlier in this chapter, it can be seen that apart from the U.K. and the U.S.A., the other major producers in order of importance are Japan, West Germany, France, Italy and the Netherlands. Several aspects of the chemical industry in these countries are already apparent from the figures and tables which were used in these comparisons.

3.5.1 *Japan*

Although Japan's chemical industry is second only to that of the U.S.A., most of its production is for home use. Even the Netherlands, whose chemicals production is far below that of Japan, outranks it both for exports and imports. Another feature of the Japanese chemical industry is its organizational structure. Although only the U.S.A. produces more chemicals, only one Japanese company is included in the top 20 companies for chemical sales, at position 13. In contrast West Germany has three and these occupy positions 1, 2, and 3. Even the U.K. has $1\frac{1}{2}$ companies in the list (Table 3.10). Thus the Japanese industry consists of a large number of medium- and small-sized companies.

3.5.2 *West Germany*

Like Japan, Germany has no indigenous oil or natural gas, and the performance of its chemical industries is therefore even better than appears at first sight. The German industry is dominated by the three giants BASF, Bayer and Hoechst who in 1986 occupied first, second and third places in the world's top 20 chemical companies. As might be expected they are multi-nationals in every sense. In common with many of the European companies, they achieved excellent levels of profitability in the last few years.

3.5.3 France

As we have already seen, the French chemical industry is similar in size to that of the U.K. Its two major companies are Rhône–Poulenc and the oil company Elf–Aquitaine, which has extensive chemical interests.

3.5.4 Italy

The Italian industry is approximately two-thirds of the size of the U.K. industry. It has only one really big chemical company—Enimont—which was formed at the beginning of 1989 by merging Enichem with 60% of Montedison.

3.5.5 Netherlands

Although it is only about one-third of the size of the U.K. industry, the Dutch industry boasted $1\frac{1}{2}$ of the top 20 companies in 1986, with Royal Dutch/Shell in 7th position and AKZO in 11th position.

3.6 World's major chemical companies

Table 3.10[7] lists the world's top 20 companies based on their 1986 sales figures. The 1984 and 1982 figures are also given. (Caution must be exercised over the detailed positions since currency conversions were carried out in arriving at some of the figures—this may favour some companies to a certain extent and disadvantage others.) The conversions were carried out at the end of the appropriate year. Analysis shows that of these 20 companies, 6 are based in the U.S.A., 3 in West Germany, 3 in Switzerland, 2 in France, $1\frac{1}{2}$ each in the U.K. and the Netherlands and one each in Belgium, Japan and France.

Note that although the recession of the early 1980s is apparent with many of the 1984 sales figures being lower than in 1982, with the exception of the U.S. companies (except Pfizer) all the others had significantly higher sales figures in 1986. The difference is because the recession hit the U.S. later than Europe.

3.7 General characteristics and future of the chemical industry

3.7.1 General characteristics

This section collects together important points which have previously been made in this chapter. Comments on the general characteristics of the chemical industry apply to a large extent to any individual national industry.

Table 3.10 World's largest chemical companies (by sales proceeds)

		(all sales figures in $ million)			
Position	Company	Country	1982	1984	1986
1	BASF	W. Germany	13 641	12 827	21 057
2	Bayer[1]	W. Germany	13 843	13 426	20 003
3	Hoechst[1]	W. Germany	13 948	12 472	18 686
4	Du Pont[1,3]	U.S.A.	14 435	15 861	15 828
5	ICI	U.K.	11 920	11 487	15 032
6	Dow Chemical	U.S.A	10 618	11 418	11 113
7	Ciba-Geigy[4]	Switzerland	6 887	6 721	9 929
8	Montedison	Italy	6 597	6 405	9 599
9	Shell[2,3]	Netherlands/U.K.	6 828	6 913	8 646
10	Rhone-Poulenc	France	5 524	5 320	8 273
11	Akzo	Netherlands	5 381	4 647	7 186
12	Monsanto	U.S.A.	6 325	6 691	6 879
13	Mitsubishi Kasei[5]	Japan	3 090	3 141	6 643
14	Union Carbide	U.S.A.	9 061	9 508	6 343
15	Elf Aquitaine[2,3]	France	2 436	4 828	6 090
16	Exxon[2,3]	U.S.A.	6 049	6 870	6 079
17	Solvay	Belgium	3 805	3 550	5 401
18	Sandoz	Switzerland	3 019	2 859	5 203
19	Roche-Sapac	Switzerland	3 543	3 179	4 868
20	Pfizer	U.S.A.	3 454	3 855	4 476

[1] Includes only 50% of sales of 50% owned affiliates
[2] Chemical interest only.
[3] Excludes intersegment transfers
[4] Data based on current values
[5] Consolidated results

The first point to make is that the industry is very research-intensive, the research and development effort being devoted to (i) development of entirely new products, (ii) better and more economical routes to existing products, and (iii) improving the efficiency of existing processes and developing new applications for existing procedures. In the 1950s and 1960s the accent was very much on (i), and many new polymers, drugs and pesticides were developed. As the need to reduce costs is now more acute than ever, research and development directed towards (iii) is becoming increasingly important. An example, already encountered, is the increasing effort being directed towards energy conservation. The figures in Table 4.1 confirm the research-intensiveness of the industry. Note that the final column expresses R & D expenditure as a percentage of *sales* (not profits). A corresponding figure for the U.K. engineering industry would be about 1.5% of sales income. Note the extremely high figures for the pharmaceutical companies, Ciba-Geigy and Glaxo.

The range of chemicals produced and of scale of operation is very wide. Despite this, the various processing plants have many similarities and have been designed on the basis of a number of common unit operations, examples of which are discussed in chapter 6.

It is important to visualize a typically high-technology, very capital-intensive industry, where an individual plant and its control equipment can cost hundreds of millions of dollars. All large plants run on a continuous basis, and complete automatic control by computer or microprocessor is quite common. Some of these giants (mainly ethylene crackers) have capacities in excess of 500 000 tonnes per annum. Development of more cost-effective routes and advances in technology have quite often meant that existing plant and processes have rapidly become redundant. For example, development of the cumene route to phenol meant that the benzenesulphonic acid route became obsolete some years ago.

The industry plays a very important part in every highly industrialized country and, excepting Japan, a limited number of giant multinational chemical companies play a key role throughout the world. Even Japanese companies are now appreciating the necessity for globalization of their activities. All countries which are in the early stages of real industrialization are developing chemical industries, e.g. Saudia Arabia, Mexico. Be prepared also, over the next few decades, for China, with its massive home market, to become a major chemicals producer.

Rapid growth has been a dominant feature of the post-war chemical industry, with growth rates typically twice those of manufacturing industry generally. However, in countries where the industry may be regarded as mature, i.e. Western Europe and North America, growth rates are falling and are likely to be much lower during the next decade.

3.7.2 The future

Looking ahead towards the turn of the century it is logical to divide the world's chemical industries into two categories, as indicated above; the well-established, mature ones such as those in North America and Western Europe, and those in countries just starting, or continuing their industrialization.

Following the major recession at the beginning of the 1980s, the world chemical industry is now buoyant with a high level of demand for its products being accompanied by good profits. This is the reward for the hard decisions which had to be taken during the recession to trim staff and rationalize businesses. For the latter the guiding principle was to strengthen the activities that you were good at and move out of activities that you were not so good at. As an example of this, ICI merged its petrochemical and plastics operations, combined this with its other heavy chemicals interests, i.e. fertilizers and general chemicals, and made them into a wholly-owned subsidiary—ICI Chemicals and Polymers Ltd.—which has over 37,000 employees and sales in 1988 of over £4 billion, yielding profits of £473 million. In terms of assets it represents one third of ICI's total operations and if it were independent it would be a major U.K. company in its own right.

Product and plant swaps with BP chemicals in the early 1980s enabled ICI to concentrate on PVC production and BP to strengthen its low-density polyethylene activities. There are many other recent European examples of inter-company agreements, and other cost-cutting moves such as moving company headquarters out of major cities[8]. Globalization of activities by takeovers has been a recent trend. Thus the structure of the chemical industry is changing and may be significantly different by the end of the century. A study by Dow Chemical Europe on the likely situation in the year 2000 has been presented by Frank P. Popoff, its president[9]. They predict that as many as 10 of today's top 30 firms will not be around as we currently know them—due to mergers, acquisitions, rationalization and nationalization. Further it is suggested that only 10 of the top 20 chemical companies will then be from Europe and the U.S.A. At present 19 are from these areas.

New chemical industries, although small at present, are growing rapidly in the industrializing countries which have an abundance of the raw materials for chemical feedstocks, namely crude oil. Saudi Arabia, for example, through its Saudi Basic Industries Corporation, is pressing ahead with a massive downstream petrochemical development in the Al Jubail area[10], with a combined capacity in excess of 1·5 million tonnes for the several plants due on stream in 1987–8. Large oil discoveries in Mexico will meant it will also follow Saudi Arabia's lead in developing a major chemical industry of its own. Other rapidly developing nations in this sphere are Korea and China, which, with its massive home market, will be a major chemical producer by the end of the century. These new developments have been considerably aided by the fact that many petrochemical processes are now so well established that the plant and process technology can readily be purchased, almost off the shelf.

Due to the difficulty in competing with these developing countries in terms of raw material costs we can expect European, Japanese and U.S. companies to move further downstream in petrochemicals, where they can use their greater technical skill, expertise, and experience. This is already evident in polymers, where the only profitable area is speciality polymers which require the above attributes for their production. Finally, the availability of feedstocks for organic chemicals as oil reserves diminish will also have an influence, but this is difficult to predict. The influence of the communist block is similarly difficult to predict but the construction of a pipeline to transport natural gas from Russia to West Germany has been completed. Reports that Russia may become a major oil exporter (as its large reserves of natural gas are tapped) are being viewed with concern by OPEC. This apparent abundance of feedstocks may lead to Russia becoming a significant exporter of basic organic intermediates, alongside Mexico and Saudi Arabia.

As ever, the difficulty for the industry is to balance capacity with expected demand, although the latter is extremely difficult to predict even 5 years ahead because it is so strongly affected by the world's economy which is very cyclical with a boom inevitably followed by a slump. The diversity of experts' views clearly illustrates the difficulties. Views on ethylene in the mid 1990s range from predictions of significant over-capacity in Europe to a worldwide shortage.

Clearly the actual outcome depends on how the market for this product expands (or contracts).

References

1. *The Liverpool Section of the Society of Chemical Industry* 1881–1981: *A Centennial History*, D. Broad, Society of Chemical Industry, 1983.
2. *Rèigel's Handbook of Industrial Chemistry*, 8th edn, ed. James A. Kent. Van Nostrand Reinhold, 1983, p. 363.
3. *Annual Review of the Chemical Industry* 1979, by the Economic Commission for Europe (ECE/CHEM/34), Table 86.
4. Derived from: *C&EN*, 1978, (May 1), 33, and *C&EN*, 1982, (Dec. 20), 48.
5. *C&EN*, 1982, (Dec. 20), 50.
6. *C&EN*, 1983, (Feb. 14), 15.
7. Taken from *Chemical Insight*.
8. See for example ref. 4, 54, 55.
9. *C&EN*, 1982, (Oct. 11), 6.
10. *European Chemical News*, 1982, (Dec. 20/27), 16.

Bibliography

A History of the Modern British Chemical Industry, D. W. F. Hardie and J. Davidson Pratt, Pergamon Press, 1966.
The Chemical Economy, B. G. Reuben and M. L. Burstall, Longmans, 1973.
UK Chemicals Industry Facts, September 1989, Chemical Industries Association Ltd.
Chemicals Information Handbook, 1988, Shell International Chemical Company Ltd.
'World Chemical Outlook', *Chemical and Engineering News*, 1982, (Dec. 20), 45–67.

Note: The data for Tables 3.4, 3.5, 3.7, 3.8 and 3.9, and Figures 3.2 to 3.7 are taken from *UK Chemical Industry Facts*. September 1989, Chemical Industries Association Ltd., London.

CHAPTER FOUR

ORGANIZATION AND FINANCE

D. G. BEW

4.1 Introduction

The objective of the present chapter is to examine, in broad outline, how a large chemical company is organized to carry out the many functions which are required to maintain an on-going business. The organization and use of labour, and aspects of labour relations, although important areas of the operation of a company, are outside the scope of this chapter.

Chemical companies must respond to changes brought about by both internal and external events. New opportunities will arise from research activities inside the company—probably stimulated by a perceived need in an external market—and threats will appear from competitive action by other companies. The company which can respond most rapidly to these stimuli and take effective, rapid action in R&D, production, marketing and other areas will have the best chance of continued, profitable existence. Many studies have been carried out, by university behavioural scientists and of existing business organizations, on the different structures adopted and the resultant effects on the operation of the company[1].

This chapter will consider the main structural units of a large chemical company and also the ways in which the company can raise money to finance its activities.

4.2 Structure of a company

The detailed structure of a company is highly individual and depends on many factors such as the fields of activity—whether growing or declining—the markets involved, the past history of the company and the people involved. Although there are as many structures as companies, the structure outlined here is typical of many large chemical companies. Much information on the organization of a company can be found in the annual report; German companies are more informative than many others. A typical organization chart for a large chemical company to the division level is shown in Fig. 4.1.

79

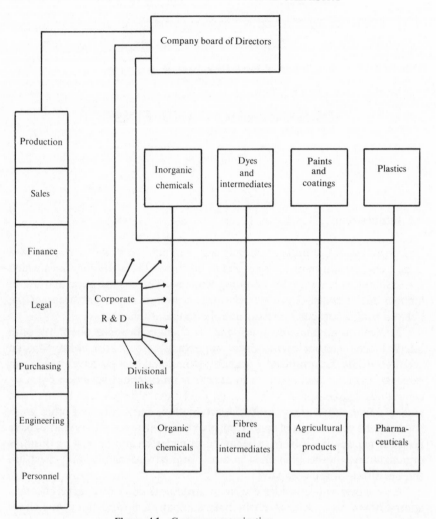

Figure 4.1 Company organization.

4.2.1 *Company board functions*

At the summit of the corporate structure is the board of directors with overall responsibility for the long-term strategic planning of the company. The work of the board can be split in various ways. In some cases individual board members have responsibility for the technical aspects of a group of products, while another board member looks after commercial aspects. Where overseas activities are an important facet of company activities, as is the case with most major chemical companies, then board members may combine their technical or commercial role with a general overall responsibility for an overseas country or group of overseas markets. Board members also combine their activities in committees which consider company-wide activities which must be brought together and matched on a central basis, such as plans for capital expenditure or personnel matters.

The function of the board is to plan the long-term growth, or at least survival, of the company, so they are concerned with overall strategy and long-term plans rather than with the detail of individual operations. To carry out this function the board needs information on company-wide activities condensed to show the main strategic elements. Supplying the board with the required data is a number of support groups or staff functions, often quite small, which coordinate activities in other parts of the company and handle those aspects which must be managed for the company as a whole. Examples of these are finance and accounting, where all parts of the company must adopt mutually compatible accounting methods and financial plans covering capital and revenue expenditure must be matched by the corporate resources. Similarly matters of taxation, legal matters, licensing and patents may be handled on a corporate basis by a central board support group. The number of such staff-function departments varies from company to company and it is not always possible to know whether some companies separate these functions into smaller units than others. Hoechst and Bayer, for example, list 8 or 9 corporate departments or coordinating departments, whereas Du Pont show 15 staff departments. However, the latter included separately such functions as transportation, information systems, marketing communications and public affairs, which are probably rolled up within other functions by the European companies quoted.

4.2.2 Operating divisions

Within a large chemical company the major production and marketing activities are separated into a number of operating divisions in order to form units of manageable size. The operating divisions are separated by product type or by technology to give units which have products with similar markets or common technological features. Each of the operating divisions of the major chemical companies is a large company in its own right and would be a world-ranking organization if it was not part of a major company. Thus the sales of Hoechst pharmaceutical division make the division the No. 2 pharmaceutical company, and those pharmaceutical sales alone would put the company in the top 40 of world chemical companies[2].

Research and development activities may be organized on a divisional basis, with the R&D staff function coordinating divisional plans and initiating studies which overlap or are outside existing operating division activities. At the other extreme, many companies operate a centralized R&D facility to concentrate their technical staff, by which they hope to obtain the benefits of cross-fertilization of ideas and also optimize the use of analytical and physical-

method (NMR, electron microscope, etc.) equipment. In this case individual divisions will have relatively small laboratory teams, associated with
the production facilities and carrying out quality control and plant
improvement/problem-solving work. The number of divisions is a result of
company policy, product range and company history. A corporate R&D
organisation is shown in Fig. 4.1 although, as indicated above, these activities
may be operated on an individual division basis.

Each division has its own management and organizational structure which
may be on a product-line or operating-function basis. The division may
operate almost as an autonomous unit with its individual headquarters and
production site; alternatively, the commercial and technical staff of the division
may work in the corporate headquarters with only the divisional production
facilities on individual sites. In the case of ICI, for largely historical reasons
divisions have their own headquarters and sites—e.g. Agricultural Division
has its headquarters, R&D, sales, technical and much of its production
facilities at Billingham in north-east England, whilst Organics Division has
headquarters, R&D, sales and technical service at Manchester
with major production facilities nearby at Huddersfield. However, under
recent reorganization most of the ICI Divisional structure has been broken up
and the company organized on a product basis. Thus Agricultural Division
activities have been combined with other commodities into ICI Chemicals and
Polymers Ltd.—a wholly owned subsidiary. By contrast, Du Pont in the
U.S.A. and BASF in Germany operate corporate R&D departments with
divisional sales, purchasing and technical staff in the company headquarters.
Production facilities are then widely dispersed to be near major markets and
minimize transport problems (Du Pont) or largely concentrated for integration and optimization, and to take advantage of a major transport facility
(BASF with Rhine barge facilities).

4.2.3 Divisional structures

Divisional organizations structured on a product-line or functional-
department basis are shown in Figs. 4.2 and 4.3. In both cases there are strong
possibilities of duplication of activities within a function or, what might be
more damaging, omission of work in an area which is thought to be the
responsibility of another group. To avoid these drawbacks, the development
of some form of matrix organization is usual to ensure coverage of all the
planned activities with minimum overlap. In the product line organization, a
linking of all the R&D group activities through a R&D line management
structure will help the optimum use of R&D resources and avoid the separate
solution of analogous problems within the individual product lines. For the
individual scientist the matrix structure maintains links with his own
discipline and provides opportunities for job movement and career development which could be lacking in an R&D group operating in a rigid product-
line structure.

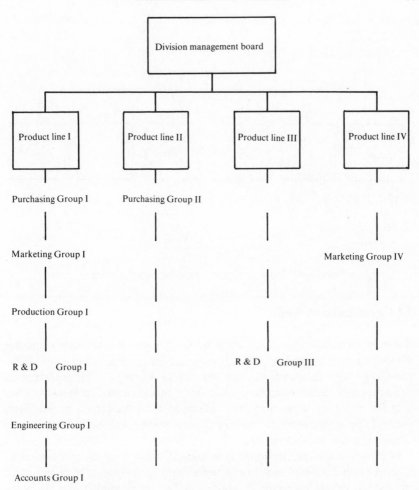

Figure 4.2 Division structure—product line basis.

In a division structured on a departmental basis the R&D scientist will have close contact with his peers in his own and other scientific disciplines and with the scientific hierarchy in the division. A matrix structure, through an interdepartmental product group team, could then have the benefit of giving the individual scientist an insight into the problems and restraints of other departments. This should then enable him to plan his work to tackle most directly the problems in hand and give him the added responsibility and involvement as the R&D specialist in the multi-disciplinary team. An example of a matrix based on a departmental structure is shown in Fig. 4.4. The further organization of some of the indicated divisional activities—in R&D, production, marketing—is examined in more detail in subsequent parts of this chapter.

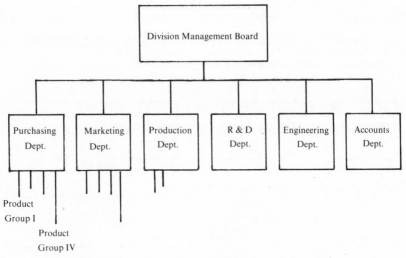

Figure 4.3 Division structure—functional department basis.

4.3 Organization of R&D

In the chemical industry, or any other industry, research is a business function like other activities such as production or marketing. The business of R&D is knowledge and information, and the R&D activity is the purchase of information to reduce uncertainty and solve problems and to provide better data on which to base decisions. Information is 'bought' and problems resolved by application of the company's money and skilled manpower resources in the R&D function.

Within an academic environment a research activity may be carried out for its own intrinsic interest and to add to the sum of knowledge. In an industrial context the primary reason for doing research is economic and the work carried out is done for the benefit of the sponsoring company. The chemical industry invests heavily in R&D, as do other high-technology industries (aircraft, computers), and much more intensively than the industrial average. This is a result of the complex processes used in the chemical industry and the fact that by proper use of science and technology, the processes can be improved, costs reduced and innovations introduced. In the highly competitive environment of the chemical industry such improvements and innovations are essential to improve the company position in the market.

4.3.1 *Long-term activities*

Work in an industrial R&D organization covers a very wide range of activities and time-spans, and the long-term nature of some R&D activities represents a major difference with respect to other company activities. The scope of

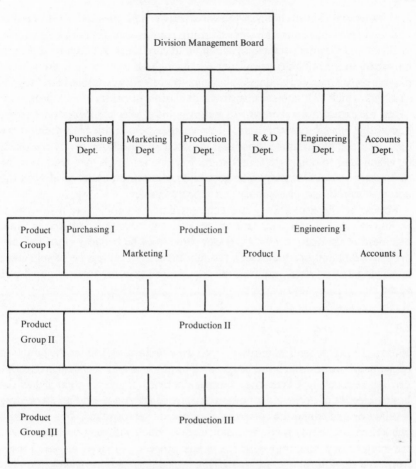

Figure 4.4   Matrix structure.

activities can range from an exploratory search for a new catalyst to the development of an existing product.

At the longer time-scale end of R&D activities, the work is often classified as fundamental or exploratory. Fundamental work usually refers to basic studies and derivation of data in areas of current or potential interest to the company. The effort is not normally aimed at any particular commercial goal but is less specifically targeted and is directed to helping ultimately to solve problems in broad areas of activity. Thus broad-ranging studies on hydrocarbon oxidation could provide information which would be useful in more specific areas such as cumene oxidation for phenol production or cyclohexane oxidation for production of nylon intermediates. Exploratory work is more specifically targeted on a commercial objective and involves the investigation, as a preliminary step, of promising lines or methods of solving a problem. The initial exploratory target is to eliminate the less attractive activities and leave a smaller number of possibilities to be followed up more intensively. Exploratory work is, in effect, a preliminary screening activity and care must be taken to achieve a balance between missing a promising line through too narrow an approach and passing on too many candidates to be followed up. As an example, searching for catalysts which will achieve the direct oxidation of propylene to propylene oxide, the exploratory chemist will make an initial selection based on existing knowledge of oxidation catalysts. These will then be examined in a preliminary oxidation test to eliminate those that produce no detectable propylene oxide or which lead to complete combustion. Further tests will then lead to a few candidates for detailed study. The possibility of missing a useful catalyst by the initial selection and preliminary test is inescapable.

Following the exploratory stage the research activities have a somewhat shorter time horizon and are more directly oriented towards a final commercial application. Activities can sometimes be broadly classified into process R&D and product R&D, though this cannot always be so split since modifying the product requirement can require process research, and vice versa.

4.3.2 Shorter term—process R&D

Within the petrochemical industry, with the emphasis on tonnage production of chemical entities which can be specified by physical and chemical properties, process research has been the dominant activity. Process research has the basic objective of optimizing the manufacture of a product—either an existing product or one which the company does not, as yet, produce. This activity is concerned not simply with the basic reaction which will have been studied at the exploratory stage but with the whole process. All steps in the process, from the handling of raw materials, through reaction stages, product separation and purification to product storage, must be covered. The process research may concentrate on one area—say the reactor where a new catalyst or reactor design is being considered for an existing process. Alternatively, the work can encompass all stages where a new process is under development by examining the effect of reactor conditions on yields and by-product formation and the subsequent separation and recycle of streams, leading to product purification. The use of mathematical models—equations representing the reaction kinetics, heat and mass balances—is well developed and a major factor in process research work.

The end product of the process R&D study is a package of information which can be used for engineering design and costing of the equipment required, and data which will assist the safe start-up, operation and shutdown of the process. In order to achieve this, the process R&D study will have to progress from the initial small-scale laboratory experiments to a larger scale which will enable the difficulties likely to be encountered on the final scale up to a production unit to be examined. There are many reasons why it is not possible to convert a laboratory reaction to a production-plant scale by a simple multiplication factor. These are considered in detail in chapter 6, but as a brief indication, problems of heat transfer affected by surface/volume ratio in stirred reactors, possible wall effects in vapour-phase reactors, choice of material of construction to avoid corrosion, concentration of trace impurities causing corrosion at phase boundaries or forming azeotropes are all areas where work on a larger-than-laboratory scale is needed to provide reliable data to ensure safe operation of the process and for design of a full-scale plant. From this brief description of process research activities it can be seen that the development of the results of an exploratory chemical study into an operable chemical process requires the examination of a complicated interacting matrix of factors. Input will be needed from a range of scientific disciplines—chemists, chemical engineers, electrical and mechanical engineers and mathematicians will certainly be involved. If problems of effluent disposal arise, biologists and biochemists will also have vital contributions to make as members of the process design team.

4.3.3 Shorter term—product R&D

The objective of product R&D is to develop new products or to improve existing products in the company range. Product research may be directed to the improvement in quality or performance characteristics of an existing product to match competition or respond to perceived market needs. It can also be involved in developing new uses to extend the market for existing products or in producing completely new products to open up new market outlets or strengthen a position in current markets.

Chemical products can be broadly classified into those sold as chemical entities to a chemical/physical property specification and those sold on performance in given applications and for the effect which is produced. Examples of the former are ammonia, ethylene, sulphuric acid (large-tonnage commodity chemicals), and aspirin, citric acid, Vitamin A (small-tonnage, fine chemicals). The latter category includes plastics, resins, elastomers, synthetic fibres (large-tonnage, pseudocommodities) and cleaning formulations, catalysts, crop protection chemicals (small-tonnage, speciality formulations). Products in the first class are essentially the same from any manufacturer and it is in products of the second class that most of the product R&D work is carried out. Products in the pseudocommodity and speciality chemical classes are sold on end use performance and each manufacturer's product will be different. Product R&D to improve the properties of the existing product range or to develop new products to meet customer needs is extremely important in maintaining market position. Product-oriented R&D is usually more manpower-intensive than process R&D. New compounds and new formulations are required to satisfy customer demands and stay ahead of competitors. Strong innovative groups are needed who know the end-use business as well as the customer, and can combine this with knowledge of product properties to produce combinations and formulations to match the needs of the customer.

Research aimed at completely new products or processes is a long-term activity since it covers the initial exploratory screening of promising leads, through the definition, optimization and scale-up steps, to the commercial production level. A considerable amount of research in industry has a much shorter time-span and is aimed at achieving improvements in the operating efficiency of existing processes or quality improvement in current products. There is also 'trouble-shooting' work undertaken to overcome malfunctioning of an operating plant and restore performance and profitability to normal levels.

In an industrial research department there will be a range of project at differing stages of development being worked on at any one time. Exploratory teams will be searching for new catalysts, new reactions, new products to feed forward. Product and process teams will be carrying out process definition, scale-up, and product optimization work to provide data for the process design and engineering groups who will design and build the production units. Other R&D teams will be working to improve the performance of existing operations either in terms of reduced raw material and energy usage to improve profitability or improved quality of product to stimulate sales. The aim of the R&D management is to maintain a balanced portfolio of projects ranging from long-term exploratory activities with relatively low probability of technical success to short-term problems with high probability of achievement. A range of activities must be maintained to give a flow of topics through the department and meet the R&D targets which will be set as part of the corporate plan. Decisions taken on new products and processes to be worked on and developed are of little value unless they are intended to lead to capital expenditure and construction of a production unit if successful. Since these decisions imply a basic commitment of the future direction of the company they must have the approval of the top company management and line up with the corporate objectives in maintaining the continuing profitable existence of the company.

4.3.4 Evaluating results of R&D

Since R&D is a business, operating in an overall business context, the project activities within the department must be critically evaluated at various stages during development to see that they still meet the desired objectives. Evaluation of a project at any early exploratory stage will, of necessity, be crude since little information is available. However, the financial implications of such a preliminary evaluation will be limited, possibly to no more than a three-month continuation of exploratory work. As more information becomes available a detailed level of evaluation should be undertaken before committing funds to expensive process and product development with the possible need for capital expenditure on semi-technical or pilot-plant units. In most industrial research activities the problems under investigation are capable of technical solution if sufficient effort and money are expended. The major risks and uncertainties in industrial R&D projects are therefore commercial rather than technical. However, the staged evaluations of the project which are carried out must aim to ensure that the costs of developing and implementing the technical solution will result in a production operation which will provide an acceptable profit. There is no commercial value to a company in developing (say) a direct oxidation of benzene to phenol if, because of low conversions and extreme conditions, the resulting production unit is so expensive that the product cannot be produced to compete on price with the current cumene-based route. This is not to say that exploratory work to find an improved catalyst which could lead to a competitive processes should not continue. The problems of selection and evaluation of R&D projects are complex and require detailed study. A brief look at the evaluation of an R&D project appears in section 5.11 but project selection will not be considered further here.

A major problem in a research department is maintaining a balance between long- and short-term activities. A long-term project has greater uncertainty due to the problems of assessing the technical and commercial situation far into the future. Money spent today will not produce profits until many years hence even if the project is successful, and in periods of high inflation and recession the present value of such future profits is greatly reduced. Even where a research programme in the past has resulted in a profitable production unit, it is difficult to determine with any certainty how much of the profit is due to the research activity. The profit may result from a novel process design reducing the capital cost of the plant, clever purchasing of raw materials at advantageous prices, or efficient, aggressive marketing, and only partly from the original research work which resulted in the process. The value of long-term research is therefore very difficult to quantify and there is a tendency to concentrate on short-term activities, such as improving performance of an existing unit. In this case the time taken to implement the results of the research will be short and the results more easily quantifiable in terms of increased output or reduced product costs. A cost:benefit analysis to determine the value of the research is then readily carried out. However it is generally the long-term projects which offer the possibility of new products and processes which could fundamentally affect the progress and long-term viability of the company. A prime objective of R&D management is therefore to achieve the optimum balance of projects with the available resources of manpower and within the available budget.

4.3.5 Financing R&D activities

The question of how much should be spent on R&D activities is a major problem in a high-technology industry like the chemical industry. In an industry where products and processes are subject to rapid improvement and replacement, a company with an inadequate research commitment will quickly find itself at a competitive disadvantage. Although processes and operating knowhow are available for licence, a company which has failed to invest in its own research activities cannot always buy the information which it needs. There is an increasing tendency for companies to want a reciprocal exchange of technology with an agreed licence fee to give a balance of benefits, rather than a one-way transfer of technology. Furthermore, even if process technology is available for licence, the prospective purchaser needs some level of competent research activity to be able to assess the technology on offer. Research is therefore not an expensive overhead luxury to be cut back at the first sign of recession, but is an integral part of the long-term activity of a profitable company and as such cannot be considered in isolation. There is unlikely to be a shortage of problems which would limit the amount of research done by a company nor, under current conditions, is non-availability of qualified staff a probable limiting factor. One common limiting factor on research is finance to carry out the research and (even more demanding) to provide the capital to exploit the R&D innovations. The rate at which the company organization can change to absorb the results of R&D innovation can also be a limiting factor.

The amount of money to be spent on R&D activities is a matter of judgement and depends on the type of business involved. Within the chemical industry the need for research and the amount spent varies over a wide range. At one end is the heavy basic industry where product and technology change slowly and spending on research activities will be low. The other extreme, where innovation is rapid and products are subject to intensive competition and may be displaced rapidly is exemplified by the pharmaceutical industry and requires major investment in research. Table 4.1 shows research expenditure for a number of companies in $m and also expressed as a

Table 4.1   Range of research expenditure 1987

Company	R&D Spending ($m)	% of sales
Du Pont	1162·0	6·3
Hoechst	1412·2	6·0
Bayer	1464·1	6·2
BASF	1028·4	4·0
ICI	870·3	4·1
Dow Chemical	670·0	5·0
Union Carbide	159·0	2·3
Ciba–Geigy	1318·0	10·6
Glaxo	394·0	11·2
Shell Chemicals	291·0	2·5
Grace	107·3	2·4

percentage of sales. Figures are largely from Chemical Week and C&EN[2a], and illustrate the range (of % on sales) spent on research. Spending on R&D activities clearly depends on the nature of the company activities but even companies apparently in similar business areas have very different levels of expenditure. Spending on R&D has been increasing in recent years both in $ terms and as a % of sales and the trend appears to be continuing in 1988 and into 1989.

The money available for R&D comes from the company pool of 'overhead cash' provided by a charge on existing production activities. How much will be allocated to R&D will depend on company policy and will result from discussion and compromise between the various groups which are funded by overhead charges. Ideally the budget should be linked to the long-term objectives of the company; in practice the budget will be subject to short-term effects and depend on the total money available. No predominant criteria for determining a research budget are apparent but Twiss[3] identifies five possible bases:

   (i) Interfirm comparisons
  (ii) Fixed relation to turnover
 (iii) Fixed relation to profit
 (iv) Reference to previous expenditure
  (v) Costing of an agreed programme

No one basis will be completely satisfactory and the final budget will contain some influence from most of them. The level of activity by competitors must also be taken into account and may influence the final budget. Research is a labour-intensive activity and although a research team can be broken up relatively quickly (albeit painfully) in hard times the rebuilding of an effective group cannot be so easily accomplished when profits improve. One aim of a research budget should therefore be to avoid major fluctuations in workload and provide some stability in manning levels.

4.3.6 Links with other functions

Some emphasis has been placed on the view that R&D activities should not be considered in isolation and brief mention will therefore be made on R&D links with other functions in the company. From what has been said earlier it is apparent that close links should exist between the R&D function and the production function. The works will be the source of many of the short-term problems being studied in the research department, such as urgent work to resolve a plant trouble giving rise to output or quality problems. Other works-connected research activities will be more planned such as studies to improve raw material efficiency or reduce energy consumption in areas where study of standard cost sheets (see section 5.2) shows a worthwhile opportunity exists. Studies of this type are often accomplished jointly by teams working in the laboratories with some members seconded to the works. A temporary transfer of this nature can help the research worker to see the works problems in context and make him aware of limitations which may make an apparently attractive laboratory-developed solution to a problem unworkable in practice.

Links between the works and the process and product research activities are also important since the former will be the ultimate users of any process successfully developed. The involvement of experienced production staff in the process development team before process and design have been too closely defined can be highly beneficial. Incorporation of hard-won experience on safety and process operability into the team can result in a process which can be brought on line more smoothly and operated with improved safety margins.

The research department should also have close, effective links with the marketing organization in order to carry out the R&D function of looking to the future of the company. Problems with product quality, potential new uses for products, market needs which might be filled by product modification and such information obtained in marketing-department contact with customers can have a profound effect on R&D activities[4]. Many studies on research activities and successful innovations have shown that close contact with the marketing function is an important factor in successful innovation. Conversely, lack of contact with the market was a feature of many of the unsuccessful cases studied. The 'Sappho' project[5] and similar studies[6] show that attention to and understanding of user needs and attention to the needs of the market are distinguishing features of successful R&D activities.

Both the functions considered—production and marketing—are, of course, key parts in the organization of a chemical company, or any other manufacturing company, and the individual organization of these groups will be discussed.

4.4 Production organization

The production activities in a large chemical company form part of the operating division structure, and the divisional manufacturing units are combined to form 'works' each of which involves the operation and management of a number of chemical plants. Scale of production in chemical plant ranges from a few tens of tonnes to hundreds of thousands of tonnes annually. Chemical processes may be operated in batch mode using equipment—reactors, filters, stills, condensers, tanks—which can be re-arranged to produce a variety of products, rather like large-scale laboratory glassware. Other processes are operated in large, complex, specifically design-ed plants which run continuously, 168 hours per week for weeks at a time, to produce one product, possibly with some co- or by-products, on a very large scale. In either case the ultimate aim of the works management team is to produce the required quantity of product, of the desired quality and as efficiently as possible, i.e. at minimal cost.

4.4.1 Management structure

A chemical works is a complex system involving buildings, plant and machinery, raw materials, energy, and human inputs and product outputs which must be carefully managed to achieve the desired targets. The plants must be operated efficiently, equipment maintained to a high standard and processes improved where practical, to produce the required outputs safely and to minimize environmental impact.

Plants in a chemical works are generally managed by chemical or chemical engineering graduates, although non-graduate staff may sometimes be responsible for the management of a well-established process. In a large chemical works a number of plant managers will respond to an intermediate level of management—a section- or product-group manager and, in turn, the intermediate managers respond to the works manager. This classical 'line management' structure provides for the delegation of authority from the works manager through to the plant manager and process operation staff. Other activities in the works cut across the line structure and serve a general staff function, such as quality control and analytical services, safety services, labour and personnel and so on. Engineering services in the works may be organized in a line structure, analogous to the process management structure, if specialized techniques are involved in the different product groups. Alternatively the engineering and maintenance staff can operate as a central pool of skilled manpower serving the whole works.

A production management structure is shown in Fig. 4.5.

The broad production plan for the works will be developed by the works manager in consultation with the marketing, purchasing and distribution departments. The plan will also take into account information on expected availability and performance of the plants on the works fed back from the plant manager through the line structure. Production requirements to meet the plan will then be passed down the line from the works manager to the individual plant managers. Since the plant manager carries the ultimate responsibility for supplying product to meet plan requirements it is worthwhile taking a closer look at his role.

4.4.2 Plant management and operation

A production plant in a chemical works manufactures product by a chemical process requiring controlled conditions of temperature, pressure, catalyst, feed rates, etc., for its successful operation. The plant manager must understand the chemistry of his process and the effect of changes in process conditions. He must also know his plant equipment and the constraints which this imposes on his ability to make changes. However, being able to handle the process

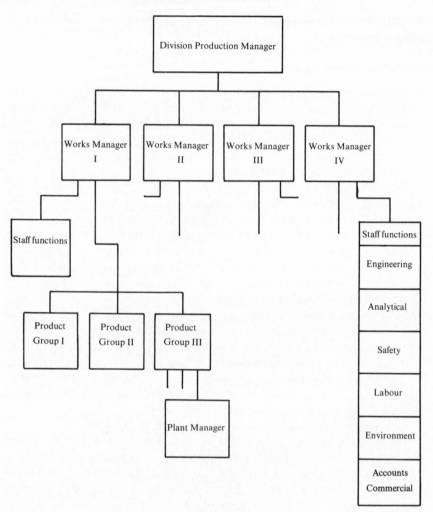

Figure 4.5 Production management.

chemistry is only part of the plant manager's job and much of his time is spent on non-chemical matters. The efficient, safe operation of a chemical plant—no matter how sophisticated the measurement and control devices—depends finally on people and a large part of the work of a plant manager involves dealing with and communicating with people. Vertical communication involves the plant manager discussing plant operation with the process shift team and passing information on the plant performance to his group manager and works manager. Similarly, information on short-term changes in the production plan is fed down to the plant manager and discussed in terms of changes required in plant operation with the process operators. The increasing sophistication of chemical plants has brought the need for process workers with more knowledge of the science behind the chemical process and the control of the units in the plant. There are now technical courses available which provide training and qualifications for process operators and it is part of the plant manager's job to ensure that his process team are fully trained and kept up to date.

The plant manager must also liaise across the line structure with other plant managers to ensure that their needs for resources do not conflict. Discussions with analytical staff on product quality, supply department on raw materials and marketing department to balance plant output with stocks, storage capacity and expected sales also form a significant part of the job. Liaison with the works engineering staff is also an important part of the plant manager's activities. The planned maintenance of equipment before breakdown can avoid considerable upset to the plant operation and process efficiencies. Major plant shutdown to change catalyst, or repair equipment which can only be worked on when the plant is off-line, must be organized well in advance to ensure that the materials and manpower are available and reduce the off-line time to the minimum. Despite all the planning, equipment breakdown will still occur and the plant manager must work with his engineering colleagues to see that such events are dealt with rapidly.

In addition to all these people/communication activities the plant manager must also monitor his process efficiencies and costs. The plant record sheets show the consumption of raw materials and services such as steam, electricity, cooling water and fuel oil and the make of on-specification product. Usages of these inputs per tonne of product must be monitored against the standards set by process design or known efficient operation; any significant discrepancy will be investigated and remedial action initiated. Product cost—at least at the plant cost level over which the manager can exert some influence—must be compared with the standard cost (see Chapter 5) to see that (say) analytical costs are not creeping up or steam being reduced at the cost of increased usage of expensive fuel oil.

A very important feature of the work of a plant manager, and the works management team in general, is concerned with the safe operation of the plants on the works. Plant inspections, checks on control systems, use of safety equipment and close cooperation with the works safety staff form part of the activities involved in maintaining a high level of awareness of safety matters in all plant staff.

The plant manager and his senior managers continually seek to improve the performance of the plants on a works. This will involve consultation with the works process improvement team and liaison with R&D staff to balance the conflicting requirements of production and experimental programmes. Any work proposed for experiments on the plant (modified reaction conditions, new catalyst, detailed sampling study) to confirm laboratory studies will almost inevitably lead to process upsets and some loss of production. The current need for product must be balanced against the short-term loss of output and potential longer-term process improvements. This summary of the factors involved in the operation of a chemical plant emphasize that, at all levels in the management structure, a knowledge and understanding of the process chemistry is only one of many skills needed.

4.4.3 Engineering function

It will be apparent from what has been said above on the process operation side of the chemical works that the engineering staff have a vital part to play in the efficient operation of the plants. Most of the work of the engineers in the chemical factory is concerned with the maintenance and repair of equipment, with plant extension and modification work forming a relatively small part. Since the maintenance activities are involved with all plants and equipment on the works the skilled personnel involved (machine fitters, pipe fitters, instrument fitters, electricians, welders and so on) are usually combined in a central works engineering group. On a large site, senior engineers—usually graduates—will supervise each group, e.g. there will be a machines engineer, instrument engineer, electrical engineer etc. There will also be a graduate engineer associated with one or more plants and responsible for maintaining the mechanical equipment on those units in good order, and working closely with the plant managers and function engineers in the planning and implementation of maintenance work. A number of plant engineers will respond to a section or group engineer who in turn will respond to a works engineer. The maintenance functions on the works operate in a closely-linked matrix organization combining a line structure, parallel to the process management line structure, combined with an engineering-skill-oriented structure operating across the works as a whole.

The smooth, efficient operating of a chemical works is critically dependent on the cooperative working of the process management and engineering management staffs. As with many aspects of industrial life, the ability to work with and through other people and develop effective personnel management skills is as important as technical ability in the works environment.

4.4.4 Links with other functions

Clearly the chemical works must have close links with many functions in the operating division to ensure that the products made contribute fully to the profits of the organization. Supplies of major feedstocks, catalysts and auxiliary chemicals are organized through the division supply function, and movement of product—by pipeline or suitably loaded in drums, sacks, road tankers or ships—arranged with the distribution department. Contact with the R&D function is also important with regard to process improvement and short-term trouble shooting to uprate plant performance. There is also a need for involvement with product and process development work as previously indicated to provide an input of operating experience and highlight potential problems at an early stage.

Links between production and marketing staff are vitally important at all levels in the management structure. At a senior level close liaison is required in developing the annual production/marketing plan and ensuring that any major fluctuations in demand can be accommodated by storage or high-rate production. For the longer term the works manager and his marketing equivalent must balance the long-term sales and market forecasts against production capacity to begin the long process of obtaining authorization, planning and building new capacity, in good time.

The individual plant manager must also keep in close touch with his colleagues in the marketing field to see that his stocks and current production rate can meet short-term demands. He must also be aware of any significant short-term changes in planned sales, arising from an urgent order (or delayed order) which could affect his production programme. Planned shutdown of the plant for maintenance work must also be covered by building up stock to meet anticipated orders. The marketing department must also be informed of such planned shut-downs well in advance so that deliveries to customers can be rescheduled if necessary and no new commitments for sales are made for the critical period.

This review of the structure and functions of the production activities in a chemical company illustrate the diversity and importance of the production function. Since the operating division, and indeed the company, primarily exists to make profits by selling chemicals, the efficient production of those chemicals may be regarded as the drive cog in the company machine. However, as with all machines, smooth meshing with other cogs is required to convert the drive into the desired machine output.

4.5 Marketing

As previously indicated the chemical industry produces a wide range of products, at scales ranging from a few tonnes each year for pharmaceuticals and some dyestuffs intermediates up to hundreds of thousands of tonnes each year of ammonia and ethylene. Most chemical sales are made to other manufacturers who will use the chemical product as feedstock (or additive, solvent, etc.) in producing another material which may be sold to the consumer or may also be an intermediate and undergo further processing. Consequently the function of marketing in the chemical industry, where the customer is technically knowledgeable, is very different from selling consumer products.

4.5.1 *Role of marketing*

In common with the marketing functions in other industries, the job of the marketing manager in the chemical industry is to maximize profits by making an optimum level of sales at the highest possible prices, bearing in mind the competition and company strategy. It may be company strategy, for example, to hold price levels down in a particular market to discourage competition or to encourage market growth. For chemicals which are produced on a large scale, or which require very complex processes for manufacture, a massive investment of capital for the production unit is required. In addition the effort of numbers of technically qualified people to manage the processes and provide technical service support to the customer is needed. As a result only large companies can undertake such activities, and for many chemicals manufacture and sales are shared between a few large producers. For many large-tonnage chemicals and petrochemicals the products from the different manufacturers are competitive on quality and price. The marketing manager must then seek other ways to obtain an order for his company rather than lose it to a competitor. Such factors as improved methods of delivery (e.g. molten product in tankers rather than flake in bags, to reduce customer's handling costs), ready availability of product and speed of delivery from local storage, and improved technical support, become important in securing an order. The marketing manager can also join forces with colleagues in other operating divisions of the company to offer a package of products to satisfy the needs of the customer.

The marketing manager must be aware of the need to recover all the direct and indirect costs of production (absorption costing) from sales in his major markets. He must also know the marginal production cost for his products so that he can grasp opportunities to make additional sales which produce an income higher than marginal cost and thus make a profit contribution (see section 5.10). However, the possible rebound effect of such marginal sales on prices in major markets must not be overlooked, and the possible depression of overall price levels following a small increase in total sales could be counterproductive.

Organization in the marketing department is usually a conventional line structure with an individual marketing manager responding to a group manager and a number of group managers linking to the division marketing manager. A marketing manager will have responsibility for one or more products depending on the complexity of the market and the number of customer accounts to be dealt with. The group manager will hold overall responsibility for a number of related products and act as a communications focus for the separate product managers and their link with higher management.

The availability of specification product to meet orders is very important and part of the job of a marketing manager is the development of production/sales balances. Marketing department, in collaboration with the works staff, produce both short-term and long-term plans and forecasts which form a basis for the optimal balancing of expected demand with production and stock levels.

4.5.2 Short-term sales plans

Short-term forecasts, for up to 12 months ahead, contain detailed sales forecasts based on close contact with the markets, and expected production plans and stock holdings. Product and stock levels contain allowances for planned shutdown of plant for maintenance work and take into account periods of limited production due to process restrictions (pump failure, partial reactor choke, etc.) to be expected based on past experience. In the case of multi-product plants, campaign lengths to match stocks and sales of individual products and lost production when changing product must also be taken into account.

Selling price and production cost forecasts are also included to provide a profit forecast and give early warning of remedial action which might be needed (improve process efficiency, reduce steam consumption, increase sales) to maintain an acceptable profit margin. These short-term production/sales plans will be used as a basis for plant operation on the works; unexpected changes (loss of catalyst activity reducing output, order from new customer) will be accommodated by regular formal meetings and by frequent informal contact between the marketing manager and his production management colleagues.

4.5.3 Long-term sales plans

Long-term forecasts, covering a period of 3–5 years, are used in the planning of long-term strategy for the operating division. Sales levels are based on analysis of individual markets and forecast growth rates over the period. These are balanced against production capacity and forecasts of availability of feedstock to show the long-term position.

Sales prices and cost forecasts are again needed to show that the product will continue to make a profit contribution in the long term. If the long-term balance indicates overcapacity for a product the marketing manager must take action to increase sales by seeking new applications, developing export markets or winning a bigger share of the existing market from his competitors. When a shortage of capacity is anticipated, consideration should be given to increasing output by modifying existing plant or building new capacity. In the latter case, with new plants taking several years to plan, design and build, it will be necessary for the marketing department to produce sales tonnage and price forecasts for ten or more years ahead. This 'crystal ball gazing' is required to show that the long-term profitability of the product is sufficient to reward the proposed capital expenditure. Producing the market support case for new capital expenditure based on long-term forecasts forms part of the division or company long-term strategy. It forms an essential part of the process of ensuring that the company will be equipped to meet sales demands in the future.

4.5.4 Market R&D

Another area of activity in which the marketing department must be future-oriented is in market research and development. Selling chemicals into technically-based markets requires an understanding of the customers' processes and his problems. The close contact with the consumer gives the technical salesman and marketing department a picture of the current needs of the customer and what it would be worth to him to have those needs satisfied.

Information on market needs and the analysis of possible market opportunities provided by the marketing department forms an important input into the R&D area of the company. Efficient transfer of market analysis and the information obtained as a result of customer contact can help direct the company R&D activities into the most relevant areas. Similarly when a new product has been produced by a company or division research activity the marketing manager and technical sales force can assist the R&D product development manager to make contact with potential users of the new product. The combined efforts of the marketing manager providing information on the commercial advantages of the new product and the product development manager on the technical merits can stimulate the customer's evaluation of the product and the development of market outlets. Growth of the market for a new product depends on the perceived need for the effect it produces, the price which a customer is willing to pay for that effect and the alternative solutions to the problem which are available. Information obtained by the marketing department on all of these factors and relayed into the producing company's R&D organization can play a vital part in the development of new products and in the development of market outlets for those products.

4.5.5 Links with other functions

The efficient functioning of a marketing department, and its influence on the division or company of which it is part, clearly depends on close links with other parts of the organization. The importance of the links with production staff and with R&D has already been mentioned in this section and earlier in the chapter. It is worth stressing again the importance of close contact between marketing and R&D since this is an area which is sometimes neglected. The very different orientations and functions of the departments, (customer/commercial—science/technical) and the different types of people who are attracted by the nature of the work in the two departments tend to make such contact difficult. Nevertheless, for the efficient operation of both functions, and the company as a whole, close contact and cooperation must be encouraged and maintained.

Having considered how the company is organized to develop, make and sell its products let us consider another important activity in the chemical (or any other) industry, namely raising the money to finance these activities.

4.6 Sources of finance

The chemical industry is a capital-intensive industry. Chemical plants are built of expensive materials to withstand the processing conditions—corrosion, high temperatures, high pressures—frequently encountered. Modern plants are also highly instrumented and automated to improve process control and efficiency and to provide better safety margins. Of the economist's 'resources for production'—labour, land and capital—the labour requirement for most processes operated in the chemical industry is relatively small and labour costs form only a small part of the total cost. Land requirements by the chemical industry are also relatively minor in terms of area. However, an appropriate site is important and a site is required with adequate facilities for transport (road and rail) and shipping (for bulk imports and exports) and facilities for handling and safely disposing of liquid and gaseous effluents. Land-related inputs—crude oil, phosphate rock, sulphur, etc., are important to the chemical industry but represent a fairly small percentage of total land usage. The resource which the chemical industry uses most is capital—the money to buy and install capital equipment to manufacture profit-generating products. These capital requirements of the chemical company can be provided in a number of ways at varying cost to the company raising the capital.

Basically a firm can provide its new capital requirements in two ways—from internal sources generated by current activities of the company, or by going to the financial market and raising money externally.

4.6.1 Internal sources of finance

To finance a capital project (such as a new production unit) using internally generated funds, some of the profits from current activities can be retained in a reserve fund instead of being distributed to shareholders as dividends. A further internal source of capital for new projects is the money set aside as the depreciation charge on existing capital assets. Use of the depreciation cash in this way carries out one of the purposes of the depreciation charge (i.e. the maintaining of the capital asset base of the company) on a continuing basis by building new capital assets. These internal sources of finance—retained profit and depreciation charge, particularly the latter—represent some of the major sources of capital for new projects in the chemical industry. The cost of such internal finance to the company is derived from the cost of the alternative uses to which the money could be applied. Possibilities are the interest rate which could be earned from a bank for short-term deposit or the net interest saved by reduction of overdraft or short-term loan borrowing.

4.6.2 External finance

In raising money from external sources the company is obtaining money to finance current capital projects by promising to repay a larger sum (the borrowed capital plus interest payments) at some time in the future. The interest rate at which an investor can be persuaded to give up current funds for future spending power is determined by many factors. A company is prepared to pay more in the future for current funds because it expects to be able to invest the borrowed capital and earn sufficient to pay back the loan plus interest charges and show a profit for its own use.

The money supply from which the company aims to obtain funds is provided by the private investor and, increasingly, the large institutions such as pension funds, insurance companies, banks and the like. Since the institutions are in turn handling the funds of large numbers of individuals— held in their pension funds, paid in insurance premiums or held in bank deposits—the individual investor represents the final source of finance for private companies. Conversion of his cash into investment in a company either as loan stock or a shareholding represents a loss of flexibility for the investor. There will be a delay and some cost involved in reconverting the investment back into cash which can be used to satisfy some other need of the investor, such as a new car or deposit on a house. There is also some uncertainty over the complete recovery of the original money purchasing power. These factors result in the necessity to offer an incentive to a potential investor in terms of an interest rate or share in expected profits to overcome the loss of present purchasing power. Assessment of the level of incentive needed—the loan interest rate or share dividend expectation—is also affected by the demand for capital by companies trying to raise money. If the total amount of money available is less than the total of all company needs, then interest rates must be increased by individual companies to obtain their needs from the limited supply available.

4.6.2.1 Loans and loan stock. If the firm decides to raise part of its new capital requirement externally it can do so on a short-term or long-term basis. Most short-term borrowing is carried out through the banking system and interest is charged at a rate about that of the prime rate or bank rate. In Britain such short-term borrowing is usually based on an overdraft facility which represents an extremely flexible method of borrowing. A bank will agree with a borrower an acceptable 'credit line' or maximum amount which can be borrowed and the company can then draw on this at any time up to the limit. Interest is charged on the actual amount borrowed at any time—on a daily or weekly basis—so the borrower does not pay for any unused part of his credit.

Short-term borrowing in this way is not really suitable for raising capital for a plant construction project as the bank can call in the loan or overdraft at any time. This represents the chief security of the banks' position in that if a company shows signs of running into financial problems the bank can quickly withdraw its loan facility and obtain repayment. This would obviously exacerbate the financial problems, and to avoid this difficulty a company should have a significant proportion of loan capital on a longer-term basis. Short-term loans or overdraft facilities are generally forms of bridging finance providing flexibility for the borrower.

Long-term borrowing is usually achieved by raising a loan from the investing public by sale of loan stock. Interest payments by the borrowing company on such stock are deducted from company income before declaring a taxable profit. The loan stock can be undated, dated or convertible. By selling undated loan stock the company is usually committed to paying a fixed rate of interest on the money raised. The rate of interest must be sufficiently high to attract investors and is payable as long as the company is in business. The investor cannot recover his investment from the company—unless it becomes bankrupt, when he will receive part of the distribution on winding up the company. The investor can, however, sell his undated loan stock to someone else and the value of the stock will be determined by the change in interest rates since the original investor bought the stock. If interest rates have risen the stock value will be reduced since the prospective buyer can get a better return elsewhere. Thus £100 undated loan stock carrying 6% interest would be worth only £75 when competitive investments offer 8% interest.

If a company issues dated loan stock it guarantees payment of the fixed interest on the borrowed money until the maturity date when the borrowed money will be repaid. With this stock the investor is certain to receive full repayment—unless the company goes bankrupt—so the selling price of dated loan stock to another investor is less directly linked to interest rates and does not fluctuate quite as much as undated stock.

Convertible loan stock carries a fixed rate of interest but at one or more specified dates the investor is offered the opportunity to convert his loan stock into ordinary shares in the company.

The aim of a convertible loan stock is to provide an initially attractive fixed income with the option to convert into ordinary (equity) shares, on advantageous terms, when earlier promises of profit increases are being realized. If the improved profits, and hence dividends on ordinary shares, have not been achieved the lender can continue to receive his fixed loan interest.

A company issuing loan stock is obliged to pay the quoted interest rate whether or not a profit is being made. Hence such stock represents a safe form of investment to the lender in terms of guaranteed income. However, no matter how much profit the company produces, the stock holder gets only his quoted interest rate. Interest rates generally have increased over the last 20–25 years and holders of undated stock in that period have lost badly. Losses have been incurred because of the lower value of the stock due to higher interest rates as indicated above. In addition, the effects of inflation have reduced the value of the money the investor receives from the sale of his stock so that his purchasing power is even further reduced.

4.6.2.2 *Shares*. As an alternative to raising money by sale of loan stock a company can sell shares to the public—either preference shares or ordinary shares (the latter are also known as equity shares). Preference shares, like loan stock, carry a fixed rate of interest or dividend but in contrast to loan stock the interest is only payable if the firm makes a profit. Buying a preference share involves the investor in a greater risk than purchase of loan stock so the preference share will have to offer a better rate of interest. In the case of so called 'blue chip' companies (such as Du Pont, Dow, ICI, Shell in the chemical industry) where the expectation is that the company is likely to continue making profits for the foreseeable future, then a preference share would need only a slightly higher rate of return than a loan stock.

The final method by which a company can raise money is by the issue of ordinary or equity shares in the business. By purchase of equity shares the investor notionally buys part of the company and becomes a part owner. Buying loan stock does not carry this status and the investor in such stock is an outsider lending money to the company. As a part owner the holder of equity shares is entitled to receive a dividend out of the company profit at the end of the year.

Equity capital (including retained reserves) accounts for more than half of the total capital for most companies—in the case of ICI the 1987 Annual Report shows 59%. Although holders of ordinary shares are entitled to a share of profits they can only receive this after all creditors, loan interests and preference share dividends have been paid.

Interest on loan stock is a charge on the company which is paid before profit is calculated, but the preference share holders have first claim on their interest rate out of the declared profit. The ordinary share holder is therefore in the position of highest risk—in times of low profit (after paying loan interest) the preference share holder has first claim and there may be no dividend declared on ordinary shares. Conversely, in good times with high profits, the preference share holder receives only his fixed rate and dividends on ordinary shares will be high. Because of the high risk involved, the average return on equity shares has been above that for other sources of finance, particularly government loan stocks—so-called 'gilt-edged' stocks. However, in periods of rapid inflation investors have decided that a share in a company, which would increase in value in money terms as inflation reduced the real value of money, was preferable to a fixed interest stock. As a result yields on ordinary shares have sometimes been lower than yields on fixed interest stocks.

4.6.3 *Financial structure*

A company will have its total capital funded by a combination of most of the sources of funds outlined above. The average cost of capital to the company will be the weighted average of the components of the total capital. This figure can be used as a minimum return which a new investment must earn in order to be considered acceptable.

The composition of the company's financial base—i.e. the relationship between loans and shareholder's funds—is referred to as the gearing of the company. Gearing is the ratio (as a percentage) of loans and other fixed interest capital to total capital invested. A 'highly geared' company has a high proportion of its invested capital carrying an obligation to pay interest whatever the profit earned. The residual income, which 'belongs' to the ordinary shareholders, then fluctuates considerably for relatively small changes in 'profit-before-interest charges' due to the high fixed-interest load.

Company gearing ratio is one of the factors which a would-be investor will consider when studying a company. A number of other financial ratios are important in assessing a company's credit base and suitability as a reliable source of income, particularly to large institutional investors. One such ratio is the 'current ratio' which is the ratio of current assets to current liabilities. Current assets are items such as stock holding, cash, debtors (accounts to be received) which can be converted to cash within a short period. Conversely, current liabilities must be cleared in a short period (usually within a year) and include trade creditors, tax liabilities and short-term loans. This ratio shows the cover available on existing short-term liabilities and is of interest to suppliers and short-term lenders. Another ratio of interest is the long-term debt/net worth at book value and gives a measure of the extent to which the assets could fall below book value when sold but still cover debt.

The information needed to assess the financial strength of a company can be found in the annual report with its 'profit and loss' and 'balance sheet' statements. The former presents the net earnings of the company during the financial year and how it is split between loan interest, taxation, reserves and shareholders. A balance sheet provides a statement of the assets which form the capital base of the company and the sources of that capital at the end of a financial year. The two accounts, balance sheet and profit/loss, are complementary showing the financial situation at a given time and the change during a given period respectively.

A study of the intricacies of these financial accounts is outside the scope of the present chapter and discussion of the principles of accountancy and their applications should be sought in the appropriate texts.

4.7 Multinationals

A significant feature of the chemical industry is the international or multinational character of the activities of many (if not all) of the major companies. It is interesting to consider how this has come about and what factors have been influential in leading to the multinational development of chemical companies.

4.7.1 Growth of multinationals

The growth of multinational companies is often regarded as a post-World War II phenomenon and, during the late 1950s and 1960s, was thought to be an American threat to the rest of the world[7]. The increasing presence of American-produced goods and American-owned manufacturing and mineral extraction companies gave rise to fears of loss of economic control and sovereignty in the governments of many small (and some not so small) countries. As a result in some countries—particularly France and Italy in Europe—small producers were encouraged, by government actions, to merge and form combines with larger companies to form a unit strong enough to match the threat of the multinationals.

However, even a cursory study of available historical data shows that multinational manufacturing and mineral exploitation companies have been a fact of industrial life for over a century, particularly in the chemical industry. The development of such activities is also seen to be by no means an exclusively one-way American-dominated process. Companies in the chemical industry were among the pioneers of multinational manufacturing activities and many had established foreign manufacturing subsidiaries before 1914. Studies of the history of the dyestuffs industry show that major European companies—particularly German and Swiss producers such as BASF, Bayer, Hoechst, Ciba-Geigy—all had numerous manufacturing units outside their national borders prior to 1914. Nevertheless, in more recent times American firms were predominant in establishing foreign manufacture in the 1950s and early 1960s although British, Western European and Japanese activity increased markedly from the early 1960s. ICI for example set up its European Council in 1960 to form the base for expansion into continental Europe, although the company had been, to some extent, multinational on being formed in 1926 with interests in several (then) Empire countries. ICI now has factories in more than 40 countries and sells in more than 150[8]. Other major chemical companies such as Du Pont, BASF, Hoechst, Bayer, Dow and many more have comparable worldwide activities.

4.7.2 Reasons for development

There are many and varied reasons for the development of multinational activities but within the chemical industry the process seems to have followed a basic pattern. Most chemical companies started the process of becoming multinational concerns by initially developing exports based on a strong position in innovation and technology. Having established a market position and developed a sales and possibly technical support organization based on export, the subsequent step to retain and protect that position is local manufacture. Again, studies on the early history of the dyestuffs industry can be used as an example. In the late 19th and early 20th century German and Swiss chemical firms raced ahead with process and product innovation and commercial production of dyestuffs. This dominant position enabled the companies to export the dyestuffs products throughout the world and then consolidate the situation by manufacture in the most important markets[9]. The international development of many American chemical companies in the late 1950s followed a similar pattern of being export-led. The large U.S. market encouraged the construction of large plants for the new plastics and petrochemicals which were being developed. When home market demand was below capacity, low-cost marginal product could be used to develop an export market at a price which someone starting local production could not match because of the much smaller production scale involved (see section 5.8). Overseas production by the export supplier could then be started when the market justified a production scale with costs that could equal U.S. marginal cost plus freight and taxes.

The switch from chemical export from a secure home base to becoming a multinational company with foreign production facilities may be triggered in many ways. As indicated above, when an export market has grown sufficiently, local manufacture can match export cost including transport and taxes. In many situations foreign manufacture of a finished product could enable continuing exports of intermediate products. For example a large European or American company exporting Nylon 6 fibre to a less developed country could set up a fibre-spinning company in that country, but continue to export Nylon 6 chip. Later, as the market continued to grow, further spinning capacity and a caprolactam polymerization plant could be built overseas with export of caprolactam monomer from the parent company. Subsequent addition of foreign caprolactam monomer production could be based on phenol or cyclohexane exports by the parent. Thus by becoming a multinational producer of downstream products a company can find its exports of intermediate and upstream products stimulated.

In many cases the transition from exporter to multinational producer resulted from a threat to the existing market rather than an opportunity to develop. The threat could come from a local chemical company showing interest in a developed market and aiming to replaced imports by indigenous production. Alternatively the threat could come from the foreign government showing concern at the outflow of currency to provide imports or the dependence of a key section of the local economy on imports. Such concern could lead to import restrictions or high tariff barriers, coupled with tax or investment grant incentives to a local manufacturer. In either case the exporter's response to the threat could be local manufacture and a widening of the multinational character of the parent company.

As indicated above, the chemical industry has developed its multinational character ever since the early days of the industry—particularly the organic and dyestuffs side. This arises from the highly technological and science-based nature of the industry which means that introduction of a new product or process innovation can lead to a significant economic or market advantage. The term 'innovation' is used to mean the commercial introduction of a product or process rather than the making of the original invention or discovery of a basic reaction. Such innovation requires a major investment in expensive research and process development activities and the widest application of the technology to gain the maximum benefit from this investment. Some return on R&D expenditure can be obtained by licensing a technological development to a competitor. However, introduction of the new technology in the company's own production facilities in a large number of market areas should prove more rewarding.

As mentioned earlier, the development of multinational companies has not met with universal approval in the countries where subsidiary companies are based. Accusations have been made of high prices charged by a parent company for intermediates purchased by an overseas branch. This would increase the profit of the parent company and minimize the tax paid by the subsidiary to the host government. Similarly, labour unions have accused the multinationals of switching production to lowest-cost areas (lower wages, highest productivity, lower fixed costs resulting from lower safety or effluent standards) resulting in job losses in other countries.

Clearly when a chemical company becomes a multinational it collects many problems along with any benefits. However, the industry is likely to continue to develop its multinational character, particularly in such areas as joint ventures in resource-rich areas of the world.

References

1. P. R. Lawrence, and J. W. Lorsch, 'Organisation and Environment' Division of Research, Graduate School of Business, Harvard University, 1967.
2. *Chemical Week* 1988, (December) 33–39.
2a. *Chemical Engineering News* 1989, (January) 15–18.
3. B. Twiss, *Managing Technical Innovation*, Longman, 1974, p. 34.
4. W. F. Madden, 'Linking Research Activities with the Market of Opportunity' in *The Chemical Industry*, published for Society of Chemical Industry, 1981.
5. SAPPHO. Freeman *et al.*, '*Success and Failure in Industrial Innovation*', Centre for Study of Industrial Innovation, University of Sussex, 1972.
6. Langrish *et al.*, *A Study of Innovation in Industry*, Macmillan, 1972.
7. J. J. Servan-Schreiber, *Le Defi American*, Paris, 1967.
8. *ICI Annual Report* 1987.
9. D. Aldcroft, *The Development of British Industry and Foreign Competition* 1875–1914, University of Toronto, 1968.

Bibliography

A. J. Merret and Allen Sykes, *The Finance and Analysis of Capital Projects*, Longman, second editon, 1973.
B. Twiss, *Managing Technological Innovation*, Longman, 1974.
F. R. Bradbury and B. G. Dutton, *Chemical Industry–Social and Economic Aspects*, Butterworths, 1972.
L. G. Franko, *The European Multinationals*, Harper and Row Ltd., 1976.
Multinational Corporations, The ECSIM Guide to Information Sources, 1977.
Major Chemical Companies of the World, ECN Chemical Data Services.

CHAPTER FIVE

TECHNOLOGICAL ECONOMICS

D. G. BEW

5.1 Introduction

The basic function of any private company which aims to continue in business is to make a profit. Without this it cannot do other socially desirable things such as providing continuing employment, providing money for local social services through rates and local taxes and providing money for national activities through corporate taxes and through the taxes paid by the employees. In a planned economy, or a nationalized industry in the mixed economies currently operating in many countries, the profit motive is not the main imperative. Unprofitable activities may be continued for social or political reasons—to maintain employment in a depressed area or to maintain an industry to avoid dependence on imports in strategically important areas of manufacture. However, any such unprofitable operations represent a drain on the national resources and must be supported and subsidized by the taxes paid by profitable activities.

In the chemical industry, companies were founded, and continue to exist, to produce profits by making and selling chemicals. The products made are those for which a need has been identified and which can be produced for sale at a price which the consumer is willing to pay and which adequately rewards the producer. The processes which are chosen to manufacture the desired products are those which are believed to offer the best profit to the producer. They may not be the most chemically elegant routes to the products or the most scientifically interesting processes but in the context these factors are unimportant. In case it may be thought that the chemical producer operates his process to maximize his profits without concern for the effects on people or the environment it must be emphasized that, today, production of chemicals is governed by numerous regulations. Construction standards and operating safety of plant, the protection of the employees from dangerous levels of chemicals and the effect of effluents and products on the environment and people outside the plant boundaries are all subject to control by local and national legislation. These factors, particularly the control of atmospheric chemical levels within the plant and the safe disposal of gaseous and liquid effluents, can have a significant influence on the choice of process. They can also significantly increase both the capital cost of the plant and the cost of operating the process, as more equipment is required to meet increasingly stringent controls. The cost of operating a plant safely with minimum impact on the environment must ultimately be carried by the consumer in the higher product price needed to keep the producer in business. These aspects of chemicals manufacture are also considered in Chapter 8.

5.2 Cost of producing a chemical

The profit which the manufacturer obtains by producing and selling his chemicals can be measured in various ways as discussed later in the chapter. However, in order to know whether or not a profit is being made from an existing plant or product—or, in the case of a project at the planning stage, whether the new product would show an acceptable profit—we need to know the actual or estimated costs involved in producing the chemical. So let us consider how the cost of producing a chemical is built up. A considerable number of factors are involved in the production of a chemical, ranging from the supply and storage of the raw materials through to the storage and selling of the finished product. In between these steps is firstly all the complex and expensive equipment for carrying out the chemical processes, separating and purifying the product, and secondly the people who operate the plant and carry out maintenance work to keep the processes in operation. A simple way of combining information on these various cost factors into a useful economic model is the cost table. This presents information on an annual or 'per unit of product' basis and is useful for indicating the relative importance of the various factors which make up the cost. Table 5.1 shows an example for the production of cumene (isopropyl benzene) by reaction of propylene with benzene using excess benzene as solvent and phosphoric acid on a support as catalyst. Published information (*Hydrocarbon Processing* **55**, 3, March 1976, pp. 91–6) has been used in building up the table.

The reaction involves passage of propylene plus some propane diluent and an excess of benzene upwards over the supported acid catalyst at 230°C and 35 atm. Propylene conversion is high and the reactor product is flashed to recover propane plus a little propylene for recycle. Liquid product is then distilled to recover unreacted benzene, which is also recycled, and the crude product then distilled to give pure cumene and a small residue of fuel-value heavy ends.

For a process plant which is currently operating, the table can be drawn up using best data from plant records of operation under steady conditions and then used as a standard cost table to monitor subsequent operation of the unit. In the case of a new project being considered to manufacture a chemical by a new process under development, a comparable table can be drawn up using available information to indicate targets to be achieved to provide an economically attractive operation.

There are several ways of sub-dividing the components which make up the

Table 5.1 Cost build-up for cumene production

Scale of Production 100 000 tonnes/year—operation at full capacity

Operating costs	£000/year	£/tonne Cumene
Benzene 0·67 tonnes/tonne of product @ £310/tonne	20 770	208
Propylene 0·38 tonnes/tonne of product @ £305/tonne (propene)	11 590	116
Phosphoric acid catalyst + chemicals	140	1
Gross materials	32 500	325
Heavy end fuel credit 0·04 tonnes/tonne of product @ £55/tonne	(220)	(2)
Net materials	32 280	323
Energy inputs	710	7
*Plant fixed costs	1 150	12
Overhead charges	780	8
Depreciation (15 year life of fixed plant)	1 533	15
Works cost	36 453	365
Target return on total capital 10%	2 800	28
Required sales income (net of packages and transport)	39 253	393

Capital involved (early 1988 basis)	
Fixed plant	£23 million
Working	£ 5 million
Total	£28 million

*Plant fixed costs are: operating and maintenance labour and supervision, maintenance materials, rates and insurance, and works overhead charges.

full cost of a chemical; the method used depends on the purpose for which the information is required. Traditionally accountants like to divide costs into variable and fixed elements and this split is useful when considering the costs of an integrated works, an individual production unit or a single product. Variable costs comprise those factors which are only consumed (and therefore only charged to the operation) as product is being manufactured. As a result the total variable cost during an operating period—day, quarter, year—will vary directly as the plant output during that period, although the cost per unit of production will remain constant. Fixed costs are those charges which have to be paid at the same annual rate whatever the rate of production, in fact even if the plant is shut down for a short period for repairs, or any other cause, the fixed charges are still incurred. Since the total fixed costs are charged whatever the annual production from the plant, the cost per unit of output will increase significantly at output rates much below 100%. This problem will be discussed later in the chapter.

5.3 Variable costs

Let us now examine these cost components in more detail and look first at the variable costs. There are different views as to which items should be regarded as variable and which fixed, but the split described here is widely accepted.

Raw material costs　　　　⎫　Variable cost elements
Energy input costs　　　　⎬　(total sum £000/year varies
Royalty and licence payments ⎭　with plant output)

5.3.1 *Raw material costs*

Where a process plant is operating, the usages of feedstocks can be obtained by measurements of process flows during periods of steady operation. In the case of a new process at the research stage, or a project under development, raw material usages can be estimated from the process stoichiometry using assumed yields or yields obtained during process research experiments. In the latter case, if yields are based on analysis of the reactor product, allowance must be made for losses which occur during the product recovery and purification stages. Information on major chemical raw material prices is available in the techno-commercial literature (e.g. *European Chemical News, Manufacturing Chemist*) on a U.K. and European basis. Data on U.S. prices for a much wider range of materials is available (e.g. *Chemical Marketing Reporter*). In an industrial situation internal company data will usually be available for important feedstocks. Company technology strengths and strategic considerations can also influence the choice of feedstock and process route which will be selected to maximize commercial advantage. Thus a process which uses an internally available feedstock or a minor modification to an existing process could be preferred over a possibly better alternative using a purchased feedstock or requiring a major investment in process development work. The cost of catalysts and materials not directly involved in the process stoichiometry (solvents, acids or alkalis for pH adjustment, etc.) must also be included in the materials cost. Catalyst costs are based on loss of catalyst per cycle for homogeneous catalysts or cost of catalyst charge divided by output during charge life for a heterogeneous catalyst. Catalyst materials recovered from purge streams or by reprocessing a spent catalyst reduce the net catalyst cost. Solvent losses can be either physical (loss in off-gas streams, pump leaks) or chemical due to decomposition or slow reaction under process conditions. Such additional material consumptions are difficult to estimate in the absence of plant operating data but are usually relatively small cost items.

5.3.2 *Energy input costs*

This item in the cost table covers the multiple energy inputs necessary to carry out the chemical reaction and to separate the desired product(s) at the level of purity demanded by the market. Steam, at various pressure levels depending on the temperature required, is used to provide heat input to reactors and distillation column reboilers. Fuel oil or gas is used to provide higher temperature heat inputs either directly by heat exchange between process streams and hot flue gases, or indirectly by heating a circulating heat transfer medium. Electricity is used for motor drives for reactor agitators, pumps and gas circulators and compressors. Lighting of plant structures, tank farms and plant control rooms is an important, but usually minor, use of electricity. Cooling water is required to remove reaction exotherms and control reactor temperature, condense and cool still overhead streams and cool process streams. Inert gas—usually nitrogen—is used for purging equipment to provide an inert atmosphere over oxidizable or flammable materials. In the case of an operating process the consumption of energy inputs can again be obtained from process measurements during steady operation. For a process under development, estimates of major energy requirements can be made from the process energy balance and information on process conditions in reactors and operation trains. Price information on energy inputs is less readily available than raw material prices, although indicative figures do appear in chemical engineering journals. The energy inputs available (e.g. steam pressure available, use of fuel oil or gas) and the prices charged for them will vary from company to company and are very much site-dependent. A large integrated site could co-generate steam and electricity at high thermal efficiency in a site power station. Use of waste streams (liquid or gaseous) could provide at least part of the fuel input, with the balance being purchased oil or gas. Steam would then be available at a number of pressure levels after staged let-down from boiler pressure through turbines to generate electricity. In contrast, a small site or isolated production unit would have to raise steam in a package boiler, using purchased fuel oil or gas, and buy electricity from the grid. The prices charged for energy inputs on the two sites would differ appreciably.

Massive rises in crude oil prices in the 1970s meant that hydrocarbon feedstock costs and energy input costs became more important factors in process costs than previously. 'Energy conservation' became important and considerable attention was paid to the integration of process energy requirements by using hot streams from one part of a process to provide heat in another area. This can often (but not always—see Lindhoff[1]) lead to increased plant capital requirements, and a balance must be struck between increased capital charges and lower energy charges. With lower oil prices in the 1980s some of the emphasis on this aspect has been reduced. However, in a highly competitive industry, and also in the interests of environmental protection and resource conservation, efficient use of energy resources is important. Chapter 7 emphasizes the importance of energy requirements and conservation.

5.3.3 Royalty/licence payments

If the process being used for production (or to be used in the case of a planned project) is based on purchased technology rather than a process developed within the company, then a royalty or process licence fee will be incurred. This may be either a variable or fixed charge depending on the nature of the licence agreement. Usually a charge is made per unit of production and would appear in a cost sheet as a variable-cost item. Alternatively an annual licence fee related to the plant size (e.g. £1 million per 100 000 tonne installed capacity) will be charged and this would then appear as a fixed charge. The size of the royalty payment or licence fee is a matter of negotiation between the licensor and licensee and will be influenced by the nature of the technology, the advantages offered over competing processes and the number of alternative processes available for licensing.

5.3.4 *Effect of production rate on variable cost*

As has been indicated earlier, the total variable cost of a product is directly proportional to plant output and the variable cost/unit of production is normally constant. In the case of a continuously-operating process—the type of process used widely in the chemical industry—operation at low output rates or, conversely, higher than design rates can lead to process inefficiencies and an increase in the variable cost/unit of production. At low rates, increased residence time in the reactor can lead to overconversion and hence to increased usage of raw materials. The increase in by-product concentration in the product system leaving the reactor increases the *relative* demands on the product separation and purification system and hence results in increased energy consumption/unit of production. Operation at very high rates can result in increased levels of partial conversion products in the stream leaving the reactor. This, combined with reduced efficiency of distillation columns and other separation equipment when overloaded, can again lead to increased consumption of raw materials and energy inputs.

5.3.5 *Packaging and transport*

The costs involved in packaging and transport of a chemical product to the consumer are largely variable costs. However, such factors are not regarded as forming part of the production cost/income comparison. When considering process economics and profitability they are usually deducted from the money paid by the consumer to leave a net sales income which forms the revenue inflow to the producer to set against the production costs.

5.4 Fixed costs

The second category in the cost table, the fixed costs, can be divided as follows:

Operating labour and supervision
Maintenance labour and supervision
Analytical and laboratory staff
Maintenance materials   Fixed-cost elements
Depreciation   (total sum £'000/year
Rates and insurance   is fixed irrespective
Overheads—works overhead charges   of plant output)
          —general company overheads

5.4.1 *Labour charges*

The first three items represent the manpower cost associated with the process and includes the team of process workers who operate the equipment. In the case of large, continuous process units the operators work in shift teams to cover the 24 hrs per day, 7 days per week running of the unit. Maintenance labour includes fitters, electricians, plumbers, instrument artificers and other engineering workers who keep the process equipment in good working order. Usually maintenance work is planned and is generally carried out during normal day hours with only a small shift team to cover emergency breakdowns. Works analytical manpower is also included—again a shift team of analytical staff is involved to carry out checks on plant operation and product quality on a continuing basis.

The item for maintenance materials covers the cost of parts, replacement tools and similar items ranging from a replacement valve or pump to a new spanner. Only items to be expected in normal maintenance are included, the replacement, for example, of a reactor as a result of accident or fire would require a special appropriation of capital.

5.4.2 *Depreciation*

Depreciation is a term used in a number of different ways in different contexts. In the context of production costs for a chemical, the depreciation charge is regarded as an operating cost in the same way as material or energy usages. It represents the fact that the capital value of the plant is 'consumed' over the operating life of the plant. Calculation of the annual depreciation charge requires an estimate of the expected life of the equipment—a figure of 10 to 15 years is generally used in the heavy chemical industry. The annual charge can then be calculated by a simple straight-line method in which the same sum of money (equal to fixed capital costs ÷ expected life) is charged each year. Another method of calculating the depreciation charge is the declining balance or fixed percentage method. In this case the depreciation charge in a given year is a fixed percentage of the remaining undepreciated capital—thus in year 1 the depreciation charge would be $20\%C$ (for a 20% rate with fixed capital C). The next year the depreciation charge would be $20\% \times 0.8C$ and so on. This method gives higher depreciation during the early years of plant life but does not give depreciation to zero value at the end of expected service life. A high depreciation charge is then required in the final year of operating life to strike the balance.

A second interpretation of depreciation is as an allowance against tax. The annual income is reduced by an annual depreciation allowance before tax is charged thus reducing the tax payable. Calculation of the allowance is governed by the appropriate tax legislation and the depreciation shown for tax purposes may be a very different figure from that charged as an operating cost.

The use of depreciation as an allowance against tax forms part of net present value and discounted cash flow measures of profitability to be considered later.

5.4.3 Rates and insurance

This item covers the local rates or local authority tax levied in the area in which the plant is situated, together with the premium required to provide insurance cover for the facility. Actual charges will vary with the plant site and nature of the process being carried out (e.g. a plant with a high fire risk will carry a high insurance premium). Typical values for the rates and insurance item of cost lie in the range 0·5–2·0% of plant capital.

5.4.4 Overhead charges

Overhead or general charges cover those items not associated with any particular product or process but which are an essential part of the functioning of an individual site or a whole company. These charges are usually divided into two broad classifications—local works overhead and general company overheads. The former category covers items such as the general management of a works (works manager, secretarial services, plant records, planning, security, safety organization, medical services, provision of offices and canteen facilities and so on). Company overheads include head-office charges, central research and development activities, legal, patent, supply and purchasing, and other company-wide activities. The allocation of company overheads to individual works and the further allocation of this charge and the works overhead to individual plants or products is something of an arbitrary process. Methods of allocation vary from company to company but are generally based on the plant capital and/or operating manning level for a plant relative to other plants in the works, and of the works in overall relation to the company.

5.5 Direct, indirect and capital related costs

An alternative way of sub-dividing the components of full cost for a chemical product is into direct, indirect and capital-related costs. Direct charges are those arising directly as a result of the production operation and cover materials, energy, process labour and supervision costs, maintenance costs (labour and supervision and materials) and royalty payments. Indirect costs cover charges associated with, but not directly resulting from the process operation. They are essentially those charges termed overheads in the previous classification. Capital-related costs are those charges which result from the fixed-plant capital associated with the process. They consist of the annual depreciation charge and the rates and insurance item previously described. Since modern chemical processes (particularly in the heavy chemical and petrochemical industries) are highly capital-intensive, the capital-related charges form a significant part of the overall process cost. The relationship

Table 5.2 Relationship between variable and fixed, and direct and indirect, costs

Variable or fixed		Direct or indirect
V	Materials cost	D
V	Energy inputs	D
V	Royalty payments	D
F	Process and maintenance labour	D
F	Process and maintenance supervision	D
F	Maintenance materials	D
F	Rates and insurance	Capital-related
F	Works overhead	I
F	Site and company overhead	I
F	Depreciation	Capital-related

between the items classified as variable and fixed and the direct-indirect cost split is shown in Table 5.2.

5.6 Profit

Up to this point on the cost table (Table 5.1)—the works cost or total production cost—the cost build-up for the chemical allows only recovery of monies spent in producing the product and in maintaining the operation of the producing works and company. We now need to look at what profit will come from the sale of the product to provide for company growth, to reward shareholders and to pay taxes.

Profit can be measured in a variety of ways but two measures which are commonly used by accountants are gross profit (or gross margin) and net profit (net margin). The gross profit is obtained by deducting the direct production costs of the chemical (or other product) from the net sales income or revenue. It can be quoted on an annual basis or on a 'per unit of product' basis and represents the total cash available to pay for activities not directly concerned with production (indirect and capital charges in Table 5.2) and for growth, taxes and payment to shareholders. The net profit is obtained by deducting the total of direct and indirect production costs (works cost in Table 5.1) from the net sales income. This net profit is what is generally thought of as 'profit' and is essentially the money on which tax is levied leaving a net profit after tax to provide for growth of the company and pay dividends to the shareholders. For the simple annualized cost table, profit is shown as a return on the total capital involved (fixed plant and working) either derived from the sales income for an existing product or set as a target return for conceptual costing of a proposed new product. Thus the final total cost for our chemical product which will provide an acceptable return for the producer, and which must be met by the price paid by the customer, is much higher than the directly attributable production cost. This is the case both in the petrochemical industry where production is highly capital-intensive and in the speciality- and fine-chemicals areas where overhead charges are high to cover the extensive product and application development work required.

The build-up of the cost of producing a chemical has been shown for a single scale of production and with the plant operating at full capacity. Great emphasis has been placed in the chemical industry on the effect of scale and more recently, in the depressed economic climate, on the effect of operating chemical plants at low rates. What are the effects on product cost of an increase in scale or operation at less than full output?

5.7 Effects of scale of operation

5.7.1 *Variable costs*

In considering the production cost build-up the classification of costs as variable or fixed was made. The total sum of money in £'000 for the variable-cost items varies directly with the quantity produced. However, in unit cost terms—£/tonne of product or cents/lb—the variable cost is practically independent of the scale of operation. The yield in the reactor and the efficiency of product separation are not significantly affected by differing scale of operation so raw material usage and cost per unit of product will not be changed. Similarly the reactor energy requirement and separation energy inputs per unit of product will be largely unaffected by scale of operation since the endothermic or exothermic nature of the reaction is unchanged and the thermal efficiency of the separation stages does not change significantly. Thus, in producing, say, cumene from propylene and benzene, the raw-material costs and the cost of services will be the same (in £/tonne or cents/lb of cumene) whether the scale of production is 50 000 tonnes/year or 150 000 tonnes/year. When operating on a larger scale it may be possible to obtain raw materials at somewhat lower prices or it may be economic to install additional energy recovery equipment which would not be economically worthwhile on the smaller scale. These changes could reduce raw-material costs and energy costs but the effects are likely to be small and, in general, variable cost items are not significantly scale-dependent.

5.7.2 *Fixed costs*

Fixed-cost items (per unit of product) do, however vary significantly with scale of production. In the case of the labour element of fixed costs the number of process operators and maintenance workers are by no means directly proportional to scale. The precise relationship will depend on the type of process involved. A typical petrochemical industry process which involves largely fluid handling and is highly automated will have a relatively small process manning requirement and will require only a small increase in manning as process scale is increased. The number of maintenance workers will be greater than the number of process workers, but here again an increase in scale does not require a proportional increase in manpower; for example, a pump handling $50\,m^3/hr$ will not require twice the maintenance man-hours of a pump handling $25\,m^3/hr$. At the other end of the process spectrum, a small-scale batch process involving solids handling—typical of the dyestuffs industry—will have a higher manning level per unit of production and show an increase in manning requirements more nearly proportional to scale.

Attempts have been made to derive relationships between scale and manpower requirements, e.g. Wessel[2] derived a relationship of the form manpower $\propto$ no. of process steps ($=$ process complexity) and $(capacity)^{0.24}$. More recent studies[3,4] suggest even less dependence on capacity, with process complexity and company management philosophy being more important.

5.7.3 Plant capital

The major effect of change of scale is, however, on the plant capital requirement and consequently on the capital dependent charges per unit of production. Fixed-plant capital can be related to the production scale by an equation of the form

$$\frac{C_1}{C_2} = \left(\frac{S_1}{S_2}\right)^n$$

where C_1 and C_2 are fixed plant capital
S_1 and S_2 are plant production scales
and n is a fractional power.

Early studies[5] in the petroleum and petrochemical industries derived a median value of n of 0·63. Subsequent studies confirmed a median value between 0·6 and 0·7 but showed values of n between 0·38 and 0·88 depending on the nature of the process and the operating scale. This is because some elements which go to build up the capital cost of a plant, such as engineering and supervision, electrical and instrument installations, are relatively unaffected by scale, whilst costs of machinery and equipment are scale-affected. In costs for these items a relationship of the above form is followed with $C_{1/2}$ being the item costs and $S_{1/2}$ being the item sizes.

For small plants or relatively small changes in scale only the cost of the plant equipment items will change significantly. Other factors, such as civil engineering work, support structures, installation costs, electrical and instrument costs, etc., will not change greatly and the overall scale factor will be below 0·6. If however, the increase in scale involves more units of equipment rather than a single larger item then the scale factor will be nearer unity. An example of this is shown in the production of ethylene by thermal cracking of hydrocarbons. Current technology limits the capacity of the cracking furnace to 30–35 000 tonnes/year of ethylene output. As the ethylene production capacity is increased, more furnace units are added and the scale-up factor for

Table 5.3   Effects of scale of production on cost

Product:		2-Ethylhexanol by carbonylation of propylene		
Basis: early 1988.	Plants at full capacity			
Scale tonnes/year		50 000	100 000	
Capital		£ million	£ million	
Fixed plant		67	102	
Working		12	19	
Total		79	121	
Operating costs	£'000	£/tonne	£'000	£/tonne
Propylene @ £305/tonne	12 688	254	25 376	254
Carbon monoxide @ £110/tonne	2 860	57	5 720	57
Hydrogen @ £700/tonne	2 800	56	5 600	56
Catalysts and chemicals	575	12	1 150	12
	18 923	379	37 846	379
Credit:				
iso-Butyraldehyde @ £280/tonne	(1 680)	(34)	(3 360)	(34)
By-product fuels @ £55/tonne	(318)	(6)	(637)	(6)
Net materials	16 925	339	33 849	339
Service costs (= energy costs)	2 150	43	4 300	43
Variable cost	19 075	382	38 149	382
Direct fixed costs	3 220	64	4 903	49
Depreciation (15 year life)	4 467	89	6 800	68
Overhead + indirect fixed	1 340	27	2 040	20
Works cost	28 102	562	51 892	519
10% return on total capital	7 900	158	12 100	121
Production cost + return	36 002	720	63 992	640

the cracking furnace section of the ethylene plant is 0·8–0·9. For comparison scale-up factors in the gas compression/treatment and distillation sections of such a cracker are 0·55.

Table 5.3 gives an example of the effect of change in scale on production costs for two plants running at full capacity. Clearly, most of the reduction in unit cost on moving to the higher scale of production derives from the capital-dependent charges. Figure 5.1 shows the cost of production per tonne of 2-ethyl-hexanol based on propylene carbonylation assuming satisfactory sales of by-products and shows the continuing reduction in cost of production as production scale is increased. Production costs for most processes in the chemical industry—particularly in the petrochemical industry where continuous processes are operated—show a similar scale-dependence. It would seem that the only factor preventing the building of ever larger plants producing lower-cost product is the capacity of the market to absorb the

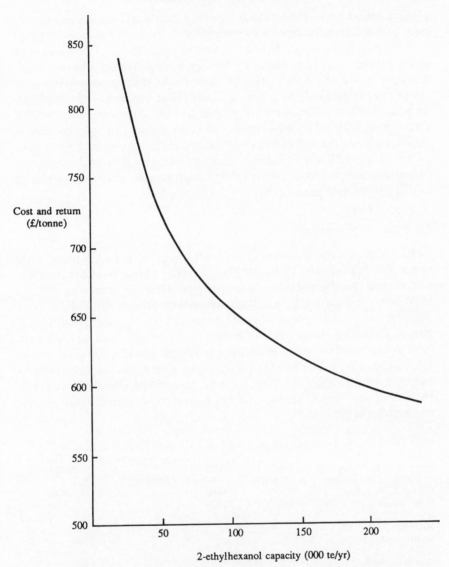

Figure 5.1 Cost vesus production scale for 2-ethylhexanol production.

products. However, by the late 1960s it was becoming apparent that the construction and operation of very large plants introduced new problems or intensified known problems. The larger plants involved more complex, sophisticated process designs and greater process integration. Also large reactors and distillation columns had to be built on site, rather than built in a manufacture's workshop and then transported to the site. Partly as a result of this, and because of the sheer size of the projects, the large plants were taking longer to build and the very large amounts of fixed capital were committed for a longer period before producing any positive cash flow. Late start-up of very large plants can mean a project never achieves the expected cash flow and profit over its operating life. In order to maximize the benefit of increased scale on the plant capital, large plants are built on a 'single stream' basis on which the capacity of a process unit (reactor, compressor, distillation column) is increased by building a single larger unit rather than duplication of equipment. As a result of this, a breakdown of a single process unit could shut down the whole plant with very serious economic consequences. Largely because of these difficulties, the size of individual plants has not increased greatly since the late 1970s and it will probably require significant breakthroughs in process technology and project construction and management techniques to initiate further increases in plant scale.

5.8 Effect of low-rate operation

Table 5.3 shows the production cost advantage of a larger plant when operated at full capacity. The larger plant provides a lower product cost whilst still meeting the desired return on total capital. However, if the unit suffers a breakdown or there is a market limitation such that the operator of the larger plant cannot sell all his potential production and has to operate at less than full output, then his cost per tonne of product will increase since his fixed costs have to be carried by the lower tonnage produced. The effect on product cost for the above 2-ethylhexanol plants operating with the larger unit achieving only 60% availability or an available market limitation of 60 000 tonnes/year is shown in Table 5.4. A sales income of £690/tonne of 2-ethylhexanol is assumed to simplify the picture.

Table 5.4 2-Ethylhexanol from propylene: effect of operation at below full capacity

	50 000		60 000	100 000	
Plant capacity tonnes/year	50 000			100 000	
Available market tonnes			60 000		
Sales possible tonnes	50 000			60 000	
Sales income @ £690/tonne (£'000)	34 500			41 400	
Operating costs	£'000	£/tonne		£'000	£/tonne
Materials	16 295	339		20 310	339
Service costs	2 150	43		2 580	43
Variable costs	19 075	382		22 890	382
Direct fixed costs	3 220	64		4 903	82
Depreciation	4 467	89		6 800	113
Overhead + indirect fixed	1 340	27		2 040	34
Works cost	28 102	562		36 633	611
Profit margin	6 398	128		4 767	79
Return on capital	8·1%			3·9%	

Cash flow = Sales income − (Variable cost + Direct fixed + Indirect fixed)
        = Profit margin + depreciation
                £10 865 000   11 567 000

Table 5.5 Break-even production rate

Sales (kilotonnes/year)	Income (£m/year)	50 kilotonnes/year plant				100 kilotonnes/year plant			
		Variable cost (£m/year)	Fixed cost (£m/year)	Total cost (£m/year)	Profit (£m/year)	Variable cost (£m/year)	Fixed cost (£m/year)	Total cost (£m/year)	Profit (£m/year)
10	6·90	3·82	9·03	12·85	−5·95	3·82	13·74	17·56	−10·66
20	13·80	7·64		16·67	−2·87	7·64		21·38	−7·58
30	20·70	11·46		20·49	+0·21	11·46		25·20	−4·50
40	27·60	15·28		24·31	+3·29	15·28		29·02	−1·42
50	34·50	19·10		28·13	+6·37	19·10		32·84	+1·66
60	41·40					22·92		36·66	+4·74
70	48·30					26·74		40·48	+7·82
80	55·20					30·56		44·30	+10·90
90	62·10					34·38		48·12	+13·98
100	69·00					38·20		51·94	+17·06

Breakeven 29·3 kilotonnes/year
Return on capital at full capacity 8·1%

Breakeven 44·6 kilotonnes/year
Return on capital at full capacity 14·1%

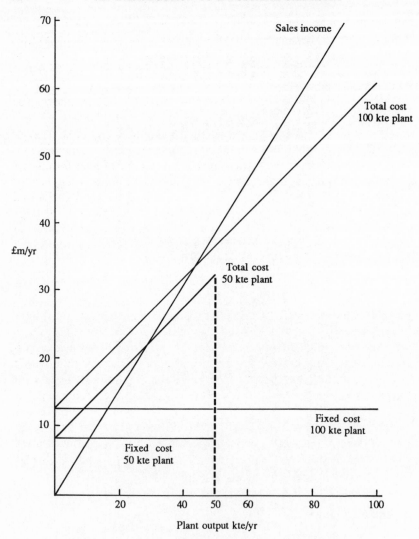

Figure 5.2 Break-even production rates for 2-ethylhexanol plants.

The operator of the smaller plant could not meet all the potential market but could operate at full output showing a profit flow of £6.37 million at the going market price for 2-ethylhexanol and a reasonable 8·1% return on his capital. The operator with the larger plant could satisfy all the available market demand but his production cost per tonne of product would be greater than that on the small plant and the profit flow and return on capital would be lower.

However, once the plant has been built the capital is regarded as 'sunk capital' and the important factor in the continuing viability of the business is the cash flow—defined as the sales income less total out-of-pocket expenses (Table 5.4). On this basis the larger plant provides more cash for the continuation of the business. Nevertheless, if the reduced market situation had been foreseen when the plants were being planned, the operator of the large plant would undoubtedly have built a unit of smaller capacity to match the lower market expectations and used the differential capital which would then be available to invest in more rewarding activities.

5.8.1 Break-even production rate

If we consider the situation, as above, where sales of 2-ethylhexanol show a net sales income of £690/tonne and the variable cost is £382/tonne, then there will be some level of production in the plants at which the sales income will cover all the costs (variable plus fixed including depreciation) but show no profit margin. This is the break-even production rate. Table 5.5 shows the build up of sales income and costs for the two plants and the figures are plotted in Fig. 5.2.

From the table and Fig. 5.2 it can be seen that the break-even production level, under the assumed conditions, for the 50 kilotonnes/year plant is 29 kilotonnes/year (58% of plant capacity) whereas the 100 kilotonnes/year plant must have sales of 45 kilotonnes/year (45% capacity) to break-even as a result of the higher fixed costs incurred by the larger plant.

5.9 Diminishing return

Although Fig. 5.1 shows that the product cost and return at full output reduces as the production scale increases, the effect diminishes with increase in scale. Table 5.6 shows some of the data used to plot Fig. 5.1, and also shows the incremental capital and reduction in cost and return for each capacity step.

Examination of the figures shows that, as capacity is increased, the reduction in cost plus return for the next increment of capacity is diminished. Looking at the incremental reduction in cost plus return against the incremental fixed capital used to produce it, it can be seen that the additional capital has less effect as the scale is increased. This is to be expected since, with increase in production scale and reduction in fixed costs per tonne of product, the scale-independent variable cost will dominate the cost plus return. Thus

Table 5.6 Effect of increased production scale on cost and return

Scale kilotonnes/year	25		50		75		100		125
Fixed capital £ million	44		67		85		102		117
ΔFC £ million		23		18		17		15	
Cost + return £/tonne	826		720		668		640		618
ΔC + R £/tonne		106		52		28		22	
$\dfrac{\Delta\text{C} + \text{R}}{\Delta\text{FC}}$		4·6		2·9		1·6		1·5	

from Table 5.3 for the 50 kilotonnes/year plant the variable cost represents 53% of the cost plus return, and fixed costs (including depreciation and return) the balance of 47%. For the 100 kilotonnes/year plant the relationship of variable costs to fixed costs is 60%:40% and the reduction in fixed capital-dependent charges as scale is increased has a diminishing effect on the overall cost plus return. The effect is enhanced if a point is reached at which the next increase in capacity requires two reactors (or furnaces or stills) instead of a single larger unit. The increase in capital will then be greater than for a comparable single-stream plant with even less reduction in cost plus return. An illustration of this is given in Table 5.7.

This diminishing effect on cost plus return resulting from a uniform incremental increase in capacity as the total capacity level is increased is usually referred to in the chemical industry as the law of diminishing returns. In classical economics the same 'law' refers to the situation in which an increase in one of the factors of production (land, labour or capital) results in a reduction in output rather than an increase.

Table 5.7 Increased capacity: effect on cost plus return of a single stream versus a twin-stream plant

	100 *kilotonnes/year plant*	125 *kilotonnes/year plant* (a) *single stream*	(b) *twinned reactors and furnaces*
Fixed capital £million	102	117	122
Variable cost £/tonne	382	382	382
Fixed capital dependent charges £/tonne	216	198	207
Other fixed costs £/tonne	42	37	37
Cost + return	640	617	626
ΔC + R £/tonne of 125-kilotonne *v.* 100-kilotonne unit.		23	14

5.10 Absorption costing and marginality

When all the costs—direct and indirect charges—together with a profit margin are recovered by the sales income, the selling price is said to have been set on a full-cost or absorption-cost basis. Recovery of all costs (at least) from the revenue provided by sales of product is essential if the producing plant and the company as a whole are to continue to function in the long term. A profit margin over and above this level is needed to persuade investors to buy shares in the company or subscribe to loans which will enable the company to continue developing.

Although the total costs must be recovered by the product sales there may be possibilities to make additional sales at a price which is less than the full product cost. Such sales can be justified for strategic reasons such as to deter a competitor from building additional capacity or to develop a foothold in a new market. In order to avoid making such sales at a disastrously low price the concept of marginality and marginal cost must be developed.

Consider a manufacturer producing a chemical with a design capacity of 50 000 tonnes/year. Fixed costs (total labour plus supervision, maintenance, materials, depreciation, rates plus insurance and overhead allocation) amount to £3·0 million per year. Variable cost based on known usages and prices for raw materials and energy inputs is £415/tonne for outputs up to plant rated capacity. Suppose experience has shown that by operating at slightly higher rates throughout the year and using a little more catalyst, production could be increased to 52 000 tonnes/year with a variable cost of £418/tonne. Further increases in rates could raise capacity to 55 000 tonnes/year but at the expense of some loss of yield and reduced efficiency of operation on product separation and recovery which raises the variable cost to £428/tonne. Table 5.8 shows how the total production cost increases with increased output and how the average cost of product (or unit cost) falls as production is increased up to nominal capacity.

The figures of Table 5.8 are plotted graphically in Fig. 5.3.

Table 5.8 Effect of increased output on total production costs and unit cost of product

Production (tonnes/year)	Fixed costs (excludes profit) (£'000/year)	Variable cost (VC) (£/tonne)	Total VC (£'000/year)	Total cost (£'000/year)	Average cost (£/tonne)	Marginal cost (£/tonne)
5 000	3 000	415	2 075	5 075	1 015	
10 000		415	4 150	7 150	715	415
15 000		415	6 225	9 225	615	415
20 000		415	8 300	11 300	565	415
30 000		415	12 450	15 450	515	415
40 000		415	16 600	19 600	490	415
50 000		415	20 750	23 750	475	415
52 000		418	21 736	24 736	476	493
55 000		428	23 540	26 540	482	601

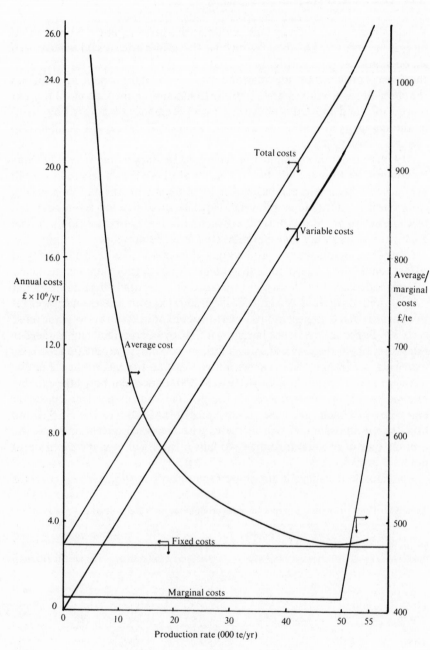

Figure 5.3 Effect of production rate on costs.

The marginal cost of production is the increase in total cost per unit increase in production: i.e. at any production level it is the cost of an additional tonne (or unit) of product. Table 5.8 and Fig. 5.3 show that up to the normal plant capacity of 50 000 tonnes/year the marginal cost is constant and is equal to the variable cost.

This can easily be confirmed:

Total cost at production rate X tonnes/year = £3·0 million fixed
$+ (X \times £415)$ variable
Total cost at production rate $X + 1$ = £3·0 million fixed
$+ ((X + 1) \times £415)$ variable
Marginal cost £415/tonne

If, however, we want to push the operation of the plant above normal capacity, to produce 52 000 tonnes this year instead of 50 000 tonnes, the average cost/tonne increases slightly but the marginal cost of the extra output increases dramatically since to achieve it all the plant output will have been produced at slightly higher variable cost. So what does this tell us about operating our existing plant to maximize profit? Consider the situation where the marketing department expect to sell 40 000 tonnes/year of product in the coming year. A look at Table 5.8 or Fig. 5.3 will tell us that if the average sales income is £490/tonne then the total of variable and fixed costs will be covered but there will be no profit. The output of 40 000 tonnes/year at an average sales income of £490/tonne represents a break-even situation. If the average sales income falls below £490/tonne at this level of output then the plant will not be making a full contribution to overhead charges. An average sales income greater than £490/tonne at our output level of 40 000 tonnes/year will generate a profit. Since the marginal cost of extra product at this level of output is equal to the variable cost (£415/tonne) then any additional sales giving a higher net sales income than £415/tonne will be beneficial. If the major sales are at a sufficiently high price to generate a profit then additional marginal sales at a sales income above marginal cost will increase profit. In the case where major sales are not producing a profit then marginal sales, made at above marginal costs, will reduce losses. To illustrate this, assume the major 40 000 tonnes/year sales are made at an average sales income of £520/tonne then total sales income is £20·80 million and total cost (Table 5.8) is £19·60 million and a profit of £1·20 million is produced. If an additional 1000 tonne could be sold (perhaps in an export market involving higher distribution costs) at a lower net sales income of £450/tonne (i.e. above marginal cost) the total income would be £21·25 million and total costs (£19·60 m + 1000 tonnes @ £415/tonne) would be £20·015 million giving a slightly larger profit of £1·235 million. Conversely, if the major 40 000 tonnes/year sales show an average sales income of £480/tonne, income would be £19·20 million against costs of £19·60 million—a loss of £0·4 million—and the manufacturing operation would not be making its full contribution to overall company costs. In this situation an addition of 1000

Table 5.9 Effect of additional sales at above marginal cost

Sales	Total sales income (£ million)	Total variable cost (£ million)	Total cost (£ million)	Profit (loss) (£ million)
(i) 40 000 tonnes @ £520/tonne	20·800	16·600	19·600	1·200
As (i) + 1 000 tonnes @ £450/tonne	21·250	17·015	20·015	1·235
(ii) 40 000 tonnes @ £480/tonne	19·200	16·600	19·600	(0·400)
As (ii) + 1 000 tonnes @ £450/tonne	19·650	17·015	20·015	(0·365)
(iii) 50 000 tonnes @ £ 520/tonne	26·000	20·750	23·750	2·250
As (iii) + 1 000 tonnes @ £450/tonne	26·450	21·318	24·318	2·132

tonnes of sales at £450/tonne would raise income to £19·65 million against costs of £20·015 million, giving a slightly reduced loss of £0·365 million. These cases are summarized in Table 5.9.

Case (iii) in Table 5.9 shows the situation with the plant scheduled to operate at full design capacity to meet major sales of 50 000 tonnes/year at £520/tonne. Under these circumstances the marginal cost of additional sales rises rapidly (Fig. 5.3) and the addition of 1000 tonnes of sales at £450/tonne— below the marginal cost at this level—is seen to reduce the total profit. To summarize this section on marginality and marginal costs: at any level of production and selling price, if marginal sales can be made at a sales income greater than marginal production costs, the marginal sales will make a positive contribution to income. However, for the overall operation of the production unit to generate a profit we must not lose sight of the fact that average sales income must be greater than average production cost at the given level of production.

5.11 Measuring profitability

In earlier sections of the chapter the term 'profit' has been used and in the present section a more detailed look will be taken at the ways in which profit is measured and the ways in which we can compare the profitability of investments.

5.11.1 Return on investment

Historically the yardstick used to measure profitability has been the percent return on capital or percent return on investment. This is defined as

$$\text{percent return on investment (ROI)} = \frac{\text{annual profit}}{\text{capital invested}} \times 100$$

The method and derived ratios are still widely used in basic accountancy, although they are being displaced by more up-to-date methods. This traditional method has the advantage of being simple and readily understood but is not very informative and can be misleading.

As used in Table 5.1 the target return on investment represents the cash income on an annual basis divided by the total capital invested. The cash income is obtained by deducting the total of direct and indirect annual production costs from the annual sales income, and the capital invested includes both fixed and working capital. Frequently the net cash income after tax is used as the numerator in the ROI ratio and this will, naturally, result in a lower return figure. One weakness of the return on capital as a measure of profitability is that the cash income generated by sales depends on the depreciation method used in the company. The annual charge for depreciation varies with the method and changes from year to year for methods other than simple straight-line depreciation. However, possibilities for confusion are multiplied by the choices available for the denominator in the ROI ratio. The capital value used can be total capital (i.e. fixed plus working) or fixed capital alone, and the fixed capital value used can be the original capital value, the current depreciated value, the average value over the life of the plant or the index-inflated current replacement value. The working capital element also depends on the methods used to value the feed and product stocks. Clearly, before using the ROI method to compare processes or companies using available published data the basis of the ratio and the definition of the factors involved must be carefully studied . Whilst all the above variations of the ROI ratio are used, the one most commonly encountered is (annual profit assuming straight-line depreciation) ÷ (total capital invested).

The traditional ROI method of expressing profitability presents a single year 'snapshot' of a process. No allowance is made for the often lengthy period when capital is being invested and no positive cash flow is generated. Distortions resulting from inflation and the effect of the time value of money are also ignored.

5.11.2 Use of inflated capital—current cost accounting

In periods of high inflation the use of the conventional ROI method to analyse profitability can seriously affect the financial strength of a company. Inflated prices for the product bring in apparent high income whilst capital charges, based on the original capital, will be low and ultimately insufficient for replacement of the plant. Apparent profits and ROI will then be high resulting in high taxes and expectations of good dividends by the shareholders. If capital charges are based on an updated valuation of assets a very different picture emerges. The example in Table 5.10 shows a plant for production of 100 kilotonnes of product with a fixed plant capital of £100 million in 1982 showing a 10% ROI at 1982 prices. For late 1988 with product prices and costs for materials, fuels and labour inflated using the appropriate indices but with

Table 5.10 Comparison of the effect of using historic and inflated values for plant capital on rate of return

Scale: 100 kilotonnes/year			
	1982	1988	1988
Plant capital £100 million		Historic plant capital £100 million	Inflated plant capital £152 million
Working capital £20 million		£26 million[3]	£26 million[3]
Operation 1982: £/tonne at full output		1988: £/tonne at full output	1988: £/tonne at full output
Raw material	100	114[1]	114[1]
Services (energy)	25	29[1]	29[1]
Labour	10	16[2]	16[2]
Overheads	15	24[2]	24[2]
Plant capital dependent	160	160	243[4]
Works cost	310	343	426
Sales income	430	563[3]	563[3]
Profit margin	120	220	137
ROI before tax	10%	17·5%	7·7%

[1]Inflated, using Index for Materials and Fuels purchased, of the chemical and allied industries.
[2]Inflated, using Index for Earnings in the chemical and allied industries.
[3]Inflated, using Index for output prices of the chemical and allied industries.
[4]Inflated, using Plant Capital Index from *Process Engineering*.

capital on an historic basis the profit margin looks good and the ROI very attractive. However if the fixed capital is also inflated to a current value the profit is reduced and the ROI not particularly attractive. Furthermore, in the 'historic capital' case, tax would be levied on a 'profit' of which approximately 40% is needed to maintain the capital base of the company.

In the situation shown in the 'historic capital' column of Table 5.10 the tax paid and, on the whole company basis, the dividend paid resulting from the apparent profit shown by the above operation would in fact be taken from cash required to maintain the capital base of the company. To correct this highly undesirable situation the 'current cost accounting' approach used in column 3 of Table 5.10, in which all factors in the cost build-up are expressed in terms of current values, is being increasingly adopted. The Institute of Chartered Accounts supports a system of this type and many major companies have worked together to develop a mutually acceptable system.

5.11.3 *Payback time*

A second long-standing method of assessing profitability—usually used at the planning stage for a new project rather than valuing an existing asset—is the payback period required. This is not so much a direct measure of profitability but rather a measure of the time taken for the positive cash flows to recover the original fixed-plant capital investment. Payback time is usually based only on the fixed plant capital and the payback time is the time taken for the cumulative positive cash flows following plant start-up to balance the earlier negative cash flows of the fixed-plant investment. Cash flows are usually taken on an 'after-tax' basis and allowance must be made for grants or allowances which offset capital expenditure.

During the plant construction period all cash flows are negative, i.e. cash is flowing out of the company 'treasury' to pay for the plant construction. When

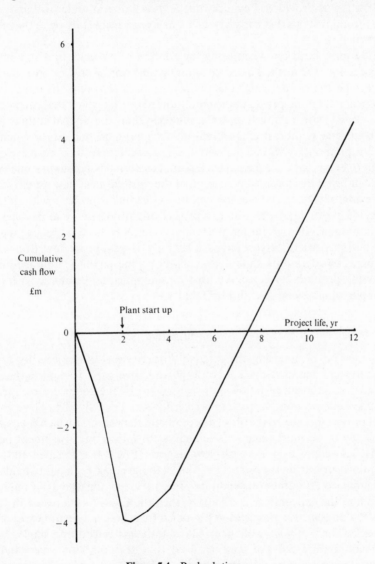

Figure 5.4   Payback time.

the plant begins to make and sell product, cash inflow begins. The positive cash flow is: cash flow = (sales income) − (total direct and indirect production costs) + (depreciation). Depreciation is part of the positive cash flow since it is cash which has gone into the general company 'treasury' although it carries a label nominally reserving it for one specific function.

Figure 5.4 shows the cash outflow during the plant construction period and start-up. Cash inflow builds up after start-up as the plant output is increased until in the third year of operation full-rate operation is reached and a steady cash inflow is provided. Payback time is shown as just under five and a half years from plant start-up or seven and a half years from start-up of the project construction.

Care must be taken when using the concept of payback time to compare projects since the method does not allow comparison of projects over the total project life. In periods of high uncertainty the method has the advantage that it is dependent on the period immediately ahead during which the situation can be assessed with relatively more confidence than the distant future. If the payback time is short it offers a reasonable guarantee, that, at the worst, the fixed capital outlay will be recovered. The method makes no estimate of the profit to be earned once the payback point has been passed but to some extent this can be inferred as a continuation of the positive cash flow established in achieving payback. The method can be misleading if projects with different expected lives are compared since a project with (say) a five-year payback and continuing profit flows for ten years could overall be more attractive than an alternative with a three-year payback but only six-year profit flow. Because the method considers only short-term cash flows the payback time method of assessing profitability is usually used as a supporting technique with other methods of assessing overall profitability.

5.11.4 *Equivalent maximum investment period*

This method of assessing the relative attractiveness of projects by a time measurement was developed at Nottingham University[6]. It requires the same cumulative after-tax cash flow data as required by the payback time method but takes into account the pattern of cash flows in the early years. These are the most important ones in the life of a project and the years for which predictions are likely to be most reliable. The equivalent maximum investment period (EMIP) is defined from the cash-flow diagram (Fig. 5.4) as the area under the cash-flow curve from the start of the project to the break-even point divided by the cumulative maximum expenditure. Since the area is measured in £ million × years and the expenditure is £ million, then the EMIP is expressed in years. EMIP represents the time period for which the whole maximum expenditure would be at risk if it were all incurred instantly and repaid by a single instant payment. In Fig. 5.4 the area involved is ∼ 16·5 £million years and the maximum cash outflow £4 million, giving an EMIP of 4.1 years. The pattern of cash flows to the break-even point (the payback time) is taken into account since the area under the curve is measured. As a result of this two projects with the same payback time can have different EMIP values. As with the payback time, the value of EMIP which distinguishes an attractive project from an unattractive one is somewhat arbitrary. It depends on the nature of the industry, the degree of risk and type of project. Analysis of data from earlier projects within a company can set guidelines which can then be used to examine new projects. Although the method is simple and easy to apply it does not seem to have been widely used.

5.12 Time value of money

The profitability measures considered so far have not considered possible charges on the capital involved nor the effect of time on money. Since a sum of money available now can be invested to earn interest it is of greater value than if the same sum were received some time in the future. Modern methods of assessing profitability take account of this time value of money and the pattern of cash flow during the life of a production unit. The most widely used methods are the Net Present Value (NPV) and Discounted Cash Flow Return (DCF) methods. These methods are generally used to aid investment decisions and in comparing the relative merits of different projects before investing money and building plant. Once capital has been invested and the chemical plant has been built and is in operation then different criteria are involved. As previously indicated, the method of operation which maximizes cash inflow and profit is to run at as high a rate as possible provided the least profitable marginal sales are made at a price which gives a sales income above the marginal production cost.

5.12.1 Net present value and discounted cash flow

When considering a new project the capital for building the plant must obviously be invested before any cash inflow or profit appears. If the money is borrowed from a bank then interest will be charged until the positive cash flow from the project repays the loan. This interest must then be included as a cash outflow in assessing the overall project. If the project capital is available from internal company funds there will be no direct interest to be paid. However, the company has lost the opportunity to invest the money and earn interest on it, possibly by lending to a bank. This loss of possible investment income represents an opportunity cost which must be charged against the project. On the other hand, when the break-even line has been crossed and the project is providing a cash inflow this money could be invested and earn interest at the rate previously charged on the borrowed capital. The rate of interest to be charged on capital outflow or credited to cash inflows will depend on the source of the finance. A company can provide capital for a project in different ways—by raising a market loan, using retained profit and depreciation, borrowing from a bank or finance house. In general a company will raise money in several ways and have a 'pool' of available funds at an agreed average company interest rate to fund possible projects.

The NPV and DCF profitability estimates are derived from cash flow forecasts covering the total life of the project. The cash flow in any year is: (net sales income) − (total production cost excluding depreciation) − (capital

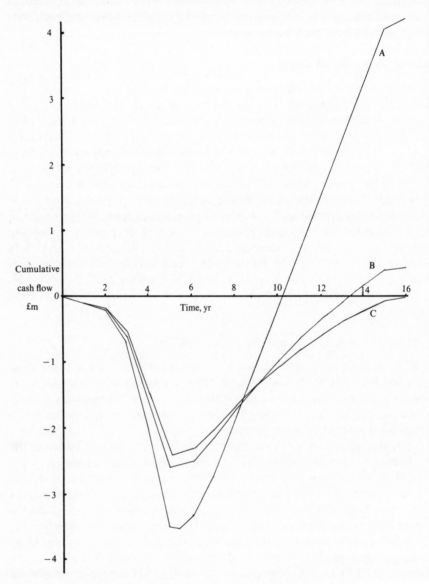

Figure 5.5 Discounted net cash flow pattern for a proposed project.

invested) after tax allowances and deductions. Depreciation is excluded since recovery of capital is implicit in the calculation and the methods thus have the advantage of being independent of depreciation methods. Since there are considerable problems in forecasting costs and sales for more than two or three years ahead, the longer the time period, the less likely the forecasts are to be right. As future sums of money have different values year by year due to the interest charged or earned they must be adjusted—or discounted—to values at a single moment in time, usually the present. For simplicity NPV and DCF calculations are normally based on the assumptions that cash flows in a given year occur at the end of that year. This is not often the case and in practice cash flows—particularly payment for items of equipment during the plant construction and income from product sales—take place throughout the year. However, the difference between annual discounting and continuous discounting will not usually be very great and it is normally much smaller than errors in the cash flow data.

If we have a present sum of money P, its value n years in the future (F) will be $F = P(1 + i)^n$ where i is the accepted compound interest rate fraction. Conversely, the present-day value (P) of a sum of money (F) to be received n years in the future is given by the expression

$$P = \frac{F}{(1 + i)^n}.$$

Thus the individual cash flows in future years can be discounted to a common base (the present) and directly compared. From the form of the expression for present value it will be seen that a sum of money received in two years time is worth more than the same sum received ten years hence. This reduces the effect of the less certain cash flows in the long-term future and emphasizes the importance of early positive cash flows.

The net present value of a project or investment is the sum of the present values of each individual cash flow. To reduce the arithmetic involved the net cash flow in each year is determined and the present value of the annual cash flows is summed. For a project with net annual cash flow C_t in year t, the net present value;

$$\text{NPV} = \frac{C_0}{(1 + i)^0} + \frac{C_1}{(1 + i)^1} + \frac{C_2}{(1 + i)^2} + \cdots \frac{C_t}{(1 + i)^t} + \cdots \frac{C_n}{(1 + i)^n}$$

or, in shorter form

$$\text{NPV} = \sum_{t=0}^{t=n} \frac{C_t}{(1 + i)^t}$$

where n is the total project life.

The 'present' is usually the time of the evaluation which may be the start of the project or may be after a period of research and development but before significant capital outlay. Figure 5.5 shows the forecast net cash flow pattern for a proposed project. The forecast shows two years of research work at a net cost of £100 000 per year followed by one year of process and market development including process design and costing £500 000. Plant construction is forecast to require two years giving a total negative cash flow of £3·5 million by the end of year 4. Some £50 000 further expenditure is incurred during start-up but then the positive cash flow from sales begins and by the end of year 5 a positive flow of £150 000 is shown. Production and sales build up in the following year, reaching full capacity in year 7. After eight years of steady operation and positive cash flow, production ceases, the plant is scrapped and working capital recovered. The undiscounted net cash flows are shown in Table 5.11 and plotted as curve A in Fig. 5.5. If the cost of capital—that is the interest rate to be charged on negative cash flows and credited to positive cash flows—is set at 10% as the average cost to the company, then the discounted cash flows are shown in Table 5.11 and as curve B in Fig. 5.5.

Tables of discount factors which provide values of $1/(1 + i)^n$ for different values of i and n are available and simplify manual calculations. Now, of course, computer programs are available to carry out discount calculations on any small computer. The project is seen to have a net present value of + £422 000 having recovered the capital and paid interest charges at the company rate of 10%. If, however, the company interest rate for raising money is 15%, then as shown in Table 5.11 the project shows a NPV of − £276 000. At this higher rate for borrowing money to carry through the project, a loss would be made and the project would be unattractive. The greater the

Table 5.11 Projected discounted cash flows

End of year	Net cash flow after tax (£000)	Discounted at 10% (£000)	15% (£000)
0	− 100	− 100	− 100
1	− 100	− 91	− 87
2	− 500	− 413	− 378
3	− 1 300	− 977	− 855
4	− 1 500	− 1 025	− 858
5	+ 150	+ 93	+ 75
6	+ 600	+ 339	+ 259
7	+ 850	+ 436	+ 320
8	+ 850	+ 397	+ 278
9	+ 850	+ 361	+ 242
10	+ 850	+ 328	+ 210
11	+ 850	+ 298	+ 183
12	+ 850	+ 271	+ 159
13	+ 850	+ 246	+ 138
14	+ 850	+ 223	+ 120
15	+ 150	+ 36	+ 18
Total	+ 4 200	+ 422	− 276

positive NPV of a project at the set discount rate the more economically
attractive it is. A project showing a negative NPV is unprofitable at the set
discount rate and would not be pursued.

5.12.2 *Discounted cash flow return*

As Fig. 5.5 and Table 5.11 show, increasing the discount rate from 0 to 10%
reduces the NPV from £4·2 million to £422 000. A further increase in discount
rate to 15% produces a negative NPV showing a loss-making situation. At a
discount rate somewhere between 10% and 15% the NPV will be zero, and this
is the DCF value. DCF, which is also referred to as 'internal rate of return' is
defined as that value of discount rate which results in a zero NPV for a project.
The higher the value of the DCF the more economically attractive the project.
A minimum acceptable level of DCF is set by the cost of capital to the
company considering an investment. If the project shows a DCF greater than
the cost of capital then the project will show a profit, conversely a DCF below
the cost of capital indicates a loss-making project.

Finding the DCF for a project is a matter of trial and error calculation.
When carrying out manual calculations a value of i (the discount rate) is
selected and the project NPV calculated. If this is positive the value of i chosen
is lower than the project DCF, so a higher value of i is selected and a new NPV
calculated. If a negative value of NPV is then obtained then the higher value of
i is above the DCF for the project. Graphical interpolation between NPV
values will enable the DCF value to be estimated. From the figures in Table
5.11 the project DCF rate lies between 10% and 15%. Interpolation gives a
DCF value of 13% and curve C in Fig. 5.5 shows the cash flows discounted at this rate. In fact, discounting the cash flows of Table 4.11 at 13%
shows a small negative NPV, discounting at 12% shows a small positive NPV
and interpolation gives a DCF rate of 12·7%. Although the arithmetic
operations can be carried out to give an even more precise value of DCF—
particularly if a computer is used—it must be remembered that the cash flows
are based on forecasts of sales, prices, raw material costs and production costs
for many years into the future. The cash flows of Table 5.11 represent an
attempt to project 15 years into the future and the uncertainties in the data do
not justify calculation of DCF rates to decimal places. This problem of
uncertainty will be examined later when considering the assessment and
evaluation of a new project.

The DCF method is not suitable for projects where a net negative cash flow
takes place late in the project taking the cumulative cash flow back below the
zero line. In most cases such a large cash outflow is a result of additional
capital investment in a planned expansion of capacity. It is unlikely that a
decision to commit funds to the expansion will be made at the same time as the
decision on the original project so the two can be treated separately and
individual cash flows evaluated.

5.12.3 Use of NPV and DCF as profitability measures

Since the same data and method of calculation are used to determine NPV and DCF it might be considered that one or other would suffice. There are, however, differences in the two measures and in their interpretation. Calculation of the project NPV by discounting the net cash flows at the company cost of capital gives a measure, in current money values, of the profit which the project will earn. This profit is produced after recovering the initial investment and paying all costs including the cost of 'borrowing' the capital from the company pool. It is not a ratio or relative value but a direct measure of the total profit expected from the project. The DCF for the project is a calculated rate of return on the invested capital at which the project breaks even, i.e. pays all expenses but shows no residual profit. It is a measure of the earning power of the project and also provides an indication of the efficiency with which the invested capital is used.

When a single project is being considered without reference to other possible projects the use of NPV or DCF will lead to the same conclusion on the economic acceptability of the project. This results from the fact that if a project shows a positive NPV, it must also show a DCF return higher than the discount rate used to calculate the NPV, that is a DCF rate above the cost of capital to the company. However, a more usual situation is one in which a choice must be made between alternative forms of a project which are mutually exclusive or in which a portfolio of projects must be selected. In this case the need for optimization arises and the company objectives and constraints must be taken into account.

In the former case, where a choice must be made between alternative forms of a project these will usually have different investments and operating costs. Thus although the income from sales will be the same the annual cash flows will differ. In order to maximize the profit from the project the proposal with the greater NPV at the cost of capital should be selected. This may not be the same as the proposal with the highest DCF since the return is measured on different levels of investment. A lower DCF on a larger capital investment could provide a higher NPV (i.e. a greater cash profit) than a higher DCF on a smaller investment.

When selecting a portfolio of projects the company aims must be considered. If the company objective in economic terms is to maximize the profit flow from the proposed list of projects, then this is an objective in money terms and can be attained by maximizing the total NPV of the projects chosen. This requires the acceptance of all independent projects which show a positive NPV at the company cost of capital. Such a solution is only possible if the resources available (supply of capital, project management, engineering facilities, etc.) are sufficient to carry out all projects with a positive NPV. If the company has more potentially suitable projects than can be financed with the available capital then the objective will be modified to maximizing the total NPV within the available capital restraint. Since the DCF is a measure of efficiency of use of capital, ranking projects in decreasing order of DCF and selecting projects in DCF order until the available capital budget has been taken up will give the desired maximum NPV. Of course, selection of projects is rarely as straightforward as implied above. Other resources besides capital could be inadequate to deal with all potentially attractive projects—more than one project might require the same technical manpower or engineering facilities. To resolve these problems, detailed network analysis and mathematical project programming techniques are required.

The NPV is generally more widely used in project evaluation studies than DCF measures. The NPV provides a direct measure in money terms of the attractiveness of a project and when dealing with more than one project NPVs are additive since they are measured at the same cost of capital. A DCF is, on the other hand, a rate of return and values cannot be combined for multiple projects since they are measured on differing capital investments.

5.13 Project evaluation

In the context of the chemical industry, a project can range from initial research studies (on a new product or new route to an existing product) to a capital project costing many millions of pounds or dollars for the installation of a new production unit. The common ground is the ultimate objective of generating profit at some time in the future by the use of present resources (skilled manpower, money). Economic evaluation of a project involves the consideration of those factors which can be measured and compared in money terms to provide information which will help decision making. Project evaluations may be carried out several times during the life of a project to assist in making decisions at the various stages. The initial decisions may involve continuation of a research programme, then to commit more resources to development of the process. Later the need to build a pilot plant to study scale-up problems may have to be studied, followed ultimately by a decision whether or not to invest the capital and build a full-scale unit to use the new process or manufacture the new product. The stages involved in project evaluation are summarized on p. 150.

5.13.1 *Comparison of process variable costs*

When considering a new route to an existing product the initial studies at an early stage must compare the proposed new route to the existing process or processes. Projects of this type are typical of the petrochemical or heavy chemical industries involving the production of bulk chemical intermediates, fertilizers and commodity polymers. Since the product is established and marketed then material from the new process must be able to compete, in quality and manufacturing cost, with that from existing processes. An initial evaluation of the proposed new route can then be made by comparing the cost plus return for the new process with that for existing technologies. It is essential that the comparison studies are made on the same basis and using the same methods to make the comparison as realistic as possible.

From the earlier part of the chapter it can be seen that, for large continuously operating units, the major factors in the build up of production cost are the variable costs (net raw materials and services) and capital-dependent charges. For smaller-scale batch production units the labour element can also become a significant factor. Hence, in order to compare our potential new process with existing methods we need to estimate the variable costs and the capital required for a plant to operate the process. As an example, consider maleic (cis-butenedioic) anhydride, the bulk of world production of which is currently manufactured by the oxidation of benzene. New catalysts are being developed by many companies to use n-butane as a feedstock. Let us assume that our research group is studying such a catalyst and the estimated production cost by the proposed butane oxidation needs to be compared with that of the benzene oxidation. Information on raw material and energy usages for the existing benzene oxidation process can be found in the literature[7], and numerous patents describe reaction conditions, yields, conversion of benzene and methods of product work-up, enabling the variable cost for maleic anhydride by benzene oxidation to be determined. Results from the research work will give indications of yield of maleic anhydride and n-butane coversion for the reaction step although data on recovery and recycle of unreacted feedstock are unlikely to be available. Purification by distillation or crystallization is normally very efficient, and in the absence of evidence for azeotropes or eutectics and making assumptions on separation efficiency (e.g. as good as in the benzene route) the overall raw material usage and cost for the new route can be estimated. If there is information on the expected life for the new catalyst this can be used to give an estimate of catalyst cost on an annual or per unit of product basis. In the absence of catalyst life data an acceptable charge for catalyst can be assumed and used to indicate a target catalyst life which must be achieved.

Estimation of the cost of energy inputs for the new process is difficult at the early research stage when process information is limited. One method which is useful at this stage is described by Marsden[8] and requires only data on the number of major process operations, process throughputs and chemical heats of reaction. It enables an estimate of the total energy requirement of the process to be made and inspection of the assumed reactor conditions and expected distillation temperatures will usually enable a split into steam, fuel and electricity consumption to be made.

5.13.2 Estimation of plant capital

Having made an estimate of the variable cost for the new process the next requirement is to produce an estimate of the plant capital requirement. Traditional methods of estimating capital, involving development of a process flowsheet and chemical engineering design, are time-consuming (expensive) and require a level of detailed information which is not available at the early research stage. Several workers have developed methods for the rapid estimation of plant capital at the predesign stage (see Bridgewater[9]) which require varying levels of input data. One method—'Process Step Scoring'[10] developed within ICI—requires only a simple flow diagram indicating the main conceptual process steps (react, filter, distill, separate etc.) with throughputs and estimated reaction times and conditions. The capital cost for a desired scale of operation can then be derived quickly at a level of accuracy sufficient for most preliminary evaluations. In order to compare the new process and the existing process on the same basis an estimate of capital requirement for the existing process should be prepared by the same method as that used for the new route. If capital estimates are available in the literature for the existing process they can serve as a cross-check on the estimate made. However, because of inflation, care must be taken to adjust earlier capital estimates to the same time base as the new estimate by use of the appropriate index. Suitable indexes showing change of plant capital cost with time are the CE index published in *Chemical Engineering* for American conditions, and the PE Index published in *Process Engineering* for UK conditions. The latter journal also publishes indexes for several other key countries, e.g. West Germany, Japan, France, Canada, and Australia.

5.13.3 *Process cost comparison*

With estimates of variable cost and capital requirement available the process cost table could be built up if an estimate of the labour requirement (process and maintenance and analytical) could be made. These items form part of the plant fixed costs and studies of a number of petrochemical processes show that an annual charge equivalent to 3–6% of the plant capital would be a suitable figure to include in a preliminary estimate for total plant fixed costs. For smaller batch operations a range of 8–12% of plant capital would be a reasonable figure to apply. A comprehensive cost table (Table 5.12) can then be built up.

This level of evaluation is sufficient at an early stage and would justify continuing research on catalyst and process developments. As more data become available the stage will be reached where a preliminary process flowsheet can be prepared and initial chemical engineering studies carried out to enable a more detailed estimate of capital and energy inputs to be made. This would involve sizing of the main items of the flowsheet using short-cut methods and building up to a full capital estimate using a factor method such as that of Miller[11], Cran[12] or Sinha[12a]. At this point, process optimization studies can be valuable in determining the energy or capital-intensive steps in the process and indicating where efforts to make process improvements would be most rewarding.

However, as previously stated, the cost table provides only an instant

Table 5.12   Process cost comparison for the production of maleic (cis-butenedioic) anhydride by benzene and butane oxidation

Process	Benzene oxidation	n-Butane oxidation
Scale	25 000 tonnes/year	
Basis	Costs at January 1989 prices. Capital at a *Process Engineering* index level of 211 (1980 = 100)	
Plant capital	£30 million	£34 million
Working capital	£ 3 million	£ 2 million
	£33 million	£36 million
Operating costs per tonne of maleic anhydride	£	£
Benzene 1·13 tonnes/tonne of product @ £310/tonne	350	—
Butane 1·17 tonnes/tonne of product @ £185/tonne	—	216
Catalysts and chemicals	10	20
Materials	360	236
Net services (large steam credits/allowed)	6	14
Variable cost	366	250
Plant fixed costs 6% plant cap.	72	82
Depreciation 6.67% plant cap.	80	91
Overhead fixed costs 3% plant cap.	36	41
Worked cost	554	464
Target ROC 10%	132	148
	686	612

picture of the process cost and profitability. As the stage is approached when capital will have to be allocated for pilot-plant studies there is a need for a view of the prospects for a project over its expected life. This requires the use of NPV and DCF methods of evaluation which in turn necessitate estimates of product sales, raw material prices and product selling prices over the life of the project.

In order to obtain the necessary information an estimate is required of the amount of the product which the market can take up and also an estimate of the share of the market which the Company carrying out the survey can expect to obtain. This will enable the size of the proposed new product unit to be fixed or set at a range of sizes to be evaluated in the NPV/DCF calculations.

5.13.4 *Estimating markets/prices*

The estimate of the future market size can be made in various ways— projection of historical data, comparison with other factors such as GNP where a relationship can be established (e.g. per capita usage of plastics and per capital GNP), and detailed user surveys.

Examination of consumption of many chemicals show that growth of consumption has three chief phases:

(a) Increasing rate of growth—an exponential curve.
(b) Constant rate of growth—a straight-line relation.
(c) Declining rate of growth—a curve with smaller slope than (b) and which may become negative.

A hypothetical product growth curve is illustrated in Fig. 5.6. Clearly if the

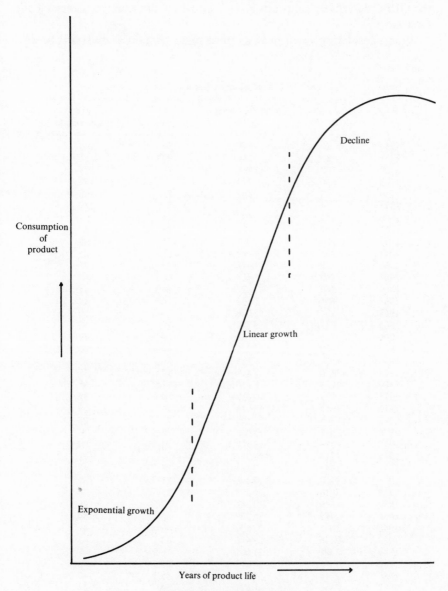

Figure 5.6 Product life cycle.

data available show exponential growth care must be taken in projecting over a long period. Techniques have been developed to modify the simple extrapolation[13] but require extensive data. The projection of past data can be improved by combining with detailed user surveys of major product outlets to provide a more rational basis for estimated demand. All methods of assessing future markets assume the absence of major perturbations, and events such as the OPEC actions on oil prices in 1973 and 1979 can make nonsense of any forecasts.

When considering a new product the product properties must first be used

Table 5.13 Acrylonitrile: U.S. production (see Fig. 5.7)

year	production × 10⁶ lb.	cumulative prod. × 10⁶ lb.	price/cents/lb.	const. 1987 price/cents/lb.
1950	25	129	36	146
1951	35	164	40	151
1952	50	214	43	159
1953	60	274	43	159
1954	70	344	31	112
1955	118	462	31	110
1956	141	603	27	93
1957	174	777	27	90
1958	180	957	27	88
1959	232	1 189	27	86
1960	229	1 418	23	72
1961	249	1 667	14·5	45
1962	360	2 027	14·5	45
1963	455	2 482	14·5	44
1964	594	3 076	17	50
1965	772	3 848	17	50
1966	716	4 564	14·5	42
1967	670	5 234	14·5	40
1968	1021	6 255	14·5	38
1969	1156	7 411	14·5	36
1970	1039	8 450	14.5	35
1971	979	9 429	14·5	33
1972	1115	10 544	13	29
1973	1354	11 898	13	28
1974	1411	13 309	14·5	28
1975	1214	14 523	23	40
1976	1518	16 041	24	39
1977	1642	17 683	27	43
1978	1752	19 435	27	39
1979	2018	21 453	25	34
1980	1830	23 283	37·5	46·3
1981	1996	25 279	44	50
1982	2041	27 320	46	49
1983	2146	29 466	43	45·7
1984	2201	31 667	45·5	47
1985	2346	34 013	45·5	46·6
1986	2314	36 327	45·5	47·2
1987	2550	38 877	34	34

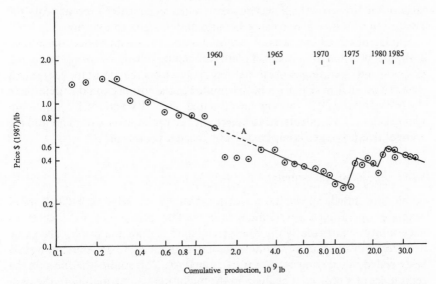

Figure 5.7   Acrylonitrile: U.S. production v. price. A: competition between established acetylene based route and Sohio propylene-based route.

to determine the market areas which will be open to penetration and the markets for competitive products used to make initial estimates for the new product. Whether the product is new or already established, forecasts of markets and market share are at best uncertain. It is impossible to forecast the changing circumstances which will affect growth of consumption and even the most refined mathematical model cannot allow for the effect of unforeseen step changes. Nevertheless, market evaluation is essential and will show trends and show the potential manufacturer the general direction if not the detailed path.

In forecasting price trends for established products the available data on prices and tonnages sold again forms the basis for extrapolation. Published information on many chemicals (e.g. US Tariff Commission reports, reports of Office Central des Statistiques de la CEE) enable a plot of log (price—adjusted to constant money values) against log (cumulative production) to be drawn. This is the so called 'Boston experience curve' (Fig. 5.7, Table 5.13) which was introduced in 1968 (Boston Consulting Group)[14]. The concept of experience curves postulates that costs (or price) in constant money value declines by a characteristic amount for each doubling of experience (production or sales). From studies of rapidly growing sectors of industry the Boston group indicated a decline in price of 20–30% for each cumulative doubling of production. A more recent study[15] confirmed the characteristic decline for products showing rapid growth. However, for products with low growth rates the relationship between price and cumulative production is much more erratic and shows a much lower rate of decline than the expected 20–30%. In view of the low growth shown by many major chemicals in recent years the projection of future prices using historic data is open to question.

Having obtained forecasts of potential market share and price, the annual cash flow position for a selected plant size can be calculated using the capital estimate and operating cost estimates prepared from the literature or research data. The cash flows can then be discounted at the cost of capital to determine the NPV and the DCF rate can be estimated. The NPV and DCF for different plant scales can be calculated to determine the optimum size of plant and the competitive processes compared over expected project life.

5.13.5 Effects of uncertainty

As already stated, all the data used to calculate the NPV or DCF (capital estimate, operating costs, annual sales, selling price, etc.) are subject to uncertainty—particularly the commercial data of sales and prices. The value of a project evaluation using single point data as an aid to decision taking can be extended by carrying out a sensitivity analysis. This studies the effect on the economics of a project of changes in the major items contributing to the cash flows. It highlights areas which are most important and where uncertainty has the greatest effect on the project. The effect of a 10% change in fixed capital, change in sales or sales price, change in raw material cost, delay in start-up and other variations on the basic project estimates can be explored and the resulting changes in NPV and DCF calculated. A sensitivity analysis does not attempt to quantify the uncertainties in the different factors but explores the effects on the project of changes in these factors.

In order to quantify the uncertainties in a project it is necessary to indicate the relative chances that a variable (e.g. selling price) will have different values. This can be done subjectively on the basis of, say, a 10% chance that the price will be as low as x, a 10% chance that the price will be as high as z against an expected mid-value of y. If it is assumed that the variables lie in a normal distribution in the range considered then the subjective probability estimates can be used to define the total distribution. Having made estimates of subjective probability distribution for each of the major variables we need a method of combining the various inputs to the project cash flows to obtain the resulting distribution of NPV or DCF. One such method is the Monte Carlo simulation. If there are a number of independent inputs to the project (capital, materials costs, etc) each represented by a probability distribution of values, then there is an infinite number of possible outcomes. Representatives of these can be calculated by selecting a value of each input from the range and calculating a value of NPV or DCF. Choosing a different value for any one input will lead to a different outcome.

The basis of the Monte Carlo technique is to carry out a large number of project evaluations with different input values selected from the individual distributions in a random way. The random selection is done in such a way

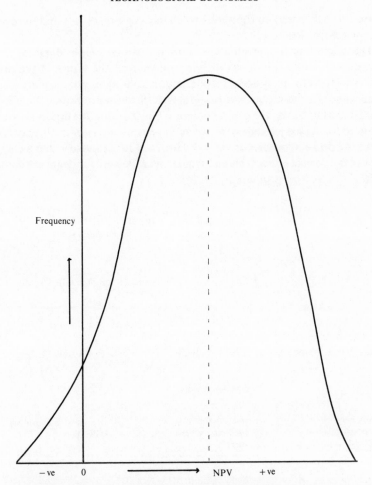

Frequency

−ve   0   NPV   +ve

Figure 5.8 Monte Carlo technique—frequency distribution of NPV values.

that the number of times a value of a variable is selected is proportional to its probability. Clearly, the large number of calculations involved requires the use of a computer to make the method feasible. The result of the many calculations is a range of NPV or DCF values and, since the same value of NPV or DCF can result from different combinations of input data, a frequency distribution of the results can be plotted (Fig. 5.8).

The example in Fig. 5.8 shows the range of possible outcomes for the project with a most likely value indicated by the dotted line. It also shows a small chance ($\sim 4\%$ based on ratio of negative area to total area) of a negative NPV at the company cost of capital. Having the distribution of possible outcomes enables the project manager to analyse the possible benefits and risks in the project in detail. Decisions can then be made by applying company policy with respect to risk, based on the indicated chances of gain or loss and the possible size of gain or loss.

The aim of project evaluation is to use the available data to provide information which will assist decision making on the future of the project. Use of sensitivity analysis and risk analysis techniques point up areas where uncertainty in the input data has greatest effect and indicates the effects of these uncertainties on the project outcome. Such evaluation does not eliminate the need for skilled judgement in the management team nor does it necessarily make the decision process any easier. However, it does ensure that a complete view of the project is available and makes clear the need for definitive company policy on risk and profitability criteria.

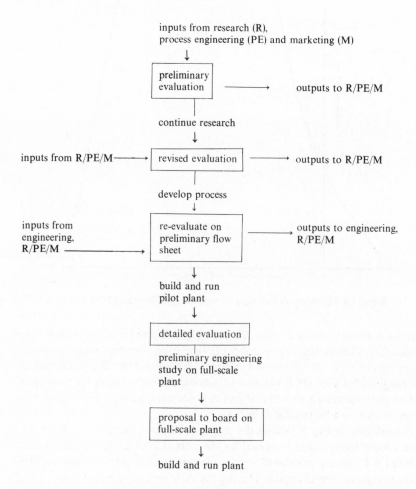

Figure 5.9   Stages involved in project evaluation.

5.14 Conclusion

The objective of all the technical economic methods discussed in this chapter is to provide information to assist management decision-making. Use of standard cost sheets helps plant managers to check the short-term performance of the units under their control and prompts corrective action if raw material usages or utilities consumptions increase. Information on absorption cost and marginal costs—particularly when marginal cost increases as plant capacity is reached—enables the marketing manager to quote for additional business at a price which will make a positive cash-flow contribution. Finally the techniques of project evaluation including sensitivity analysis and risk analysis enable decisions on new projects to be taken based on the fullest information on the likely consequences interpreted in the light of company policy on risk taking and reward seeking.

Appendix

Examples of discounted cash flow (D.C.F.) calculations

1. *Cash flow comparison of benzene and n-butane routes to maleic anhydride*

Consider a 25 kilotonnes/year plant.
Sales forecasts: 10 kilotonnes in first year of operation, 20 kilotonnes in second year
　　　　　　　25 kilotonnes/year in subsequent years.
Assuming plants take 2 years to build, with sales income estimates shown below the project revenue inflow is shown in Table A for a ten year production life.

(1) *Benzene oxidation* (based on the data in Table 5.12).

Plant capital of £30 million is spent in two years with £1 million of second year expenditure on plant buildings. Plant equipment is allowable against tax on other company profits at 52% in the year following expenditure. Allowance for buildings is spread over the project life with balancing adjustment in the final year.

Working capital is spent as sales develop and is recovered by clearance of stocks the year after plant shutdown.

Variable cost is £366/tonne in the initial years of operation, declining slightly over the project life due to minor savings in service usages at constant benzene and service prices.

Fixed cost excluding depreciation totals £2·70 million/year during most of the project life. No reduction is shown for this old, well developed process. Charges increase during the final three years life as increased maintenance is required to maintain full rate operation.

Table A

Year of Project Life	0	1	2	3	4	5	6	7	8	9	10	11
Sales Forecast kilotonnes/year	—	—	10	20	25	25	25	25	25	25	25	25
Sales Income £/tonne	—	—	690	690	685	685	680	680	675	670	670	665
Total Revenue Inflow £ thousands	—	—	6900	13 800	17 125	17 125	17 100	17 100	16 875	16 750	16 750	16 625

(Money values are all in £ of year 0)

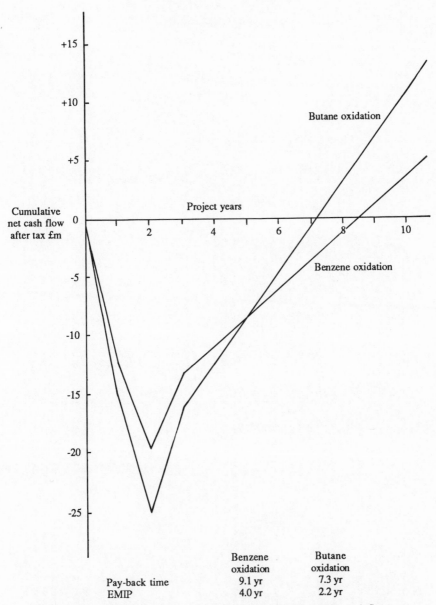

	Benzene oxidation	Butane oxidation
Pay-back time	9.1 yr	7.3 yr
EMIP	4.0 yr	2.2 yr

Figure A.1 Benzene and butane oxidation to maleic anhydride—cash flows.

The build up of annual cash flows and NPV and DCF is shown in Table B. Discounting the net cash flows after tax at the assumed cost of capital (10%) the project shows a NPV of − £3·86 million. At the lower discount rate of 5% a NPV of + £0·62 million is shown and by interpolation the DCF rate

Table B Benzene oxidation

Year	0	1	2	3	4	5	6	7	8	9	10	11	12	Total/NPV
Fixed capital £000	-14000	-16000												
Depreciation allowance £000		+7280	+8080	+21	+21	+21	+21	+21	+21	+21	+21	+21	+51	
Working capital £000			-1200	-1200	-600								+3000	
x Net capital flow £000	-14000	-8720	+6880	-1179	-579	+21	+21	+21	+21	+21	+21	+21	+3051	
Annual variable cost £000			-3660	-7320	-9150	-9100	-9100	-9050	-9050	-9050	-9050	-9050		
Fixed costs £000			-2700	-2700	-2700	-2700	-2700	-2700	-2700	-2725	-2725	-2725		
Revenue outflow £000			-6360	-10020	-11850	-11800	-11800	-11750	-11750	-11775	-11775	-11775		
Revenue inflow Table A £000			+6900	+13800	+17125	+17125	+17000	+17000	+16875	+16750	+16750	+16625		
y Net revenue inflow £000			+540	+3780	+5275	+5325	+5200	+5250	+5125	+4975	+4975	+4850		
Tax at 52% NRI £000				-281	-1966	-2743	-2769	-2704	-2730	-2665	-2587	-2587	-2522	
NRI after tax £000			+540	+3499	+3309	+2582	+2431	+2546	+2395	+2310	+2388	+2263	-2522	
z = x + y Net cash flow after tax £000	-14000	-8720	+7420	+2320	+2730	+2603	+2452	+2567	+2416	+2331	+2409	+2284	+529	Total +7341
Discount z at 10% £000	-14000	-7927	+6132	+1743	+1865	+1616	+1384	+1317	+1127	+988	+929	+800	+168	NPV -3858
Discount z at 5% £000	-14000	-8301	+6730	+2004	+2247	+2041	+1829	+1825	+1635	+1503	+1479	+1336	+295	NPV +623

DCF 6%

Table C n-Butane oxidation

Year	0	1	2	3	4	5	6	7	8	9	10	11	12	Total/NPV
Fixed capital £000	-16000	-18000												
Depreciation allowance £000		+8320	+9120	+21	+21	+21	+21	+21	+21	+21	+21	+21	+51	
Working capital £000			-1200	-1200	-600								+3000	
x' Net capital flow £000	-16000	-9680	+7920	-1179	-579	+21	+21	+21	+21	+21	+21	+21	+3051	
Annual variable cost £000			-2500	-5000	-6250	-6300	-6300	-6375	-6375	-6375	-6425	-6425		
Fixed costs £000			-3075	-3075	-3075	-3075	-3075	-3075	-3075	-3100	-3120	-3120		
Revenue outflow £000			-5575	-8075	-9325	-9375	-9375	-9450	-9450	-9475	-9545	-9545		
Revenue inflow Table A £000			+6900	+13800	+17125	+17125	+17000	+17000	+16875	+16750	+16750	+16625		
y' Net revenue inflow £000			+1325	+5725	+7800	+7750	+7625	+7550	+7425	+7275	+7205	+7080		
Tax at 52% NRI £000				-689	-2977	-4056	-4030	-3965	-3926	-3861	-3783	-3747	-3682	
NRI after tax £000			+1325	+5036	+4823	+3694	+3595	+3585	+3499	+3414	+3422	+3333	-3682	
z' = x' + y' Net cash flow after tax £000	-16000	-9680	+9245	+3857	+4244	+3715	+3616	+3606	+3520	+3435	+3443	+3354	+631	Total +15724
Discount z' at 10% £000	-16000	-8799	+7636	+2897	+2899	+2307	+2039	+1850	+1644	+1456	+1329	+1174	+201	NPV +231
Discount z' at 12% £000	-16000	-8644	+7368	+2746	+2699	+2106	+1833	+1630	+1422	+1240	+1109	+963	-162	NPV -1690

DCF 10%

Table D Cumene-project—cash flow build up

	Year	0	1	2	3	4	5	6	7	8	9	10	11	12	
	Fixed capital £000	−6080	−9120											+2500	
	Depreciation allowance £000		+3162	+4742											
	Working capital £000			−1250	−625	−375	−250								
A	Net capital flow £000	−6080	−5958	+3492	−625	−375	−250							+2500	
	Production/sales tonnes			25 000	37 500	45 000	50 000	50 000	50 000	50 000	50 000	50 000	50 000		
B	Sales income £000			+10 625	+15 938	+19 125	+21 250	+21 250	+21 250	+21 250	+21 250	+21 250	+21 250		
	Variable cost £000			8 250	12 375	14 850	16 500	16 500	16 500	16 500	16 500	16 500	16 500		
	Fixed costs £000			1 274	1 274	1 274	1 274	1 274	1 274	1 274	1 274	1 274	1 274		
C	Revenue outflow £000			9 524	13 649	16 124	17 774	17 774	17 774	17 774	17 774	17 774	17 774		
B−C	Net revenue inflow £000			+1 101	+2 289	+3 001	+3 476	+3 476	+3 476	+3 476	+3 476	+3 476	+3 476		
	Tax @ 52% NRI				−573	−1 190	−1 561	−1 808	−1 808	−1 808	−1 808	−1 808	−1 808	−1 808	
D	NRI after tax		+1 101	+1 101	+1 716	+1 811	+1 915	+1 668	+1 668	+1 668	+1 668	+1 668	+1 668	−1 808	
A+D	Net cash flow after tax £000	−6080	−5958	+4 593	+1 091	+1 436	+1 665	+1 668	+1 668	+1 668	+1 668	+1 668	+1 668	+692	NPW 7447 } DCF
	Discount at 15% £000	−6080	−5181	+3 472	+718	+821	+828	+721	+627	+545	+474	+412	+359	+129	−2155 } ~12%

is 6%. Since the project shows a negative cash flow at the cost of capital it is unattractive.

(2) *n-Butane oxidation* (based on the data in Table 5.12).

Plant capital of £34 million is spent in two years with £1 million of second year expenditure on buildings. Depreciation allowance as above.

Working capital is spent as sales develop and recovered after plant closure.

Variable cost is £250/tonne in the initial years of operation, increasing subsequently as improvements in the efficiency of the new process are offset by a rise (in constant money) in n-butane price.

Fixed cost excluding depreciation totals £3·075 million during most of the project life. No reduction is shown as a result of assumed operation problems with a new process. An increase is shown in the final three years as increased maintenance is needed.

The build-up of annual cash flows and NPV and DCF is shown in Table C. Discounting the net cash flows after tax at the cost of capital (10%) shows a NPV of + £0·23 million. At the higher rate of 12% the NPV is − £1·69 million and by interpolation the DCF rate is 10%. This represents a marginally attractive project i.e. DCF only just above the cost of capital. Further work to carry out sensitivity and risk analysis would be needed before a decision to go ahead could be made.

Undiscounted Cash Flows for each year for each process are plotted in Fig. A.1.

2. *Cumene production*

A company is planning to build a cumene plant with capacity 50 000 tonnes/year. Plant constructions will take two years with 40% of the fixed capital spent in year 0. The plant starts up in year 2 and operates at 50% capacity, building up to 75%, 90% and 100% in subsequent years.

The following information is available (all monies are in £ of year 0)

Fixed capital	£15 200 000
Working capital	£2 500 000 at full output
Sales income	£425/tonne cumene
Variable cost	£330/tonne cumene
Plant fixed costs	£760 000/yr
Overhead fixed costs	£514 000/yr

The plant is expected to have a ten-year production life and scrap value will just cover the cost of demolition and site clearance. Working capital expenditure builds up to the maximum in proportion to production rate and is recoverable after plant ceases production.

If depreciation allowance and tax on net revenue are 52% payable a year in arrears, what is the payback period and EMIP for the project? If the company cost of capital is 15% does the proposed cumene project represent an attractive investment?

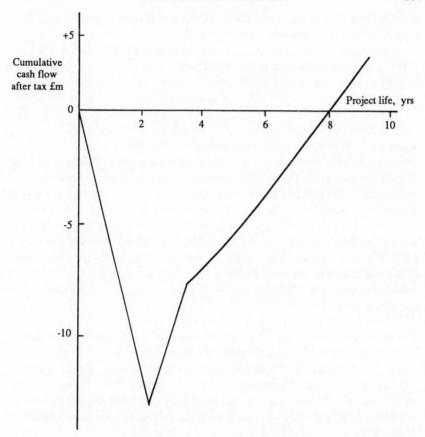

Figure A.2 Cumene project cash flows.

Table D shows the development of the cash flows. Fixed capital expenditure takes place in years 0 and 1 with depreciation allowance received in years 1 and 2. Working capital expenditure starts in year 2 and additional sums are required in years 3, 4 and 5 as production and sales build up. Sales income builds up as sales increase at the fixed sales income of £425/tonne cumene. Variable cost of production also increases as production builds up, and together with the fixed costs, gives the revenue outflow. The difference between sales income and revenue outflow provides the net revenue inflow. Tax at 52% is deducted from the net revenue inflow one year in arrears to give the net revenue inflow after tax. Summations of this and the net capital flow gives the net cash flow after tax.

The cumulative net cash flow after tax is plotted against project life in Fig. A.2. This shows a payback time of almost 7 years from the start of the project. Measurement of the area under the curve to break-even shows just under 42 £ million years and with a maximum cumulative expenditure of £12·04 million the EMIP is almost 3·5 years. These values would have to be compared with target values set by known successes within the company.

Discounting the net cash flow after tax at the cost of capital shows a negative NPV. The proposed project would therefore be unattractive and ways must be sought to reduce cost or increase sales income.

3. Practice examples

Example 1. A company proposes to invest £500 000 this year in a plant to make a speciality chemical on a scale of 500 tonnes/year. Working capital is estimated to be £100 000 at full output and would be spent proportionally to output. Sales are forecast as 300 tonnes in year 1 rising to 400 tonnes and 500 tonnes in subsequent years. Product market life is expected to be five years before being replaced. A scrap value of £100 000 is expected for the plant after shutdown. The variable cost of the product is £350/tonne and fixed costs are £155 000/year. Depreciation allowance and tax rate are 52% payable one year in arrears and sales income is £1100/tonne. The company needs a 15% return; is the project viable? What is the DCF rate?

Yes; 16·5%.

Example 2. A study is being carried out on a proposal to produce a hydrocarbon solvent on a scale of 100 000 tonnes/year. The plant will take two years to construct with 45% expenditure in the initial year and the remaining 55% in the next year. Plant capital is estimated at £15·0 million in £ of the initial year. Production starts in the third year and build up is shown below; working capital is estimated at £3·0 million at full output, spent proportionally to production.

Variable cost of the solvent is £275/tonne and fixed costs £3·0 million/year. Sales income of £350/tonne solvent is expected to remain constant (in constant money values) over the plant life. A production life of ten years is forecast and there will be no residual scrap value. Depreciation allowance and tax rate are 52% paid one year in arrears.

The project is seen as a long-term strategic investment and the company is prepared to accept a rate of return as low as 12%. Should the project go ahead? What is the payback time and DCF?

Year of production	1	2	3	4
Production (tonnes)	60 000	75 000	90 000	100 000

Yes; 6.9 *years from start of construction*, 13.8%

Example 3. Hycar Resins plc is considering the introduction of a new resin. Fixed plant capital of £1·0 million will be spent this year for a plant to produce 3000 tonnes/year of product. Working capital at full output would be £300 000 and would be spent as sales build up. Variable cost of production is £300/tonne and fixed costs of £100 000/year. A ten-year production life is expected and net scrap value will be nil. Depreciation allowance and tax on net revenue inflow are 52% paid one year in arrears.

Market studies have shown that, at the target sales income of £470/tonne build-up of sales will be slow as shown:

Year	1	2	3	4	5	10
Sales/production (tonnes)	500	1000	1500	2300	3000	3000

The company requires a 15% return and wants to see a payback time of under six years from the start of construction. Is the project attractive? What DCF and payback time would be achieved?

No; DCF just under 15% (14·3%) payback 6·7 years

Example 4. In the above study market research shows that selling price is critical. If a slightly lower sales income of £450/tonne is accepted sales build-up would be much quicker as shown:

Year	1	2	3	10
Sales/production (tonnes)	1000	2000	3000	3000

How does the project look in terms of DCF and payback time in this situation?

Attractive; DCF 16·4%, payback 5·9 years.

Example 5. A project is under consideration for production of a detergent product on a scale of 100 000 tonnes/year. Market studies show sales build-up as follows:

Year of production	1	2	3	4
Sales (kilotonnes)	60	80	90	100

A sales income of £530/tonne is expected. Plant construction will take two years with fixed capital spent equally in each of the two years. Depreciation allowance and tax are 52% payable one year in arrears. Working capital outlay is proportional to production. Residual plant scrap value after ten years' production will be nil.

Two schemes are being studied:

(i) An available intermediate is purchased and reacted in a single step to give the desired product. Fixed capital in this case is £60 million and working capital at full output £8·0 million. Variable costs are £325/tonne and fixed costs £4·77 million/year.

(ii) A basic chemical is purchased and reacted in two stages to give the desired detergent product. Fixed capital requirement is greater at £85 million and working capital at full output increased to £11·0 million due to need for addition stocks and first stage catalyst. Variable cost is, however, only £250/tonne and fixed costs £5·00 million/year.

Capital is available internally at 13% interest. How do the schemes compare?

(i) *NPV @ 13% + £619 000, DCF 13·4%*
(ii) *NPV @ 13% + £3 188 000 DCF 14·3%*

160   AN INTRODUCTION TO INDUSTRIAL CHEMISTRY

If internal capital is limited to £60 million and the additional capital for scheme (ii) has to be borrowed at 15% interest rate, would the extra expenditure on scheme (ii) be justified?

Yes; differential DCF 16·2%.

References

1. B. Linnhoff, and J. A. Turner, *The Chemical Engineer*, 1980, 742.
2. H. E. Wessel, *Chemical Engineering*, 1952, **59** (July), 209.
3. W. L. Nelson, *Oil and Gas Journal*, 1977 (8 August), 61.
4. Stanford Research Institute, Process Economics Program report, 140.
5. C. H. Chilton, *Chemical Engineering*, 1949, **56** (June), 49.
6. D. H. Allen, *Chemical Engineering*, 1961, **74** (3 July), 75.
7. *Hydrocarbon Processing*, 1981, **60** (11), 179.
8. R. S. Marsden, P. J. Craven, and J. H. Taylor, *Transactions of the 7th International Cost Engineering Congress*, London, 1982, B10-1–B10-8.
9. A. V. Bridgewater, *Cost Engineering*, 1981, **23** (5), 293.
10. J. H. Taylor, *Engineering and Process Economics*, 1977, **2**, 259.
11. C. A. Miller, *Chemical Engineering*, 1965, **72** (13 September), 226.
12. J. Cran, *Chemical Engineering*, 1981, **88** (6 April), 65.
12a. V. T. Sinha, *Engineering Costs and Production Economics*, 1988, **14**, 259.
13. W. W. Twaddle and J. B. Malloy, *Chemical Engineering Progr.*, 1966, **62**, 90.
14. *Perspectives on Experience*, The Boston Consulting Group Inc. U.S.A., 1968.
15. J. H. Taylor and P. J. Craven, *Process Economics Internat.*, 1979, **1**, 13.

Bibliography

Introduction to Process Economics, F. A. Holland, F. R. Watson and J. N. Wilkinson, Wiley, 1974.
The Chemical Economy, B. G. Reuben and M. L. Burstall, Longmans, 1973.
Institute Francais du Pétrole, *Manual of Economic Evaluation of Chemical Processes*, McGraw-Hill, 1981.
A Guide to the Economic Evaluation of Projects, 2nd Edition, D. H. Allen, Institute of Chemical Engineers, London, 1980.
Managing Technological Innovation, B. Twiss, Longmans, 1974.
Assessing Projects: A Programme for Learning, ICI Methuen, 1972.
Preliminary Chemical Engineering Plant Design, W. D. Baasel, Elsevier, 1976.
Modern Decision Analysis, G. M. Kaufmann and H. Thomas (eds.), Penguin Modern Management Readings, Penguin, 1977.

CHAPTER SIX

CHEMICAL ENGINEERING

R. SZCZEPANSKI

6.1 Introduction

The aim of this chapter is to provide an introduction to the basic principles of chemical engineering and some of the techniques used by engineers in the analysis and design of chemical processes. It is impossible to be comprehensive within the bounds of a single chapter, and some important topics such as safety, materials and chemical reaction engineering have been omitted. The reader is referred to the general bibliography at the end of this chapter which lists books covering the whole of the subject.

The role of the chemical engineer is not just to carry out chemistry on a large scale. Typically, the engineer is involved in the design of a chemical plant, its construction, commissioning and its subsequent operation. All this requires a wide range of skills and knowledge and hence chemical engineering is a very broadly-based discipline. The practising engineer must be able to apply basic principles of mathematics, physics and chemistry to solve complex problems on a large scale. Information available may be unreliable or incomplete and there are usually many constraints including those of time, money and legal requirements. The ability to analyse such problems and render them tractable by appropriate simplifying assumptions is one of the most important skills of the engineer.

The analysis of chemical engineering systems is based on the principles of conservation of mass, energy and momentum. The following sections are mostly concerned with the quantitative application of these simple principles. Many small-scale problems are used as illustrative examples and these provide an idea of the routine calculations carried out by chemical engineers. It must be remembered, however, that basic technical skills are only one component in tackling engineering problems.

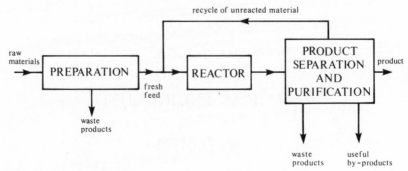

Figure 6.1   Prototype chemical process.

6.2 Material balances

6.2.1 *The flowsheet*

A process flowsheet is a schematic representation of a process which shows the equipment used and its interconnections. The flowsheet of a prototype chemical process is shown in Fig. 6.1. The stages involving preparation of feed and separation of products usually occupy most of the equipment in a plant. The reaction stage is of course crucial but is generally confined to a single vessel. An example of an actual process flowsheet is shown in Fig. 6.2. The process for production of vinyl acetate (ethenyl ethanoate) is typical in having lengthy and complex separation stages.

The type of information shown on a flowsheet depends on its purpose. Figure 6.2 is a useful aid to understanding a process description. In order to analyse process performance or specify equipment more information is necessary. It is convenient to have the flowrate, composition, temperature and pressure of each stream displayed on the flowsheet. For a process of any complexity this information would be in tabular form next to the drawing— for an example see Coulson and Richardson[1]. These authors also list the flowsheet symbols recommended by B.S.I.

6.2.2 *General balance equation*

The material balance establishes relationships between flows and compositions in different parts of a plant. It provides the principal tool for analysing a process.

Consider the system shown in Fig. 6.3. Input and output of material are related by the equation

$$\begin{bmatrix} \text{input of} \\ \text{material} \end{bmatrix} + \begin{bmatrix} \text{generation of material} \\ \text{within system} \end{bmatrix} - \begin{bmatrix} \text{output of} \\ \text{material} \end{bmatrix}$$

$$- \begin{bmatrix} \text{consumption of material} \\ \text{within system} \end{bmatrix} = \begin{bmatrix} \text{accumulation of material} \\ \text{within system} \end{bmatrix} \qquad (6.2.1)$$

SYNTHESIS

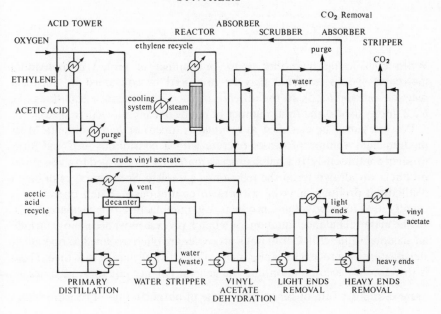

PURIFICATION

Figure 6.2 Vinyl acetate process.

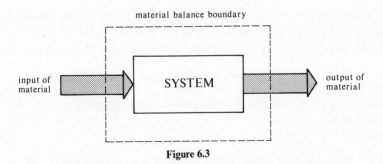

Figure 6.3

Equation 6.2.1 may be applied in different ways according to the precise definition of 'material' and the way in which the process is operated.

If the balance is applied to the total mass or to the mass of each element entering and leaving the system, the generation and consumption terms are zero (excluding nuclear reactions) and equation 6.2.1 becomes

$$\begin{bmatrix} \text{total mass} \\ \text{in} \end{bmatrix} - \begin{bmatrix} \text{total mass} \\ \text{out} \end{bmatrix} = \begin{bmatrix} \text{accumulation} \\ \text{of mass} \end{bmatrix} \qquad (6.2.2)$$

or

$$\begin{bmatrix} \text{mass of element} \\ i \text{ in} \end{bmatrix} - \begin{bmatrix} \text{mass of element} \\ i \text{ out} \end{bmatrix} = \begin{bmatrix} \text{accumulation of mass} \\ \text{of element } i \end{bmatrix} \quad (6.2.3)$$

When considering quantities (mass or number of moles) of individual molecular species, material may be produced or consumed by chemical reaction (section 6.2.5). In the absence of chemical reactions equations like 6.2.2 apply also to the total number of moles of each molecular species.

Processes may be classified as *continuous* (open) or *batch* (closed). Most modern high volume processes operate with a continuous feed and form product continuously. In a batch process, materials are charged to a vessel and products withdrawn when the reaction is complete. Sometimes, as in batch distillation, products may be withdrawn continuously. Batch operation is usually used for low volume products, e.g. manufacture of pharmaceuticals.

The material balance equation for a batch process must necessarily include an accumulation term. Continuous processes are often assumed to operate at *steady-state*, i.e. process variables such as flows do not change with time. There is therefore no accumulation and the general balance equation becomes

rate of input + rate of generation = rate of output + rate of consumption

$$(6.2.4)$$

or, for the total mass,

mass flow in = mass flow out   (6.2.5)

Alternatively, when considering the operation of a continuous steady-state process for a fixed period of time, each of the terms in equation 6.2.4 may be expressed simply as a mass or number of moles. Equation 6.2.2 thus becomes

total mass in = total mass out

The following is an example of a material balance calculation.

Example 6.1

$10 \, \text{kg s}^{-1}$ of a 10% (by mass) NaCl solution is concentrated to 50% in a continuous evaporator. Calculate the production rate of concentrated solution (C) and the rate of water removal (W) from the evaporator.

Solution

The flowsheet is shown below (Fig. 6.4). The steady state mass balance is given by equation 6.2.5. Balances can be made for individual species or total mass. Because no chemical reaction takes place we can write balances for molecular species rather than considering individual elements.

For each species, flowrate = mass fraction × total flowrate.

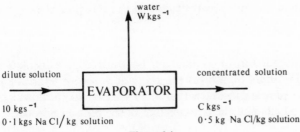

Figure 6.4

Balance on NaCl:

$$0.1 \times 10 = 0.5 \times C$$
$$\therefore C = 2\,\mathrm{kg\,s^{-1}}.$$

Total balance:

$$10 = W + C$$
$$\therefore W = 8\,\mathrm{kg\,s^{-1}}.$$

6.2.3 *Material balance techniques*

It is usually helpful to follow a systematic procedure when tackling material balance problems. One possibility is outlined below.

1. *Draw and label the process flowsheet*—organize information into an easy to understand form. If possible show problem specifications on the flowsheet. Label unknowns with algebraic symbols.
2. *Select a basis for the calculation*—the *basis* is an amount or flowrate of a particular stream or component in a stream. Other quantities are determined in terms of the basis. E.g. in Example 6.1 the flowrates of product and water were obtained on a basis of $10\,\mathrm{kg\,s^{-1}}$ of feed. It is usually most convenient to choose an amount of feed to the process as a basis. Molar units are preferable if chemical reactions occur, otherwise the units in the problem statement (mass or molar) are probably best.
3. *Convert units/amounts*—as necessary to be consistent with the basis.
4. *Write material balance equations*—for each unit in the process or for the overall process. In the absence of chemical reactions the number of independent equations for each balance is equal to the number of components.
5. *Solve equations*—for unknown quantities. This can be difficult, particularly if non-linear equations are involved. Overall balances usually give simpler equations. For complex flowsheets computer methods offer the only practical solution.
6. *Scale the results*—if the basis selected is not one of the flowrates in the problem specification the results must be scaled appropriately.

The following example illustrates the use of this procedure.

Example 6.2

An equimolar mixture of propane and butane is fed to a distillation column at the rate of $67\,\mathrm{mol\,s^{-1}}$. 90% of the propane is recovered in the top product which has a propane mole fraction of 0·95. Calculate the flowrates of top and bottom products and the composition of the bottom product.

Solution

1. Flowsheet (Fig. 6.5)
2. Basis: 10 mol of feed, i.e. $F = 10\,\mathrm{mol}$. This basis is somewhat easier to work with than the specified feed flowrate. By selecting a *quantity* of feed (rather than a flowrate) we are, implicitly, considering the operation of the process for a fixed period of time.
3. Conversion of units—not necessary, all units are molar.
4. Equations—there are no reactions and the steady-state balance equation for total or component flows is

$$\mathrm{input} = \mathrm{output}$$

    (i) Total material balance

$$F = D + B$$

    (ii) Component balances
      propane: $Fx_F = Dx_D + Bx_B$
      butane: $F(1 - x_F) = D(1 - x_D) + B(1 - x_B)$

It is evident that these equations are not independent. The total balance can be obtained by summing the component balances. Inserting specified

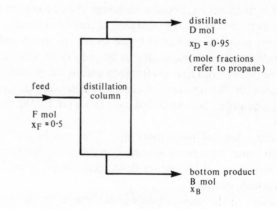

Figure 6.5

values into the first two equations gives

$$10 = D + B \qquad (1)$$

$$5 = 0.95D + Bx_B \qquad (2)$$

A third independent equation is provided by the specification on propane recovery:

$$x_D D = 0.90 F x_F$$

$$\text{or} \quad 0.95D = 0.90 \times 10 \times 0.50$$

$$\therefore 0.95D = 4.5 \qquad (3)$$

5. Solution
From (3) $D = 4.74\,\text{mol}$

Substituting in (1), $B = 5.26\,\text{mol}$

and equation (2) is now solved for the bottom-stream composition, $x_B = 0.095$.

6. Scaling—compositions are unchanged by the choice of basis. The quantities obtained above must be scaled by

$$\frac{67\,\text{mol}\,\text{s}^{-1}}{10\,\text{mol}} = 6.7\,\text{s}^{-1}$$

to solve the problem as stated, i.e.

$$\text{Bottom flowrate} = 5.26 \times 6.7 = 35.2\,\text{mol}\,\text{s}^{-1}$$
$$\text{Top flowrate} \quad = 4.74 \times 6.7 = 31.8\,\text{mol}\,\text{s}^{-1}$$

6.2.4 *Multiple unit balances*

Most processes consist of many interconnected units. In analysing such processes material balance equations may be written for each unit, for groups of units, or for the whole plant. To obtain a unique solution the number of equations describing a process must be equal to the number of unknown variables. If the analysis leads to fewer equations it is necessary to specify extra *design variables*. In the case of an actual plant where values of process variables are obtained by direct measurements the number of equations may exceed the number of unknowns. In such circumstances calculations should be based on the most reliable measurements.

Example 6.3

The flowsheet (Fig. 6.6) shows part of a process for the production of pure ethanol from aqueous solution by azeotropic distillation. Known quantities are shown on the flowsheet. Determine the remaining flowrates and compositions.

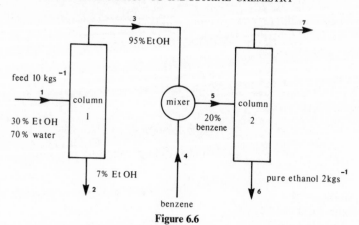

Figure 6.6

Solution

Basis: 1 second, i.e. 10 kg of feed.

Four possible material balance boundaries are shown in Fig. 6.7. The equations and unknowns associated with each balance will now be considered.

A: Overall balance

The notation F_j will be used to represent the total (mass) flowrate of stream j and $x_{i,j}$ denotes the mass fraction of component i in stream j. Component 1 is ethanol, 2, water, and 3, benzene.

Equations:

$$x_{1,2}F_2 + x_{1,7}F_7$$
$$= x_{1,1}F_1 - x_{1,6}F_6 = 3 - 2 = 1 \text{ (ethanol)}$$
$$x_{2,2}F_2 + x_{2,7}F_7 = 7 \text{ (water)}$$
$$F_4 = x_{3,7}F_7 \text{ (benzene)}$$
$$x_{1,2} + x_{2,2} = 1$$
$$x_{1,7} + x_{2,7} + x_{2,7} = 1$$

Variables:

$F_2 \; x_{1,2} \; F_7 \; x_{1,7}$
$x_{2,2} \; x_{2,7}$
$F_4 \; x_{3,7}$

total 5   total 8

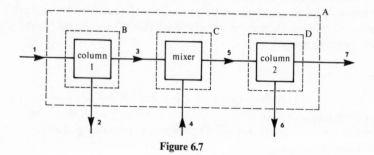

Figure 6.7

The overall balance does not give enough equations to solve for all the variables. Three more unknowns would have to be specified before this balance could be used.

B: Balance on column 1

Equations: Variables:

$0.95F_3 + 0.07F_2 = 3$ (ethanol) $F_3\ F_2$

$0.05F_3 + 0.93F_2 = 7$ (water)

---

total 2   total 2

Equations resulting from balance B may be solved immediately to give $F_2 = 7.39$ kg, $F_3 = 2.61$ kg.

As an exercise confirm that the numbers of equations and variables involved for the other unit are as follows:

C: Balance on mixer

4 equations   5 variables: $F_3\ F_4\ F_5\ x_{1,5}\ x_{2,5}$

D: Balance on column 2

5 equations   7 variables: $F_5\ x_{1,5}\ x_{2,5}$
$F_7\ x_{1,7}\ x_{2,7}\ x_{3,7}$

Only balance B yields enough equations to solve directly for the unknowns. However, solving B for F_2 and F_3 eliminates F_3 from the unknowns in balance C and leaves 4 equations in 4 unknowns. Solving these gives $F_4 = 0.65$ kg, $F_5 = 3.26$ kg, $x_{1,5} = 0.75$ and $x_{2,5} = 0.04$. Balance D equations can now be solved to give $F_7 = 1.26$ kg, $x_{1,7} = 0.48$, $x_{2,7} = 0.10$, $x_{3,7} = 0.52$.

Example 6.3 illustrated the technique of working forwards through a process, solving balance equations unit by unit. This is usually possible for processes without recycle streams provided that the feed is specified. If no individual balance yields enough equations it is necessary to solve simultaneously equations arising from balances on two or more units. Most multiple unit processes involve recycle streams but the treatment of these will be postponed until section 6.2.5.3.

When no reactions occur the number of independent equations contributed by each balance is equal to the number of components in the streams. Balances in terms of total flows and composition fractions (as in Example 6.3) must also satisfy a constraint equation on the sum of fractions in each stream j,

$$\sum_i x_{i,j} = 1 \tag{6.2.6}$$

Overall balances are simply the sum of all unit balances and do not introduce extra independent equations. However, it is often advantageous to use overall balances and discard some of the unit balances. This is because most information is usually known about feed and product streams.

6.2.5 Chemical reactions

Many examples of industrially important reactions are described in other parts of this book, particularly in Chapters 2, 10 and 11.

The treatment of material balances for systems in which chemical reactions take place involves some new considerations. Generation and consumption terms must be included for molecular species and the stoichiometric constraints must be observed.

6.2.5.1 *Stoichiometry*. The *stoichiometric equation* of a reaction defines the ratios in which molecules of different species are consumed or formed in the reaction, e.g.

$$C_2H_6 + \tfrac{7}{2}O_2 \rightarrow 2CO_2 + 3H_2O$$

For the purposes of defining *stoichiometric coefficients* v_i, it is convenient to write the stoichiometric equation with all species, i, on the right-hand side, i.e.

$$0 = 2CO_2 + 3H_2O - C_2H_6 - \tfrac{7}{2}O_2$$

The stoichiometric coefficients are

$$v_{CO_2} = 2 \qquad v_{C_2H_6} = -1$$
$$v_{H_2O} = 3 \qquad v_{O_2} = -\tfrac{7}{2}$$

A general stoichiometric equation may be written as

$$\sum_i v_i A_i = 0 \tag{6.2.7}$$

where A_i are the participating species and v_i is negative for reactants, positive for products, and zero for inerts (substances unchanged in the reaction).

6.2.5.2 *Extent of reaction*. It is useful to have a measure of the amount of material consumed or produced in a chemical reaction. The most convenient quantity is the *extent*. Consider a material balance for a chemical reactor (Fig. 6.8). The extent ξ is defined by the equation

$$f_{i,out} = f_{i,in} + v_i \xi \tag{6.2.8}$$

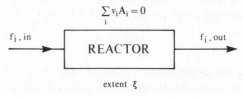

$$\sum_i v_i A_i = 0$$

f_i, in   REACTOR f_i, out

extent ξ

Figure 6.8   Material balance for a chemical reactor.

where $f_{i,out}$ is the number of moles of species i leaving the reactor
$f_{i,in}$ is the number of moles of i entering
v_i is the stoichiometric coefficient.

i.e. $$\xi = \frac{f_{i,out} - f_{i,in}}{v_i} \qquad (6.2.9)$$

The extent has the same units as f_i, i.e. mol or mol/unit time. It is always a positive quantity because of the sign convention for v_i and has the same value for all species because $(f_{i,in} - f_{i,out})$ is proportional to v_i.

Most chemical reactions do not result in complete conversion of reactants to products. The thermodynamic relationships which govern the extent of a reaction are discussed in section 10.3. The maximum possible extent depends on the equilibrium constant but it is often more convenient to define a *fractional conversion* based on the quantities present in the feed:

$$\text{fractional conversion of species } i \text{ (a reactant)} = \frac{f_{i,in} - f_{i,out}}{f_{i,in}} \qquad (6.2.10)$$

Unless the feed composition is stoichiometric the fractional conversion will be different for each species. Even for a reaction which goes to completion the fractional conversion will not be unity except for the limiting reactant.

The following is a simple example of a material balance calculation for a reactive system.

Example 6.4

200 mol of ethane are burned in a furnace with 50% excess air. A conversion of 95% is achieved. Calculate the composition of the stack gases.

Solution

Flowsheet: (Fig. 6.9)
The stated quantity of ethane will be used as a basis. The stoichiometric

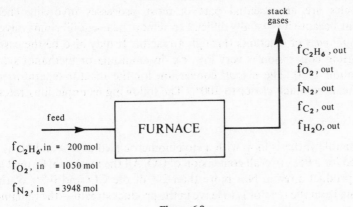

stack
gases

$f_{C_2H_6, out}$
$f_{O_2, out}$
$f_{N_2, out}$
$f_{C_2, out}$
$f_{H_2O, out}$

feed

FURNACE

$f_{C_2H_6, in} = 200\,mol$
$f_{O_2, in} = 1050\,mol$
$f_{N_2, in} = 3948\,mol$

Figure 6.9

equation is

$$C_2H_6 + \tfrac{7}{2}O_2 \rightarrow 2CO_2 + 3H_2O$$

A stoichiometric feed would require $200 \times \tfrac{7}{2} = 700$ mol O_2, thus 50% excess air provides 1050 mol O_2. Taking air to be 21% (molar) O_2 and 79% N_2 gives 3.76 mol $N_2/$mol O_2, i.e. 3948 mol N_2 in the feed.

The conversion is 95%; since no reactant is specified we assume that it refers to the limiting reactant—ethane in this case.

$$\text{conversion} = \frac{f_{C_2H_6,in} - f_{C_2H_6,out}}{f_{C_2H_6,in}} = 0.95$$

$$\therefore f_{C_2H_6,out} = 10 \text{ mol.}$$

$$\text{Extent } \xi = \frac{f_{C_2H_6,out} - f_{C_2H_6,in}}{v_{C_2H_6}} = \frac{10 - 200}{-1} = 190 \text{ mol}$$

The material balance is summarized in the table below

	IN mol	OUT mol		mole fraction
C_2H_6	200	$200 - \xi$	10	0·0019
O_2	1050	$1050 - \tfrac{7}{2}\xi$	385	0·0727
N_2	3948	3948	3948	0·7459
CO_2	0	2ξ	380	0·0718
H_2O	0	3ξ	570	0·1077
		$5198 + \tfrac{1}{2}\xi$	5293	1·0

6.2.5.3 *Recycles.* Processes in which part of a product stream is separated and recycled back to the feed are very common in the chemical industry. The prototype chemical process in Fig. 6.1 shows a recycle stream. Material balance techniques are, in principle, the same as for non-recycle processes. However, because the recycle stream is usually unspecified, a large number of equations may have to be solved simultaneously.

Recycles are an essential part of most processes involving chemical reactions because it is usually difficult to achieve near-equilibrium conversion in a *single pass* of reactants through a reactor. It may also be the case that equilibrium conversion is very low, e.g. in ammonia or methanol synthesis (see section 11.4). The *overall* conversion for the reactor-separator-recycle system can be much closer to 100%. The following example illustrates this.

Example 6.5

A methanol synthesis loop with a stoichiometric feed of CO and H_2 is to be designed for a 95% overall conversion of CO. All the methanol formed leaves in the product stream. Not more than 2% of the CO and 0·5% of the H_2 emerging from the reactor is to leave in the product stream—the remainder is recycled. Calculate the single pass conversion, the recycle ratio and the composition of the product.

Solution

Flowsheet: (Fig. 6.10)
A basis of 100 mol CO in the feed will be used.

$$\text{Reaction: } CO + 2H_2 \rightarrow CH_3OH, \qquad \text{extent } \xi \text{ mol.}$$

Let $f_{i,j}$ represent the number of moles of component i in stream j. Components are numbered as follows: 1, CO; 2, H_2; 3, CH_3OH.

Overall balances
The overall balance equations have the same form as the reactor balance (equation 6.2.8) because the only consumption and generation terms are due to the chemical reaction.

$$CO: f_{1,4} = f_{1,1} - \xi$$

$$H_2: f_{2,4} = f_{2,1} - 2\xi$$

$$CH_3OH: f_{3,4} = \xi$$

Inserting basis quantities for the feed gives

$$f_{1,4} = 100 - \xi \tag{1}$$

$$f_{2,4} = 200 - 2\xi \tag{2}$$

$$f_{3,4} = \xi \tag{3}$$

Mixer

$$100 + f_{1,5} = f_{1,2} \tag{4}$$

$$200 + f_{2,5} = f_{2,2} \tag{5}$$

$$f_{3,5} = f_{3,2} = 0$$

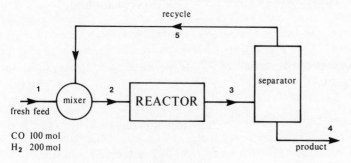

Figure 6.10

Separator

$$f_{1,3} = f_{1,5} + f_{1,4} \tag{6}$$

$$f_{2,3} = f_{2,5} + f_{2,4} \tag{7}$$

$$f_{3,3} = f_{3,4} \tag{8}$$

Specifications

$$\text{Overall conversion of CO} = \frac{\begin{array}{c}\text{moles in} \\ \text{fresh feed}\end{array} - \begin{array}{c}\text{moles in} \\ \text{product}\end{array}}{\text{moles in fresh feed}} = 0.95$$

i.e.

$$\frac{100 - f_{1,4}}{100} = 0.95 \tag{9}$$

product purity:

$$f_{1,4} = 0.02 f_{1,3} \tag{10}$$

$$f_{2,4} = 0.005 f_{2,3} \tag{11}$$

Solution of equations

Note that the overall balances replace balances over the reactor.

Equation (9) gives $f_{1,4} = 5$ mol and thus from (1) $\xi = 95$ mol. The remaining equations are solved in the order (2), (3), (10), (11), (6), (7), (8), (4), (5). The following table summarizes the results.

Component	Stream (quantities are in mol)				
	1	2	3	4	5
1. CO	100	345	250	5	245
2. H_2	200	2190	2000	10	1990
3. CH_3OH	0	0	95	95	0
total	300	2535	2345	110	2235

$$\frac{\text{Single pass}}{\text{conversion of CO}} = \frac{\text{CO in reactor feed} - \text{CO in reactor outlet}}{\text{CO in reactor feed}}$$

$$= \frac{345 - 250}{345} = 0.275$$

The single pass conversion of 27% is increased to 95% overall by recycling unreacted CO.

$$\text{Recycle ratio} = \frac{\text{recycle flowrate}}{\text{fresh feed flowrate}}$$

$$= \frac{2235}{300} = 7.45 \text{ mol recycle/mol feed}$$

Composition of product stream is 4·5% CO, 9·1% H_2 and 86·4% CH_3OH.

6.2.5.4 *Multiple reactions.* If more than one reaction occurs, equation 6.2.8 becomes

$$f_{i,out} = f_{i,in} + \sum_k v_{i,k}\xi_k \qquad (6.2.11)$$

where $v_{i,k}$ is the stoichiometric coefficient of component i in reaction k and ξ_k is the extent of reaction k.

The terms *selectivity* and *yield* are used to describe how far a particular (desired) reaction proceeds relative to other (undesired) reactions.

$$\text{Selectivity} = \frac{\text{moles desired product formed}}{\text{moles undesired product formed}}$$

$$\text{Yield} = \frac{\text{moles desired product formed}}{\text{moles of specified reactant fed or consumed}}$$

The following example includes two competing reactions and also introduces the concept of a *purge* stream. A fraction of the recycle leaves the process in the purge—it may be treated or simply vented. The purge is a means of removing substances which do not leave the system elsewhere. If a purge was not used in Example 6.6 methane would accumulate in the system and eventually prevent any useful reaction. It is desirable to keep the purge flowrate as small as possible to minimize loss of useful recycled materials.

Example 6.6

The catalytic dealkylation of toluene to benzene involves recycling of unreacted toluene after removal of by-product phenylbenzene. Using the information shown on the process flowsheet (Fig. 6.11) determine

    (a) quantities of recycle streams,
    (b) quantity and composition of purge stream,
    (c) quantity of hydrogen make up.

Solution

Basis: 100 mol toluene in reactor feed—this is the most convenient basis since most information is provided about stream 2.

Reactions:
$$\text{toluene} + H_2 \rightarrow \text{benzene} + CH_4, \text{extent } \xi_A$$

$$2 \text{ benzene} \rightarrow \text{phenylbenzene} + H_2, \text{extent } \xi_B$$

Balance on reactor
Let $f_{i,j}$ denote the number of moles of component i in stream j. Components are numbered: 1, toluene; 2, hydrogen; 3, benzene; 4, methane; 5, phenylbenzene.

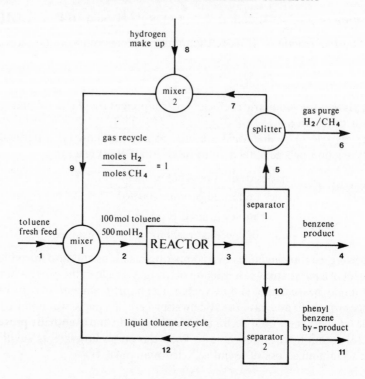

Conversion of toluene per pass = 25%

Yield of benzene (based on toluene consumed) per pass = 75%

Figure 6.11

$$f_{1,3} = 100 - \xi_A$$

$$f_{2,3} = 500 - \xi_A + \xi_B$$

$$f_{3,3} = \xi_A - 2\xi_B$$

$$f_{4,3} = 500 + \xi_A \quad \text{(quantity of } CH_4 \text{ into reactor is}$$

$$f_{5,3} = \xi_B \qquad\qquad \text{equal to quantity of } H_2)$$

$$\text{Conversion} = \frac{\xi_A}{100} = 0.25$$

$$\therefore \xi_A = 25 \text{ mol.}$$

$$\text{Yield} = \frac{f_{3,3}}{100 - f_{1,3}} = \frac{\xi_A - 2\xi_B}{\xi_A} = 0.75$$

$$\therefore \xi_B = 3.12 \text{ mol.}$$

Overall balances

$$0 = f_{1,1} - \xi_A$$

$$f_{2,6} = f_{2,8} - \xi_A + \xi_B$$

$$f_{3,4} = \xi_A - 2\xi_B$$

$$f_{4,6} = \xi_A \qquad (1)$$

$$f_{5,11} = \xi_B$$

Equation (1) expresses the requirement that all methane formed in the reactor must leave in the purge. Substituting for the extents gives

$$f_{1,1} = 25 \, \text{mol}$$

$$f_{2,6} = f_{2,8} - 21 \cdot 88 \, \text{mol} \qquad (2)$$

$$f_{3,4} = 18 \cdot 75 \, \text{mol}$$

$$f_{4,6} = 25 \, \text{mol}$$

$$f_{5,11} = 3 \cdot 12 \, \text{mol}$$

Purge

All the hydrogen and methane leaving the reactor are in stream 5, thus

$$f_{2,5} = f_{2,3} = 478 \cdot 12 \, \text{mol}$$

$$f_{4,5} = f_{4,3} = 525 \, \text{mol}.$$

The fraction of this methane leaving in the purge (stream 6) is

$$\frac{f_{4,6}}{f_{4,5}} = \frac{25}{525} = 0 \cdot 0476$$

The same fraction of the H_2 in stream 5 leaves in the purge

$$\therefore \frac{f_{2,6}}{f_{2,5}} = \frac{f_{2,6}}{478 \cdot 12} = 0 \cdot 0476$$

$$\Rightarrow f_{2,6} = 22 \cdot 77 \, \text{mol}$$

Answers

(a) Gas recycle (stream 9) = 1000 mol
(500 mol H_2, 500 mol CH_4)
Liquid toluene recycle (stream 12):
25 mol of fresh feed ($f_{1,1}$) make up the 100 mol of toluene fed to the reactor. Thus, recycle = 75 mol.

(b) Purge = 47·44 mol (52% methane, 48% hydrogen).

(c) Hydrogen make up: from the overall balance (equation (2))
$$f_{2,8} = f_{2,6} + 21.88. \text{ i.e. make up } = 44 \cdot 65 \, \text{mol } H_2.$$

6.3 Energy balances

The energy balance is based on the principle of conservation of energy. It provides an important additional technique for analysing processes. An energy balance is used to determine the energy requirements of a process or unit in terms of heating, cooling or work (pumps, compressors, etc.). Although for many purposes it is possible to carry out material and energy balances independently this is not always the case, e.g. in the design of a chemical reactor the extent is strongly dependent on the temperature and the two balances must be solved simultaneously.

6.3.1 *Energy balance equations*

The general balance equation (6.2.1) may be applied to energy but the generation and consumption terms are always zero. Energy is neither generated nor consumed in chemical reactions—there is merely a difference in energy associated with chemical bonds in reactants and products.

$$\begin{bmatrix} \text{energy} \\ \text{in} \end{bmatrix} - \begin{bmatrix} \text{energy} \\ \text{out} \end{bmatrix} = \begin{bmatrix} \text{accumulation of} \\ \text{energy within system} \end{bmatrix} \tag{6.3.1}$$

6.3.1.1 *Steady-state flow systems.* Energy flows for a steady-state system are represented in Fig. 6.12. The forms of energy important in chemical processes and their relative magnitudes are discussed in more detail in section 7.2. The internal energy term includes both 'chemical' energy of bonding and 'thermal' energy due to molecular motion and intermolecular interactions. The 'flow work' term represents the work done by the surroundings or system in transferring material across system boundaries—see Himmelblau[2]. In addition to energy associated with streams, energy may also be transferred between the system and its surroundings in the form of heat or by doing work. We use the IUPAC convention that *heat transferred to the system and work done on the system is defined as positive.*

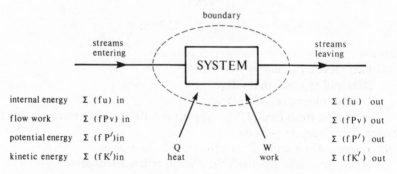

Figure 6.12 Energy flows for a continuous chemical process.

For a system at steady state

$$[\text{energy in}] = [\text{energy out}]$$

hence

$$\sum_{\substack{\text{inlet} \\ \text{streams}}} f(u + Pv + P' + K') + Q + W$$

$$= \sum_{\substack{\text{outlet} \\ \text{streams}}} f(u + Pv + P' + K') \qquad (6.3.2)$$

where f is the flowrate, P is the pressure and u, v, P' and K' are specific internal energy, volume, potential energy and kinetic energy. The difference between potential and kinetic energy terms in inlet and outlet streams is usually small compared with the remaining terms. In such cases

$$\sum_{\text{outlet}} f(u + Pv) - \sum_{\text{inlet}} f(u + Pv) = Q + W$$

or in terms of the enthalpy $h = u + Pv$,

$$\left.\begin{array}{c} \sum_{\text{outlet}} f_i h_i - \sum_{\text{inlet}} f_i h_i = Q + W \\[2mm] \Delta H = Q + W \end{array}\right\} \qquad (6.3.3)$$

This form of the energy balance equation is the one most frequently used.

Note that lower-case symbols are used for specific quantities and upper case for total quantities, i.e. h has units of $J\,mol^{-1}$ or $J\,kg^{-1}$ whilst H has units of J.

6.3.1.2 *Closed systems.* A closed system has no flow of matter in or out. Batch processes are usually operated in this way. Changes in the internal energy of the system must be due solely to heat flow and work done. Hence

$$U_{\text{final}} - U_{\text{initial}} = Q + W$$

or

$$\Delta U = Q + W \qquad (6.3.4)$$

where U is the total internal energy of the system.

6.3.2 *Estimation of enthalpy changes*

To apply equation 6.3.3 the enthalpy of inlet and outlet streams must be obtained. For a few common substances tables of thermodynamic properties are available (see reference 3 for a reasonably up-to-date list). In most cases a correlation or estimation method has to be used. The thermodynamic background is outside the scope of this chapter but is well covered in standard texts[3]. Reid et al.[4] have produced a compendium of correlation/estimation methods for a wide range of thermophysical properties. Simple correlations for use in energy balances have been compiled by Himmelblau[2]. Most of the data used in this section are taken from that source.

It is necessary to evaluate enthalpy changes for the following elementary processes:

   (i) change of temperature at constant pressure
  (ii) change of pressure at constant temperature
 (iii) change of phase
 (iv) mixing of pure substances
  (v) chemical reaction (section 6.3.3.)

Only the first of these will be elaborated on in this section—for details of (ii) to (v) see references 2–5.

6.3.2.1 *Use of heat capacities.* The change in enthalpy at constant pressure due to a change in temperature from T_0 to T_1 is given by

$$\Delta h = \int_{T_0}^{T_1} c_p \, dT \tag{6.3.5}$$

where c_p is the isobaric heat capacity. Heat capacity data for many substances have been correlated as functions of temperature by simple empirical equations.

Example 6.7

Determine the heat load required to heat a stream of nitrogen, flowing at $50 \, \text{mol} \, \text{min}^{-1}$, from 20°C to 100°C.

Solution

This is a steady-state flow system, hence $\Delta H = Q + W$. Assuming that work done is negligible in comparison with the heating

$$\Delta H \approx Q$$

The heat capacity of nitrogen gas at low pressures is represented[2] by

$$c_p = 29 \cdot 0 + 2 \cdot 2 \times 10^{-3} \, T + 5 \cdot 7 \times 10^{-6} \, T^2$$
$$- 2 \cdot 87 \times 10^{-9} \, T^3 \, \text{J} \, \text{K}^{-1} \, \text{mol}^{-1} \tag{6.3.6}$$

where T is in °C.

$$\Delta h = \int_{20}^{100} c_p(T) \, dT$$

Performing the integration gives $\Delta h = 2332 \, \text{J} \, \text{mol}^{-1}$. The flowrate is

$50\,mol\,min^{-1}$.

$$\therefore Q = 50 \times 2332 = 116\cdot6\,kJ\,min^{-1}$$

$$= 1\cdot94\,kW$$

The pressure was not specified in Example 6.7 but it has little effect on the answer. Experimental data for nitrogen are tabulated below.

P/MPa	$\Delta h(20°C \rightarrow 100°C)/J\,mol^{-1}$
10^{-6}	2330
10^{-1}	2333
1	2358
10	2594
100	2922

At low pressures the heat capacity changes little with pressure. Up to about 1 MPa the use of low pressure equations like 6.3.6 involves little error.

6.3.3 Reactive systems

Energy changes which accompany chemical reactions are usually large and therefore form an important component in the energy balance. Equation 6.3.3 is applicable to reactive systems but care must be taken to use appropriate and consistent expressions for the enthalpy.

6.3.3.1 Enthalpy.
For the purposes of energy balances the enthalpy of a substance i at a pressure P and temperature T may be evaluated by expressing it in the form

$$h_i(P, T) = (h_i(P, T) - h_i^\circ) + \Delta h_{f,i}^\circ \qquad (6.3.7)$$

where $\Delta h_{f,i}^\circ$ is the standard enthalpy change of formation of i from its elements and $h_i^\circ = h_i$ (10 1325 Pa, 298 K) i.e. the enthalpy at standard conditions. Equation 6.3.7 is based on the convention that enthalpies of elements at 10 1325 Pa and 298 K are zero. The total enthalpy is thus defined as the sum of two enthalpy *changes* which are separately evaluated. Enthalpies of formation are tabulated[2,3]. The enthalpy change between standard conditions and process conditions may be evaluated by the methods referred to in section 6.3.2.

6.3.3.2 Balance equations.
The energy balance equation for the continuous steady state process shown below (Fig. 6.13) is

$$\Delta H = H_{out} - H_{in} = Q + W$$

Using equation 6.3.7 and assuming ideal mixing the total enthalpy of the input stream(s) is given by

$$H_{in} = \sum_{input} f_{i,in}[(h_i(P, T)_{in} - h_i^\circ) + \Delta h_{f,i}^\circ],$$

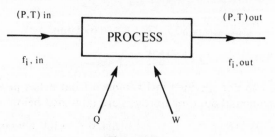

Figure 6.13

where f_i is the flowrate of component i and the sum is over all components in all input streams. A similar equation may be written for the output stream(s). The energy balance equation thus becomes

$$\sum_{\text{output}} f_{i,\text{out}}[(h_i(P, T)_{\text{out}} - h_i^\ominus) + \Delta h_{f,i}^\ominus]$$

$$- \sum_{\text{input}} f_{i,\text{in}}[(h_i(P, T)_{\text{in}} - h_i^\ominus) + \Delta h_{f,i}^\ominus] = Q + W \qquad (6.3.8)$$

Equation 6.3.8 is applicable to both reactive and non-reactive systems. For the latter the standard state enthalpies and enthalpies of formation cancel because input and output flowrates are equal. A more convenient form of equation 6.3.8 for reactive systems is obtained by introducing the enthalpy change for the reaction. For a reaction described by the general stoichiometric equation (equation 6.2.7), the standard enthalpy change is defined as

$$\Delta h_r^\ominus = \sum_i v_i \Delta h_{f,i}^\ominus \qquad (6.3.9)$$

Using the material balance (equation 6.2.8) and equation 6.3.9 to eliminate $\Delta h_{f,i}^\ominus$ terms from equation 6.3.8 gives

$$\sum f_{i,\text{out}}[h_i(P, T)_{\text{out}} - h_i^\ominus] + \xi \Delta h_r^\ominus$$

$$- \sum f_{i,\text{in}}[h_i(P, T)_{\text{in}} - h_i^\ominus] = Q + W \qquad (6.3.10)$$

Note that the above equations apply to reactions taking place at *any* pressure and temperature—they are not restricted to standard conditions.

6.3.4 Energy balance techniques

The following steps are involved in carrying out an energy balance:

(i) Determine flowrates from a material balance.
(ii) Find values of $\Delta h_{f,i}^\ominus$ and evaluate $\Delta h_r^\ominus$.
(iii) Evaluate enthalpy difference between process conditions and standard conditions for each component i.e. $[h_i(P, T) - h_i^\ominus]$.
(iv) Use above information to solve energy balance equation for the desired quantity.

Example 6.8

$100 \, \text{mol min}^{-1}$ of methane are mixed with air in stoichiometric proportions and used to fuel a boiler. The methane is at 25°C and the air is preheated to 100°C. The reaction products leave at 500°C. What is the rate of heat generation in the boiler? A 90% conversion of methane is achieved.

Solution

Flowsheet: (Fig. 6.14)
Basis: 100 mol methane in feed (i.e. 1 min of operation)
Reaction: $CH_4(g) + 2O_2(g) \to CO_2(g) + 2H_2O(g)$

Material balance

There is a 90% conversion of CH_4, hence $\xi = 90 \, \text{mol}$. The quantity of each substance leaving is obtained from the equation

$$f_{i,out} = f_{i,in} + v_i \xi$$

It is convenient to set up a table containing all required flowrates and enthalpy differences. As quantities are evaluated they should be entered in the table; the completed table is shown below.

Substance	$\Delta h_f^{\ominus}$	IN		OUT	
		f_i	$h_i - h_i^{\ominus}$	f_i	$h_i - h_i^{\ominus}$
$CH_{4(g)}$	−74·85	100	0	10	23·10
$O_{2(g)}$	0	200	2·23	20	15·03
$N_{2(g)}$	0	752	2·19	752	14·24
$CO_{2(g)}$	−393·5	—	—	90	21·34
$H_2O_{(g)}$	−241·8	—	—	180	17·00

Enthalpies are in $kJ \, mol^{-1}$; values of f_i are in mol.

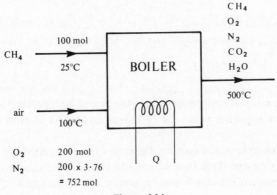

Figure 6.14

Enthalpy of reaction

$$\Delta h_r^\circ = \Delta h_{f,CO_2}^\circ + 2\Delta h_{f,H_2O}^\circ - \Delta h_{f,CH_4}^\circ - 2\Delta h_{f,O_2}^\circ$$

From Δh_r° values in the table[2], $\Delta h_r^\circ = -802\cdot3\,\text{kJ mol}^{-1}$

Enthalpy changes
Using heat capacity equations[2] of the form of equation 6.3.6 the following values are obtained.

CH_4

$$h(500^\circ\text{C}) - h(25^\circ\text{C}) = 23\cdot10\,\text{kJ mol}^{-1}$$

O_2

$$h(100^\circ\text{C}) - h(25^\circ\text{C}) = 2\cdot23\,\text{kJ mol}^{-1}$$
$$h(500^\circ C) - h(25^\circ\text{C}) = 15\cdot03\,\text{kJ mol}^{-1}$$

N_2

$$h(100^\circ\text{C}) - h(25^\circ\text{C}) = 2.19\,\text{kJ mol}^{-1}$$
$$h(500^\circ\text{C}) - h(25^\circ\text{C}) = 14\cdot24\,\text{lkJ mol}^{-1}$$

CO_2

$$h(500^\circ\text{C}) - h(25^\circ\text{C}) = 21\cdot34\,\text{kJ mol}^{-1}$$

$H_2O_{(g)}$

$$h(500^\circ\text{C}) - h(25^\circ\text{C}) = 17\cdot00\,\text{kJ mol}^{-1}$$

Energy balance
Assuming work done is negligible

$$Q = \sum f_{i,out}[h_{i,out} - h_i^\circ] + \xi\Delta h_r^\circ - \sum f_{i,in}[h_{i,in} - h_i^\circ]$$
$$= 16\,220\cdot7 + 90(-803\cdot3) - 2092\cdot9\,\text{kJ}$$
$$= \underline{-58\,080\,\text{kJ}}$$

This is for one minute's operation, hence the rate of heat production is 968 kW.

Note that a negative value for Q indicates heat is transferred from the system.

The final example of the section illustrates the calculation of an adiabatic reaction temperature. This type of calculation is often used to estimate the maximum temperature which can be attained in a reaction, e.g. maximum flame temperature.

Example 6.9

$10\,000\,\text{mol}\,\text{hr}^{-1}$ of limestone (pure $CaCO_3$) are calcined continuously in a kiln by burning $12\,000\,\text{mol}\,\text{hr}^{-1}$ of a fuel gas, with $18\,000\,\text{mol}\,\text{hr}^{-1}$ of air, in direct contact with the limestone. The limestone enters the kiln at 25°C, the fuel gas at 400°C and the air at 25°C. The gaseous products leave the kiln at 200°C and consist of CO_2, N_2 and O_2 only. Assuming there is no heat loss, determine the temperature of the solid product (pure CaO) leaving the kiln.

The molar composition of the fuel gas is CO 60%, CO_2 13% and N_2 27%. The mean heat capacity of the CaO product is $47\,\text{J}\,\text{mol}^{-1}\text{K}^{-1}$.

Solution

Flowsheet: (Fig. 6.15)
Basis: 10 mol limestone in feed.

Material balance
The following reactions take place,

calcination: $CaCO_3 \rightarrow CaO + CO_2$ (1)
combustion: $CO + \tfrac{1}{2}O_2 \rightarrow CO_2$ (2)

The fuel gas contains $2\cdot7\,\text{mol}\,CO$, $1\cdot56\,\text{mol}\,CO_2$ and $3\cdot24\,\text{mol}\,N_2$. Air contains $3\cdot76\,\text{mol}\,N_2/\text{mol}\,O_2$ i.e. $3\cdot78\,\text{mol}\,O_2$ and $14\cdot22\,\text{mol}\,N_2$ are supplied in the air stream. Both reactions go to completion thus $\xi_1 = 10\,\text{mol}$ and $\xi_2 = 7\cdot2\,\text{mol}$. Quantities pertaining to inlet and outlet streams are entered in the table below. Temperature T in °C, flowquantities f in mol, $\Delta h_f^\ominus$ in $\text{kJ}\,\text{mol}^{-1}$, $\Delta h = h(P, T) - h^\ominus$ in $\text{kJ}\,\text{mol}^{-1}$.

Substance	$\Delta h_f^\ominus$	IN			OUT		
		T	f	Δh	T	f	Δh
$CaCO_3$	$-1206\cdot9$	25	10	0	—	—	—
CaO	$-635\cdot6$	—	—	—	T_s	10	see text
CO	$-110\cdot5$	400	7·2	11·25	—	—	—
CO_2	$-393\cdot5$	400	1·56	16·35	200	18·76	7·08
N_2 (fuel gas)	0	400	3·24	11·35 ⎱			
				⎰	200	17·46	5·13
N_2 (air)	0	25	14·22	0			
O_2	0	25	3·78	0	200	0·18	5·13

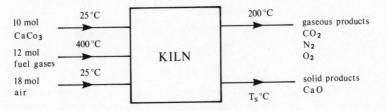

Figure 6.15

Enthalpies of reaction
Reaction 1 (calcination)

$$\Delta h_{r,1}^{\circ} = -393 \cdot 5 - 635 \cdot 6 - (-1206 \cdot 9)$$
$$= 177 \cdot 7 \, \text{kJ mol}^{-1}$$

Reaction 2 (combustion)

$$\Delta h_{r,2}^{\circ} = -393 \cdot 5 - (-110 \cdot 5)$$
$$= -283 \cdot 0 \, \text{kJ mol}^{-1}$$

Enthalpy differences
Enthalpy differences between process conditions and standard conditions have been entered in the table. These are evaluated in the usual way from heat capacity correlations. Note that nitrogen in the fuel gas (at 400°C) and nitrogen in air (at 25°C) must be treated separately when calculating the input enthalpy.

For CaO in the outlet stream the temperature is not known. Using the mean heat capacity we obtain the following expression

$$\Delta h_{\text{CaO}} = 0 \cdot 047 (T_{s} - 25 °C) \, \text{kJ mol}^{-1}$$

where T_s is the outlet temperature of the CaO in °C.

Energy balance
The counterpart of equation 6.3.10 for the case of multiple reactions is

$$\sum f_{i,\text{out}}[h_{i,\text{out}} - h_i^{\circ}] + \sum_{\substack{\text{reactions} \\ k}} \xi_k \Delta h_{r,k}^{\circ}$$

$$- \sum f_{i,\text{in}}[h_{i,\text{in}} - h_i^{\circ}] = Q + W$$

Assuming that the reactor operates adiabatically, $Q = 0$. The work term is taken to be negligible.

Using data from the energy balance table

$$\sum f_{i,\text{in}}[h_{i,\text{in}} - h_i^{\circ}] = 7 \cdot 2 \times 11 \cdot 25 + 1 \cdot 56 \times 16 \cdot 35 + 3 \cdot 42 \times 11 \cdot 15$$
$$= 144 \cdot 64 \, \text{kJ}$$

$$\sum f_{i,\text{out}}[h_{i,\text{out}} - h_i^{\circ}] = 18 \cdot 76 \times 7 \cdot 08 + 17 \cdot 44 \times 5 \cdot 13 + 0 \cdot 18$$
$$\times 5 \cdot 13 + 10 \times 0 \cdot 047 (T_s - 25)$$
$$= 211 \cdot 46 + 0 \cdot 47 \, T_s \, \text{kJ}$$

$$\sum_k \xi_k \Delta h_{r,k}^{\circ} = 10 \times 177 \cdot 8 - 7 \cdot 2 \times 283 \cdot 0$$
$$= -259 \cdot 6 \, \text{kJ}$$

Inserting terms in the energy balance equation

$$211{\cdot}46 + 0{\cdot}47\,T_s - 259{\cdot}6 - 144{\cdot}64 = 0$$

$\therefore$ Temperature of solid product $\underline{T_s = 410°C}$

6.4 Fluid flow

Materials are usually transferred between different parts of a chemical plant in the fluid state. Large quantities of gases, liquids and fluidized solids may be easily and economically transported by pumping along a network of pipes. In process design and analysis it is necessary to determine pump duties, pipe sizes, pressure drops, flowrates etc. This section reviews the basic principles of fluid mechanics on which such calculations depend.

6.4.1 *Types of fluid*

The term 'fluid' covers a wide range of materials—from gases and simple liquids to polymeric materials and semi-solid slurries. Fluids may be classified as either *compressible* or *incompressible*. The density of a compressible fluid depends on the pressure. Although this is true for all real fluids, the compressibility of liquids is very small under most conditions and they may be considered incompressible. The flow of gases must usually be treated as compressible unless pressure changes are small.

A second type of classification may be made according to the behaviour of a fluid subject to shear stress. Consider a fluid contained between parallel plates as shown in Fig. 6.16. The upper plate is moved in the x direction at a constant velocity V by a constant force F. The shear stress exerted on the fluid by the moving plate of area A is $\tau = F/A$. For a *Newtonian fluid* in laminar flow* the velocity profile in the y direction is linear and the shear stress is proportional to the velocity gradient,

$$\tau = \mu \frac{dv}{dy} \tag{6.4.1}$$

The constant of proportionality μ is the *viscosity* of the fluid and equation 6.4.1 is known as *Newton's law of viscosity*. The viscosity of a fluid is a measure of how easily momentum is transferred through the fluid. A high viscosity indicates rapid momentum transfer from the moving plate to the stationary plate and hence a high resistance to shear. The units of shear stress are the same as those of pressure, i.e. Nm^{-2} or Pa, therefore viscosity has units of Pa s. The viscosity of water at room temperature is approximately 10^{-3} Pa s whereas that of air is about 2×10^{-5} Pa s.

---

*see section 6.4.2.1

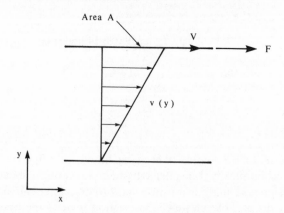

Figure 6.16   Velocity profile for a Newtonian fluid subject to shear.

Gases and most simple liquids (e.g. water) are Newtonian fluids. Non-Newtonian fluids do not obey equation 6.4.1; they include many materials of industrial importance. Examples are polymeric substances, slurries and suspensions. For non-Newtonian fluids the viscosity depends on the shear stress and may also depend on the previous deformation history of the fluid.

The treatment of fluid flow in this chapter is restricted to incompressible flow of Newtonian fluids. For a wider coverage see textbooks detailed in references 6, 7 and 8.

6.4.2 Flow regimes

6.4.2.1 *Laminar and turbulent flow.* The flow pattern which develops when a given fluid flows through a pipe or channel of fixed dimensions depends on the velocity. At low velocities the pattern tends to be *laminar* or *streamline*—adjacent layers of fluid move past each other with no mixing in the direction normal to the flow. At higher velocities the flow becomes *turbulent*—smooth laminar flow breaks up and eddies form. These eddies are of varying sizes and move rapidly in random directions. Turbulent flow results in a high degree of mixing between fluid elements and hence leads to high rates of convective heat transfer (see section 6.5.1.2). In laminar flow, by contrast, mixing only occurs by diffusion and heat transfer by conduction—both of which are relatively slow processes.

6.4.2.2 *Reynolds number.* Experiments show that the velocity at which the transition from laminar to turbulent flow occurs depends on the physical properties of the fluid and the geometry of the flow. The nature of the flow is indicated by a dimensionless group[9] known as the *Reynolds number Re*. The Reynolds number represents a ratio of inertial forces (rate of change of momentum of fluid elements) to viscous shear forces acting in a fluid. For flow in a pipe the Reynolds number is defined as

$$Re = \frac{\bar{v}\,d\rho}{\mu}$$ (6.4.2)

where $\bar{v}$ is the mean velocity, d is the pipe diameter, ρ is the density of the fluid and μ its viscosity. When these quantities are expressed in a consistent set of units Re is a pure number, i.e. dimensionless.

The following criteria for flow regimes within smooth pipes have been established

$Re < 2000$ laminar flow
$2000 < Re < 4000$ transition between periods of laminar and turbulent flow
$Re > 4000$ turbulent flow

Large values of Re correspond to highly turbulent flow with inertial forces dominant: because of high velocity, high density, large diameter or low viscosity. At low Re viscous forces are dominant and hence the flow is laminar.

6.4.2.3 *Boundary layer*. For all flow regimes, whether laminar or turbulent, the effects of viscous shear forces are greatest close to solid boundaries. Fluid actually in contact with a surface usually has no relative motion; the so called *no-slip* condition. There is therefore a region extending from the surface to the bulk of the fluid within which the velocity changes from zero to the bulk value. This region is known as the *boundary layer*.

The velocity distribution for flow over a large, smooth, flat plate is shown schematically in Fig. 6.17. The boundary layer thickness δ is conventionally (arbitrarily) defined as the distance from the surface at which the velocity is 0·99 of the bulk velocity.

In air δ would be of the order of several mm. The boundary layer is an important concept in fluid mechanics because it allows a fluid to be separated

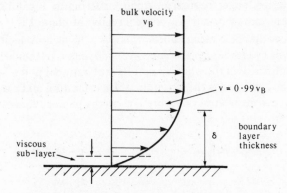

Figure 6.17 Velocity profile in flow over a flat plate.

into two regions; (a) the boundary layer, which contains the whole of the velocity gradient and all viscous effects; (b) the bulk fluid in which viscous forces are small compared with other forces. Flow in the boundary layer may be laminar or turbulent; however, very close to the wall viscous forces always dominate in a thin region called the *viscous sub-layer*.

When a fluid flows through a pipe, boundary layers form at the entry point and grow in thickness along the length of the pipe until they meet in the centre. The boundary layer thus fills the entire pipe and the flow is termed *fully developed*. If the boundary layer is still laminar then laminar flow persists.

6.4.3 *Balance equations*

The important concepts of material and energy balances were introduced in previous sections. They are restated below in forms more appropriate for fluid flow. The momentum balance gives an additional conservation equation. Together these three balance equations provide the basis for studying fluid flow phenomena.

6.4.3.1 *Material balance—continuity.* In fluid mechanics the material balance is usually known as the continuity equation. Figure 6.18 shows fluid flowing into a region through plane 1, area A_1, with a mean velocity $\bar{v}_1$, and density ρ_1. Fluid leaves across plane 2. For steady flow there is no accumulation of mass in the region and hence

$$\dot{m} = \rho_1 \bar{v}_1 A_1 = \rho_2 \bar{v}_2 A_2 \tag{6.4.3}$$

where $\dot{m}$ is the mass flow rate. For a fluid of constant density equation 6.4.3 reduces to

$$\bar{v}_1 A_1 = \bar{v}_2 A_2 \tag{6.4.4}$$

6.4.3.2 *Energy balance.* The energy balance applied to fluid flow must include the kinetic and potential energy terms of equation 6.3.2. Changes in these quantities are no longer negligible. Again referring to Fig. 6.18, the kinetic energy per unit mass entering the control region at plane 1 is $(\frac{1}{2}\bar{v}_1^2)\alpha_1$, where $\alpha = \bar{v^3}/\bar{v}^3$, the ratio of the mean cubed velocity to the mean velocity cubed. α is usually referred to as the *kinetic energy correction factor*. It is equal to unity if the velocity is uniform over the whole area. The gravitational potential energy per unit mass is gz_1, where g is the gravitational acceleration and z represents the height (relative to an arbitrary datum). The energy balance equation for steady flow becomes

$$(\bar{v}_1 A_1 \rho_1)\left[u_1 + \frac{P_1}{\rho_1} + gz_1 + \frac{\alpha_1}{2}\bar{v}_1^2 \right] + Q + W$$

$$= (\bar{v}_2 A_2 \rho_2)\left[u_2 + \frac{P_2}{\rho_2} + gz_2 + \frac{\alpha_2}{2}\bar{v}_2^2 \right] \tag{6.4.5}$$

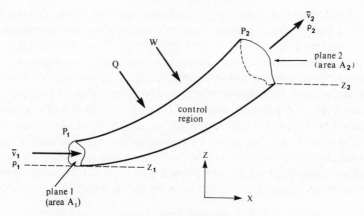

Figure 6.18 Control volume for deriving balance equations.

Note that Pv in equation 6.3.2 has been written as P/ρ. By the continuity equation $\bar{v}_1 A_1 \rho_1 = \bar{v}_2 A_2 \rho_2$, thus

$$u_1 + \frac{P_1}{\rho_1} + gz_1 + \frac{\alpha_1}{2}\bar{v}_1^2 + Q' + W'$$

$$= u_2 + \frac{P_2}{\rho_2} + gz_2 + \frac{\alpha_2}{2}\bar{v}_2^2 \qquad (6.4.6)$$

where Q' and W' are heat transfer and work per unit mass.

Energy required to overcome viscous forces within the fluid and friction at the walls is dissipated as heat. Some of this energy goes out through the walls (Q') and the rest goes into raising the internal energy of the fluid ($u_2 - u_1$). The above energy changes may be grouped together in a 'loss' term

$$l = u_2 - u_1 - Q'$$

where l represents the loss of mechanical energy by the system (transformed into thermal energy). Equation 6.4.6 may therefore be written as

$$\frac{P_2}{\rho_2} + gz_2 + \frac{\alpha_2}{2}\bar{v}_2^2 = \frac{P_1}{\rho_1} + gz + \frac{\alpha_1}{2}\bar{v}_1^2 + W' - l \qquad (6.4.7)$$

This is one of the many forms of the *Bernoulli equation*.

Energy associated with fluid flow is often discussed in terms of the *equivalent head* of liquid. Dividing equation 6.4.7 by g gives

$$\frac{P_2}{\rho_2 g} + z_2 + \frac{\alpha_2}{2g}\bar{v}_2^2 = \frac{P_1}{\rho_1 g} + z_1 + \frac{\alpha_1}{2g}\bar{v}_1^2 + \frac{W'}{g} - h_l \qquad (6.4.8)$$

where each term now has units of length. The mechanical energy losses are expressed as a head loss h_l.

6.4.3.3 *Momentum equation.* The momentum equation introduces the force acting on a body or fluid element. Newton's second law states that the sum of forces acting on a body is equal to the rate of change of momentum of the body. Force and momentum are vector quantities—we shall treat each coordinate direction separately in the following derivation of the momentum equation.

Consider a steady flow through the control volume in Fig. 6.18 with uniform velocities over each plane at positions 1 and 2. We shall first carry out a momentum balance in the x direction. Let v_{1x} denote the x component of the velocity at plane 1, etc. The mass flowrate into the control volume is $v_1\rho_1 A_1$ and hence the rate at which x momentum enters is $v_{1x}(v_1\rho_1 A_1)$. The rate at which x momentum leaves through plane 2 is $v_{2x}(v_2\rho_2 A_2)$. The rate of change of x momentum is equal to the net force on the fluid in the x direction:

$$F_x = v_{2x}(v_2\rho_2 A_2) - v_{1x}(v_1\rho_1 A_1)$$

Using the continuity equation (equation 6.4.3) gives

$$F_x = \dot{m}(v_{2x} - v_{1x}) \qquad (6.4.9)$$

with corresponding equations for the y and z directions.

In general the velocity across any sizeable area is not uniform and a momentum correction factor β must be introduced to account for this; $\beta = \overline{v^2}/\bar{v}^2$. The general momentum equation is

$$F_x = \dot{m}(\beta_2\bar{v}_{2x} - \beta_1\bar{v}_{1x}) \qquad (6.4.10)$$

Values of the momentum and kinetic-energy correction factors depend on the details of the velocity distribution for a particular flow. For flow in circular pipes the following values are obtained[7]:

	α	β
laminar	2	4/3
turbulent	≈ 1	≈ 1

In the majority of practical cases flows are turbulent and the corrections may then be omitted.

6.4.4 *Flow in pipes*

In this section we shall apply the balance equations to a selection of problems concerned with flow of fluids in pipes.

6.4.4.1 *Laminar flow.* We shall derive equations for flow in a horizontal straight pipe of circular cross-section. It will be assumed that the flow is steady and fully developed, i.e. the velocity distribution does not change with time or distance along the pipe.

Figure 6.19 shows a cylinder of fluid, radius r and length L, within a pipe of radius R. Consider a momentum balance in the axial (x) direction. Because the flow is fully developed $\bar{v}_{1x} = \bar{v}_{2x}$ and $\beta_1 = \beta_2$; hence from equation 6.4.10 the

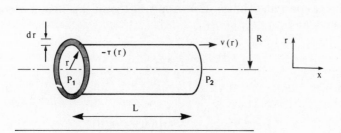

Figure 6.19   Notation used in pipe-flow equations.

total force on the cylinder in the x direction is zero. That part of the force due to the pressure difference acts on the ends of the cylinder and is given by

$$\pi r^2 P_1 - \pi r^2 P_2$$

acting in the $+x$ direction.

There is also a viscous shear force which acts on the fluid at the outer surface of the cylinder and is given by a product of the shear stress and surface area, i.e.

$$- \tau(r)2\pi rL$$

where $\tau(r)$ is the shear stress at radius r. The negative sign indicates that the force exerted on the cylinder by the fluid outside is in the $-x$ direction. The total force on the cylinder is, therefore,

$$\pi r^2 (P_1 - P_2) - \tau(r)2\pi rL = 0$$

$$\therefore \tau(r) = \frac{r(P_1 - P_2)}{2L} \tag{6.4.11}$$

But for laminar flow the shear stress is given by equation 6.4.1 which, for flow in a pipe, may be written as

$$\tau(r) = - \mu \frac{dv(r)}{dr}$$

where $v(r)$ is the velocity in the x direction at radius r (the sign is changed because here r is measured in the direction *towards* the wall in contrast to Fig. 6.16).

Substituting in equation 6.4.11,

$$\frac{dv}{dr} = - \frac{(P_1 - P_2)r}{2L\mu}$$

Integrating the above equation gives

$$v = - \frac{(P_1 - P_2)r^2}{4L\mu} + c_1$$

where c_1 is a constant of integration. Applying the no-slip boundary condition, $v = 0$ at $r = R$

$$\therefore 0 = -\frac{(P_1 - P_2)R^2}{4L\mu} + c_1$$

hence

$$v(r) = \left(\frac{P_1 - P_2}{4\mu L}\right)R^2(1 - (r/R)^2) \tag{6.4.12}$$

The parabolic velocity profile represented by equation 6.4.12 is shown in Fig. 6.20. It will now be used to evaluate the flowrate through the pipe. The volumetric flowrate through the shaded annular area in Fig. 6.19 is given by the product of the velocity at radius r and the area,

$$dQ = v(r)2\pi r dr$$

$$= \left(\frac{P_1 - P_2}{4\mu L}\right)2\pi(R^2 r - r^3)dr$$

The total flowrate is obtained by integrating over the entire cross-section:

$$Q = \int_0^R dQ = \frac{\pi(P_1 - P_2)}{2\mu L}\int_0^R (R^2 r - r^3)dr$$

$$\therefore Q = \frac{\pi R^4(P_1 - P_2)}{8\mu L} \tag{6.4.13}$$

This result is known as the *Hagen-Poiseuille equation*.

6.4.4.2 *Friction in pipes*. For straight horizontal pipes resistance to flow arises because of viscous shear or friction at the wall. Dimensional analysis[8] leads to the conclusion that for smooth pipes the pressure drop (scaled to make it dimensionless) is a function only of the Reynolds number. The dimensionless group containing the pressure drop is known as the *friction factor*:

$$f = \frac{\Delta P d}{2L\rho\bar{v}^2} \tag{6.4.14}$$

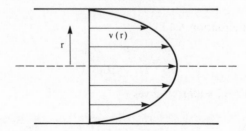

Figure 6.20 Velocity profile for laminar flow within a circular pipe.

where ΔP is the magnitude of the pressure drop, L is the pipe length and d is the diameter. The friction factor defined in equation 6.4.14 is the *Fanning friction factor*—friction factors defined in other ways are sometimes encountered, e.g. $4f$ or $f/2$.

For laminar flow the mean velocity may be obtained from equation 6.4.13:

$$\bar{v} = Q/\pi R^2$$

$$= \frac{d^2 \Delta P}{32\mu L}$$

$$\therefore f = \frac{16\mu}{\rho \bar{v} d}$$

or using equation 6.4.2

$$f = 16/Re \qquad (6.4.15)$$

There is no simple analytic treatment for turbulent flow. Experimental measurements of pressure drops have been used to determine the relationship between f and Re as shown in Fig. 6.21. For $Re \leqq 10^5$ the data are well represented by the *Blasius equation*:

$$f = 0.079 \, Re^{-1/4} \qquad (6.4.16)$$

In practice pipes are not smooth and for turbulent flow it is found that f also depends on surface roughness. Friction factor charts for rough pipes are available[7].

A relationship between the friction factor and mechanical energy losses in a flowing fluid may be obtained from the energy balance in equation 6.4.7. For a horizontal pipe

$$\frac{P_2}{\rho} + \frac{\alpha_2 \bar{v}_2}{2} = \frac{P_1}{\rho} + \frac{\alpha_1 \bar{v}^2}{2} - l_f$$

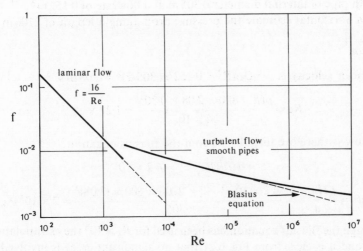

Figure 6.21 Fanning friction factor as a functioning Reynolds number for circular pipes.

If the flow is steady and fully established

$$\bar{v}_1 = \bar{v}_2 \quad \text{and} \quad \alpha_1 = \alpha_2$$

$$\frac{P_1 - P_2}{\rho} = \frac{\Delta P}{\rho} = l_f$$

From the definition of the friction factor (Equation 6.4.14)

$$l_f = \frac{f 2\bar{v}^2 L}{d} = \frac{\bar{v}^2}{2}\left(\frac{4fL}{d}\right) \tag{6.4.17}$$

The head loss due to friction is therefore

$$h_f = \frac{f 2\bar{v}^2 L}{gd} = \frac{\bar{v}^2}{2g}\left(\frac{4fL}{d}\right) \tag{6.4.18}$$

Pipe flow problems are typically of two types:

(i) Pipe size given—determine pressure drop for specified flow rate. Re can be evaluated immediately and hence f is found from the chart or correlating equations.

(ii) Pipe size given—estimate flow for specified pressure drop. An iterative solution is necessary because the flowrate is unknown; guess Re, estimate f, calculate $\bar{v}$ from energy equation and repeat until convergence.

The following examples illustrate these procedures.

Example 6.10

Crude oil (density $800\,\text{kg}\,\text{m}^{-3}$, viscosity $4 \times 10^{-3}\,\text{Pa s}$) is pumped through a smooth pipe of internal diameter $0.305\,\text{m}$ at a flowrate of $0.152\,\text{m}^3\,\text{s}^{-1}$. If the pipe is horizontal estimate the pressure drop along a length of $2100\,\text{m}$.

Solution

The mean velocity is $\bar{v} = Q/\pi R^2 = 0.152/\pi(.305/2)^2 = 2.08\,\text{m s}^{-1}$

$$Re = \frac{\rho \bar{v} d}{\mu} = \frac{800 \times 2.08 \times 0.305}{4 \times 10^{-3}} = 1.27 \times 10^5$$

The flow is therefore turbulent. From the Blasius equation

$$f = 0.079\,Re^{-1/4} = 4.2 \times 10^{-3}$$

$$\Delta P = \frac{f 2 L \rho \bar{v}^2}{d} = \frac{4.2 \times 10^{-3} \times 2 \times 2100 \times 800 \times (2.08)^2}{.305} = 2 \times 10^5\,\text{Pa}.$$

Although the Blasius equation has been used for $R_e > 10^5$ the extrapolation is small and it is clear from Fig. 6.21 that no significant error is involved.

Example 6.11

Water at a temperature of 20°C is flowing through a smooth pipe of internal diameter 0·2 m. The pressure drop over a horizontal test section of length 305 m is measured as $0·21 \times 10^5$ Pa. Estimate the flowrate. (Density of water at $20°C = 988$ kg m^{-3}, viscosity $= 10^{-3}$ Pa s.)

Solution

From equation 6.4.14,

$$\bar{v} = \left(\frac{\Delta Pd}{2L\rho f}\right)^{1/2}$$

hence, inserting values of known quantities,

$$\bar{v} = \left(\frac{6·96 \times 10^{-3}}{f}\right)^{1/2} \text{ m s}^{-1} \quad \text{and} \quad Re = 1·976 \times 10^5 \bar{v}.$$

Iteration 1
Assume flow is turbulent, guess $Re = 10^5$. From the Blasius equation $f = 4·44 \times 10^3$, thus $\bar{v} = 0·8$ m s^{-1}.

Iteration 2
$Re = 1·976 \times 10^5 \times 0·8 = 1·5 \times 10^5$, $\qquad f = 0·079Re^{-1/4} = 3·96 \times 10^{-3}$,
$\bar{v} = 0·75$ m s^{-1}

Iteration 3
$Re = 1·49 \times 10^5$, $f = 4·02 \times 10^{-3}$, $\bar{v} = 0·76$ m s^{-1}

Iteration 4
$Re = 1·50 \times 10^5$, $f = 4·01 \times 10^{-3}$, $\bar{v} = 0·76$ m s^{-1}
The calculation has converged. $Q = \pi R^2 \bar{v} = 2·3 \times 10^{-2}$ m^3 s^{-1} or $\dot{m} = \rho Q = 23·6$ kg s^{-1}.

6.4.4.3 *Bends and fittings.* Piping systems seldom consist of straight pipes of constant diameter. It is therefore necessary to account for pressure drops and mechanical energy losses associated with bends, changes in diameter, valves and other fittings.

For turbulent flow, losses are proportional to the square of the mean velocity. It is therefore customary to express pipeline losses in terms of *velocity heads* $\bar{v}^2/2g$. The loss associated with a particular fitting may be written as

$$h_f = K_f \frac{\bar{v}^2}{2g} \tag{6.4.19}$$

where K_f is the *loss coefficient*. Some typical values of loss coefficients are given

Table 6.1 Loss coefficients for bends and fittings in turbulent flow

Type	K_f
45° elbow	0·3
90° elbow	0·8
90° square elbow	1·2
coupling, no diameters change	~0
side outlet of T piece	1·8
gate valve—fully open	~0·2
gate valve—$\frac{1}{2}$ open	~5
globe valve—fully open	~5
globe valve—$\frac{1}{2}$ open	~10
entry to pipe from large vessel	~0·5
sudden expansion, cross-section $A_1 \rightarrow A_2$	$(1 - A_1/A_2)^2$

in Table 6.1. Although these are mostly determined by experiment an approximate analytical treatment is possible in some simple cases; this is illustrated in the next example.

Example 6.12

Use the conservation equations to determine the loss coefficient for a sudden expansion in a horizontal pipe.

Solution

Consider the flow arrangement shown in Fig. 6.22. We shall first apply the momentum equation to find the pressure difference between planes 1 and 2,

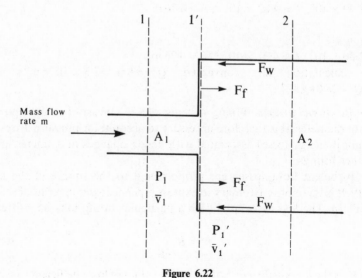

Figure 6.22

followed by the energy equation to obtain the losses. Taking components of force and velocity in the direction of flow

$$F = \dot{m}(\beta_2 \bar{v}_2 - \beta_1 \bar{v}_1)$$

where F is the total force on the fluid between planes 1 and 2. This force is made up of three contributions:

   (i) Pressure/area difference, $F_p = P_1 A_1 - P_2 A_2$
   (ii) Shear stress at wall, F_w
   (iii) Force exerted on fluid by fitting, F_f

Because planes 1 and 2 are close together the stresses at the wall are small compared with other terms, thus $F_w \simeq 0$. If required, an approximate value for F_w may be evaluated using equations 6.4.11 and 6.4.14.

   The force exerted by the fitting (at plane 1') is equal and opposite to the force exerted by the fluid, i.e. $F_f = P_1'(A_2 - A_1)$. At plane 1', just inside the expansion, the flow is still confined to an area A_1. Plane 2 is chosen to be at a point where the flow is fully expanded. Thus by continuity $\bar{v}_1' = \bar{v}_1$, and from the energy equation 6.4.7 $P_1' = P_1$ (assuming friction losses are negligible). It follows that $F_f = P_1(A_2 - A_1)$. Substituting in the momentum equation and setting $\beta_1 = \beta_2 = 1$ for turbulent flow gives

$$(P_1 A_1 - P_2 A_2) + P_1(A_2 - A_1) = \dot{m}(\bar{v}_2 - \bar{v}_1)$$

$$\Rightarrow (P_1 - P_2) = \frac{\dot{m}}{A_2}(\bar{v}_2 - \bar{v}_1) \tag{1}$$

For turbulent flow in a horizontal pipe the energy equation gives

$$h_f = \frac{(P_1 - P_2)}{\rho g} + \frac{(\bar{v}_1^2 - \bar{v}_2^2)}{2g} \tag{2}$$

From the continuity equation

$$\dot{m} = \rho \bar{v}_1 A_1 = \rho \bar{v}_2 A_2 \tag{3}$$

$$\therefore \bar{v}_2 = \bar{v}_1 A_1 / A_2 \tag{4}$$

Substituting (1), (3) and (4) in (2)

$$h_f = \frac{\bar{v}_1^2}{2g}\left(1 - \frac{A_1}{A_2}\right)^2$$

hence from the definition of the loss coefficient in equation 6.4.19

$$K_f = \left(1 - \frac{A_1}{A_2}\right)^2$$

Note that this value is based on the upstream velocity $\bar{v}_1$. A different expression is obtained if the coefficient is based on $\bar{v}_2$.

In the final example the methods of this section are applied to estimate a pump duty.

Example 6.13

Water is pumped at the rate of $35\,\mathrm{kg\,s^{-1}}$ from the collection basin of a cooling tower and delivered to the storage tank of a chemical plant at a higher level. The flow arrangement and process parameters are shown in Fig. 6.23. Estimate the power input required for a pump which has an efficiency of 60%. ($\rho = 10^3\,\mathrm{kg\,m^{-3}}$, $\mu = 10^{-3}\,\mathrm{Pa\,s}$.)

Solution

The energy equation will be applied between the two control points shown in the diagram. From equation 6.4.7 the work done (per unit mass) is

$$W' = \frac{P_2 - P_1}{\rho} + \left(\frac{\alpha_2\bar{v}_2^2}{2} - \frac{\alpha_1\bar{v}_1^2}{2}\right) + g(z_2 - z_1) + l \tag{1}$$

$P_1 \simeq 10^5$ Pa since the collection basin is open to the atmosphere. $\bar{v}_1 = \bar{v}_2 = 0$. The velocity in the pipeline is

$$\bar{v} = \frac{\dot{m}}{\rho\pi R^2} = \frac{35}{10^3\pi(0\cdot1)^2} = 1\cdot1\,\mathrm{m\,s^{-1}}$$

$Re = \dfrac{\bar{v}\rho d}{\mu} = 2\cdot23 \times 10^5$, the flow is turbulent.

Losses are made up of frictional losses in the pipe and 'minor losses'

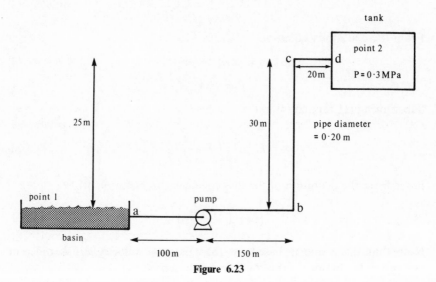

Figure 6.23

associated with fittings.

$$l = \frac{\bar{v}^2}{2}\left(\frac{4fL}{d} + K_a + K_b + K_c + K_d\right)$$

Loss coefficients are taken from Table 5.1. Point a is a pipe entry, $K_a = 0.5$; b and c are 90° elbows, $K_b = K_c = 0.8$; d is a sudden expansion with $A_1/A_2 \simeq 0$, $K_d = 1$.

Assuming that the pipe is smooth $f = 0.079 Re^{-1/4} = 3.6 \times 10^{-3}$, hence

$$\frac{4fL}{d} = \frac{4 \times 3.6 \times 10^{-3} \times 300}{0.2} = 21.6$$

$$\therefore l = \frac{\bar{v}^2}{2}(21.6 + 0.5 + 0.8 + 0.8 + 1)$$

$$= 24.7\bar{v}^2/2$$

Substituting values in equation (1):

$$W' = \frac{(3 \times 10^5 - 1 \times 10^5)}{10^3} + 0 + 9.8(25) + \frac{1.24}{2}(24.7)$$

$= 200$	$+ 0$	$+ 245$	$+ 15.3$
pressure	kinetic	height	losses
difference	energy	difference	

$$= 461 \, \mathrm{J\,kg^{-1}}$$

The mass flow rate is $35 \, \mathrm{kg\,s^{-1}}$ thus the power requirement is $16.1 \, \mathrm{kW}$. For a pump with 60% efficiency the power input is $\underline{26.8 \, \mathrm{kW}}$.

6.5 Heat transfer

Energy transfer in the form of heat is part of almost all chemical processes. Estimating and controlling the rate of heat flows in chemical reactors, separation processes, furnaces and boilers represent important problems in plant design and operation. Efficient energy utilization is also of some importance and this too requires a basic understanding of heat transfer. The scope of this section is restricted to simple heat transfer without change of phase in the heating or cooling medium. More comprehensive treatments are available elsewhere[10-12].

6.5.1 Mechanisms

The driving force for heat transfer is temperature difference: heat will only flow from a hotter to a colder part of a system. The mechanisms of heat transfer may be conveniently classified as conduction, convection and radiation.

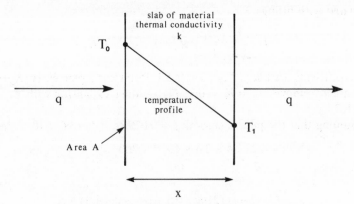

Figure 6.24   Conductive heat transfer through a slab.

6.5.1.1 *Conduction.* Heat flow by conduction is a result of transfer of kinetic and/or internal energy between molecules in a fluid or solid. The basic equation of conductive heat transfer is *Fourier's law*

$$q = -k\frac{dT}{dx} \tag{6.5.1}$$

where q is the heat flux (rate of heat transfer per unit area), dT/dx is the temperature gradient in the x direction and k is the thermal conductivity of the material through which heat is flowing. Like viscosity, thermal conductivity is an intrinsic thermophysical property: values at 300 K range from 400 W m^{-1} K^{-1} for copper, through 0·6 W m^{-1} K^{-1} for water, to 0·03 W m^{-1} K^{-1} for air.

Steady-state conduction through a slab of material is represented in Fig. 6.24. For a material of constant thermal conductivity the temperature profile must be linear because the heat flux at all points is the same. Equation 6.5.1 may therefore be written as

$$q = k\frac{(T_0 - T_1)}{x} \tag{6.5.2}$$

The rate of heat transfer is equal to the product of the heat flux and the area

$$Q = qA = \frac{kA}{x}(T_0 - T_1) \tag{6.5.3}$$

Materials of different thermal conductivity are frequently used in adjacent layers to provide thermal insulation. Consider a flat composite slab made of layers as shown in Fig. 6.25. The heat flux through each layer and the whole slab is equal, hence

$$q = h_1(T_0 - T_1) = h_2(T_1 - T_2) = h_3(T_2 - T_3)$$
$$= U(T_0 - T_3) \tag{6.5.4}$$

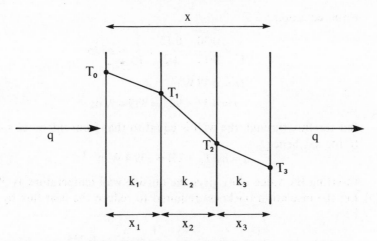

Figure 6.25 Temperature profile for heat transfer through a composite slab.

where $h_1 = k_1/x_1$ is the *heat transfer coefficient* of layer 1 etc. and U is the *overall heat transfer coefficient* of the slab. From equation 6.5.4 the temperature difference across layer i is given by

$$\Delta T_i = q/h_i$$

and the overall temperature difference is $\Delta T = q/U$
since $\Delta T = \Delta T_1 + \Delta T_2 + \Delta T_3$

$$\Rightarrow \frac{1}{U} = \frac{1}{h_1} + \frac{1}{h_2} + \frac{1}{h_3} \qquad (6.5.5)$$

i.e. the reciprocals of heat transfer coefficients (thermal resistances) are additive.

Example 6.14

A composite furnace wall consists of 0·30 m hot face insulating brick and 0·15 m of building brick. The thermal conductivities are 0·12 and 1·2 W m^{-1} K^{-1} respectively. The inside wall of the furnace is at 950°C and the surrounding atmospheric temperature is 25°C. The heat transfer coefficient from the brick surface to air is 10 W m^{-2} K^{-1}.

Estimate

(i) the surface temperature of the outside wall of the furnace
(ii) the thickness of insulating brick required to reduce heat loss from the furnace by 10%.

Solution

The problem statement is summarized in Fig. 6.26.

(i) The heat flux is $q = U\Delta T = U(950-25)$

From equation 6.5.5.

$$\frac{1}{U} = \frac{0.30}{0.12} + \frac{0.15}{1.2} + \frac{1}{10} = 2.725$$

$$\therefore U = 0.37 \, \text{W m}^{-2} \, \text{K}^{-1}$$

$$q = 0.37 \times 925 = 339.4 \, \text{W m}^{-2}$$

But the flux through the wall is equal to that from the outer surface to the air, hence

$$q = h_3(T_2 - 25) = 339.4 \, \text{W m}^{-2}$$

Inserting the value of h_3 gives the outside wall temperature as $\underline{59^\circ \text{C}}$.

(ii) Let the insulation thickness required to reduce the heat flux by 10% be x_1.

$$\frac{1}{U'} = \frac{x_1}{0.12} + \frac{0.15}{1.2} + \frac{1}{10} = \frac{x_1}{0.12} + 0.225$$

$$q' = 0.9q = 305.5 = U'925$$

$$\frac{1}{U'} = 3.028 = \frac{x_1}{0.12} + 0.225$$

hence $\underline{x_1 = 0.336 \, \text{m}}$

6.5.1.2 *Convection.* Convective heat transfer is a result of fluid motion on a macroscopic scale. A fluid element moving across a boundary carries with it a quantity of energy and hence gives rise to a heat flux. The phenomenon is termed *natural convection* or *forced convection* depending on the nature of the forces which are responsible. In natural convection, currents are generated in the fluid by differences in density which themselves result from temperature differences. In forced convection, currents and eddies are

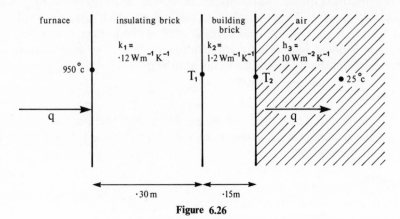

Figure 6.26

produced by work done on the system, e.g. stirring a vessel or pumping a fluid in turbulent flow through a pipe. Rates of heat transfer by forced convection are generally much higher than in natural convection.

Heat transfer by convection is a complex process but the analysis is simplified by the boundary layer concept. All resistance to heat transfer on the fluid side of a hot surface is supposed to be concentrated in a thin film of fluid close to the solid surface. Transfer within the film is by conduction. The temperature profile for such a process is shown in Fig. 6.27. The thickness of the thermal boundary layer is not generally equal to that of the hydrodynamic boundary layer. The heat flux could be expressed as

$$q = \frac{k_b}{x} (T_w - T_B)$$

where k_b is the thermal conductivity in the boundary layer. However, because the film thickness is usually not known, it is more common to express the heat flux in terms of a *film heat transfer coefficient h*, thus

$$q = h(T_w - T_B).$$

Values of film transfer coefficients are usually obtained from correlations between dimensionless groups. For forced convection (turbulent flow) within circular tubes the relevant groups are:

$$\left. \begin{array}{l} \text{Nusselt number, } Nu = \dfrac{hd}{k} \\[2em] \text{Prandtl number, } Pr = \dfrac{c_p \mu}{k} \\[2em] \text{Reynolds number, } Re = \dfrac{\rho \bar{v} d}{\mu} \end{array} \right\} \qquad (6.5.6)$$

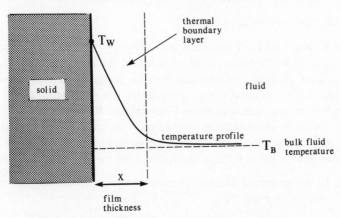

Figure 6.27 Convective heat transfer from a solid wall.

where h is the heat transfer coefficient, d is the tube diameter, k is the fluid thermal conductivity, c_p is the heat capacity (mass units) and μ is the viscosity. Experimental data have been correlated by the equation

$$Nu = 0.023\, Re^{0.8}\, Pr^{0.33} \tag{6.5.7}$$

Equations relevant to other situations such as flow across tubes are given in textbooks.[10,11,12]

6.5.1.3 *Radiation.* Radiant heat transfer occurs in the form of electromagnetic radiation with wavelengths from about 10^{-7} to 10^{-4} m; this is known as *thermal radiation*. The visible spectrum is centred on wavelengths of 5×10^{-7} m. The total energy flux emitted by a body at an absolute temperature T is given by

$$E = \varepsilon \sigma T^4 \tag{6.5.8}$$

where σ is the Stefan–Boltzmann constant ($5.67 \times 10^{-8}\, \mathrm{W\, m^{-2}\, K^{-4}}$) and ε is the *emissivity* of the body. The emissivity lies in the range 0 to 1. For polished metals it is low ($\simeq 0.05$) whereas most non-metals and oxidized metal surfaces have emissivities of about 0.8.

The net heat flux from a body to its surroundings is equal to the emitted energy (equation 6.5.8) minus the energy absorbed from the surroundings. Assuming (i) emitted energy is totally absorbed by the surroundings; and (ii) the fraction of incident radiant energy absorbed by a body is equal to ε (a reasonable assumption for metals), the net flux is given by

$$q = \varepsilon \sigma (T^4 - T_0^4) \tag{6.5.9}$$

where T_0 is the temperature of the surroundings.

It is evident that radiative heat transfer is not an important contribution to the total heat flux for small temperature differences. When large temperature differences exist, e.g. in furnaces, radiation may be the dominant mechanism of heat transfer.

6.5.2 *Shell and tube heat exchangers*

Shell and tube heat exchangers are probably the most common type of heat transfer equipment used in the process industries. They provide a high heat transfer area per unit volume and can be designed to fulfil most duties. Thermal and mechanical design procedures are well established—a fairly comprehensive account is given in Coulson and Richardson[13]. The internal layout of a simple shell and tube exchanger is shown in Fig. 6.28. One process fluid flows through the tubes and the other inside the cylindrical shell. The shell side fluid is made to flow across the tube bundle by a series of baffles.

6.5.2.1 *Basic equations.* The rate of heat transfer in an exchanger with a heat transfer area A is given by

$$Q = qA = UA\Delta T_{\mathrm{m}} \tag{6.5.10}$$

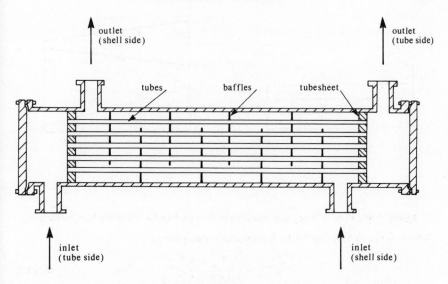

Figure 6.28

where U is the overall heat transfer coefficient and ΔT_m is a mean temperature difference between the two fluids. The definition of ΔT_m depends on the flow configuration within the exchanger.

Figure 6.29 (*a*) shows a *countercurrent* flow arrangement. Temperature profiles for the hot and cold streams are plotted against heat transferred in Fig. 6.29 (*b*): it is assumed that the overall heat transfer coefficient and heat capacities do not vary significantly with temperature. It can be shown[10] that the heat transferred is given by

$$Q = U A \Delta T_{lm} \qquad (6.5.11)$$

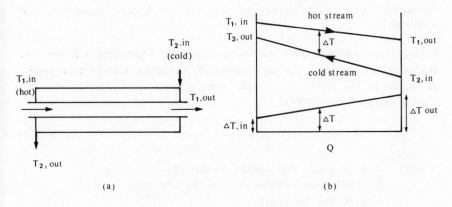

Figure 6.29 Flow arrangement and temperature profiles for countercurrent heat exchanger.

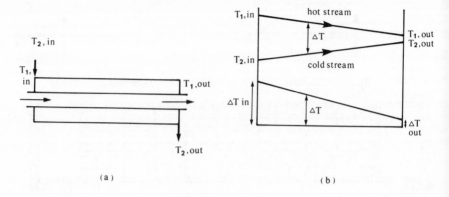

Figure 6.30 Flow arrangement and temperature profiles for cocurrent heat exchanger.

where ΔT_{lm} is the *log mean temperature difference*.

$$\Delta T_{lm} = \frac{\Delta T_{in} - \Delta T_{out}}{\ln(\Delta T_{in}/\Delta T_{out})} \qquad (6.5.12)$$

If the two fluids flow in the same direction the arrangement is termed *cocurrent* (Fig. 6.30). The temperatures of the two streams approach each other along the exchanger but the outlet temperature of the cold stream cannot exceed the inlet temperature of the hot stream. The appropriate mean temperature difference for cocurrent flow is also ΔT_{lm}. However, for a given set of inlet and outlet temperatures for hot and cold streams ΔT_{lm} is larger in countercurrent flow and hence the exchanger will be smaller.

Flow arrangements in industrial exchangers are rarely of the simple 1 tube pass, 1 shell pass type shown in Fig. 6.28. Fluids are usually passed through the shell and tubes more than once (in different directions) in order to increase the flow velocity and hence the heat transfer for a given area. In some sections of these *multipass* exchangers the flow is countercurrent and in others co-current. Correction factors to ΔT_{lm} for multipass exchangers are available[13].

6.5.2.2 *Overall heat transfer coefficient.* In an exchanger heat is transferred from one fluid to another through a tube wall. An idealized temperature profile is shown in Fig. 6.31. The overall heat transfer coefficient is obtained by summing the resistances.

$$\frac{1}{U} = \frac{1}{h_i} + \frac{x}{k} + \frac{1}{h_0} \qquad (6.5.13)$$

where h_i is the inside film transfer coefficient,
k is the thermal conductivity of the tube wall,
x is the wall thickness,
h_0 is the outside film transfer coefficient.

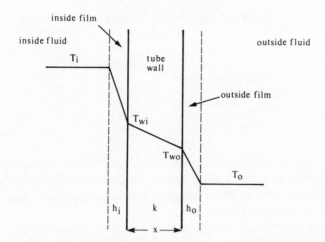

inside film

inside fluid

outside fluid

T_i

tube wall

outside film

T_{wi}

T_{wo}

T_o

h_i k h_o

x

Figure 6.31 Temperature profile for heat transfer across a tube wall.

In Equation 6.5.13 it is assumed that inside and outside surface areas are approximately equal, i.e. the tube is thin-walled.

The inside transfer coefficient may be estimated from Equation 6.5.7. The shell side coefficient is difficult to estimate because of the complex flow patterns within a baffled shell. For further details see Coulson and Richardson[13].

Some typical values of overall heat transfer coefficients are given in Table 6.2. These values are very approximate and should only be used for order-of-magnitude estimates.

6.5.2.3 *Fouling factors.* In use, heat transfer surfaces tend to become 'fouled' by deposits of scale, dirt or other solids. These deposits represent an additional resistance to heat transfer and allowance for them must be made in the design of exchangers. A modified expression for the overall heat transfer coefficient is generally used

$$\frac{1}{U} = \frac{1}{h_i} + \frac{1}{h_{d,i}} + \frac{x}{k} + \frac{1}{h_{d,o}} + \frac{1}{h_0} \tag{6.5.14}$$

Table 6.2 Typical values of overall heat transfer coefficients for shell and tube heat exchangers.

Fluids	Process	$U/Wm^{-2}K^{-1}$
water/water	heating/cooling	600–1800
oil/oil	heating/cooling	100–400
gas/gas (0·1 MPa)	heating/cooling	25–50
gas/gas (5 MPa)	heating/cooling	250–500
steam/water	heating	1400–4000
steam/aqueous solution	evaporation	1000–3000
water/organic vapours	condensation	200–1000

Table 6.3   Typical values of fouling factors

Fluid	$h_d/W\,m^{-2}\,K^{-1}$
River water	2000–10 000
Cooling tower water	5000
Boiler feed water	6000
Condensing steam	6000
Clean gases	5000–10 000
Heavy oils	2000

where $h_{d,i}$ and $h_{d,o}$ are inside and outside dirt coefficients or fouling factors.

Some typical values of fouling factors are given in Table 6.3. The values are subject to considerable uncertainty and accurate figures can only be obtained from operating experience.

The following example illustrates the calculation of heat transfer area for different flow arrangements.

Example 6.15

A shell and tube heat exchanger is to be used to cool $100\,kg\,s^{-1}$ of 98% sulphuric acid from 60°C to 40°C. Cooling water is available at 10°C and a flowrate of $50\,kg\,s^{-1}$. The overall heat transfer coefficient is $500\,W\,m^{-2}\,K^{-1}$. Determine the required surface area for (a) countercurrent and (b) cocurrent flow. Assume that the heat transfer coefficient is the same in both cases. (Heat capacities: water, $4200\,J\,kg^{-1}\,K^{-1}$; 98% sulphuric acid, $1500\,J\,kg^{-1}\,K^{-1}$.)

Solution

(a) The exit temperature of the water is obtained from an energy balance:

$$Q = F_1 c_{p_1}(T_{1,in} - T_{1,out}) = F_2 c_{p_2}(T_{2,out} - T_{2,in})$$

where Q is the heat transfered, F_1 the flowrate of sulphuric acid and F_2 the flowrate of water.

$$\Rightarrow Q = 100 \times 1500\,(60-40) = 3 \times 10^6\,W$$
$$= 50 \times 4200\,(T_{2,out} - 10)$$
$$\underline{T_{2,out} = 24\cdot3°C}$$

The flow arrangement is

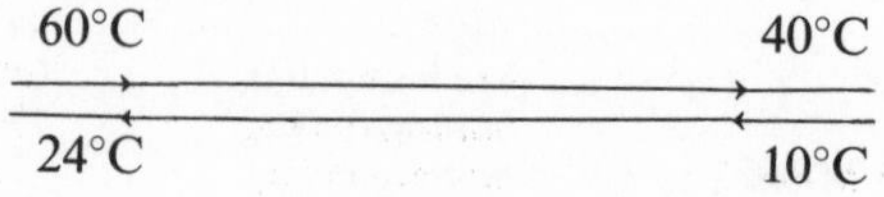

60°C   40°C

24°C   10°C

From equation 6.5.12 the mean temperature difference is

$$\Delta T_{lm} = \frac{(60-40) - (40-10)}{\ln\left((60-24)/(40-10)\right)} = 33°C$$

The heat transfer area is therefore

$$A = \frac{Q}{U\Delta T_{lm}} = \frac{3 \times 10^6}{500 \times 33} = 182\,m^2$$

(b) For cocurrent flow the arrangement is

60°C   40°C

10°C   24°C

$$\Delta T_{lm} = \frac{(60 - 10) - (40 - 24)}{\ln((60 - 10)/(40 - 24))} = 30°C$$

$$A = \frac{3 \times 10^6}{500 \times 30} = 200\,m^2$$

In this example the difference in area is only 10%. However, if the acid were cooled to 32°C the areas would differ by over 70%.

6.6 Separation processes

Separation processes are of central importance in the chemical industry. Appropriate choice and efficient design of separation equipment is essential because it represents a large proportion of the capital and operating costs of a plant. The vinyl acetate flowsheet in Fig. 6.2 is almost entirely concerned with separation of the reaction products—this is typical of very many processes. Whenever there is incomplete conversion in a reactor a separation followed by recycle may be necessary. Toxic substances in waste streams must often be removed in order to meet legal requirements. The availability of suitable separation processes is, therefore, crucial when considering the feasibility of any new chemical process.

The subject of separation processes is so extensive that the scope of this section must be extremely limited. We shall therefore concentrate on the general characteristics of separation processes and only examine simple distillation in any detail. Several excellent textbooks on the subject are available.[14, 15, 16]

6.6.1 Characteristics of separation processes

A separation process takes one or more feed streams containing mixtures and transforms these into products streams which differ in composition (Fig. 6.32). This is achieved by the addition of a separating agent which may be energy or another stream of material.

Separation usually involves the formation of more than one phase by the addition of the separating agent. The components of a mixture will be distributed unequally between the phases. Separation of phases of different density is easily accomplished by mechanical means. For example, a liquid feed

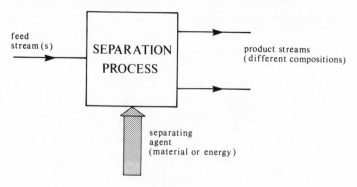

feed stream(s)

SEPARATION PROCESS

product streams (different compositions)

separating agent (material or energy)

Figure 6.32

is passed through a pressure-reducing valve into a vessel (flash drum) where it separates into two phases of different composition. Vapour is taken off at the top of the drum and liquid at the bottom. The separating agent is energy— introduced as work of compression of the feed.

The driving force for interphase mass transfer is always the thermodynamic chemical potential but in some processes thermodynamic equilibrium is not approached and the degree of separation is dependent on the rate of mass transfer. Some of the more common processes are listed in Table 6.4; all of these are equilibrium separation processes. Mechanical separation processes such as filtration, centrifuging and settling are also widely used.

6.6.2 Phase equilibria

A full treatment of the thermodynamics of phase equilibria may be found elsewhere[3]. Only the terminology and some essential concepts will be reviewed in this section.

Table 6.4 Common separation processes

Process	Phases[a] Feed	Product	Separating agent[b]
Flash evaporation	L	L + V	E pressure reduction
Distillation	L/V	L + V	E heat
Stripping	L	L + V	M stripping vapour
Absorption	V	L + V	M liquid absorbent
Extraction	L	$L_1 + L_2$	M liquid solvent
Crystallization	L	L + S + V	E heating or cooling
Evaporation	L	L + V	E heat
Drying	S/L	S + V	E heat
Leaching	S	S + L	M liquid solvent

[a]V, vapour; L, liquid; S, solid
[b]E, Energy-separating agent
   M, material-separating agent

6.6.2.1 *K-values*. The equilibrium compositions in a vapour-liquid system are related by *equilibrium K-values*:

$$K_i = \frac{y_i}{x_i} \qquad (6.6.1)$$

where y_i is the mole fraction of component i in the vapour phase and x_i is the mole fraction in the liquid phase. In general K-values are functions of temperature, pressure and composition but for ideal mixtures they are independent of composition.

The separability of two species i and j is indicated by their *relative volatility*.

$$\alpha_{ij} = \frac{K_i}{K_j} = \frac{y_i x_j}{y_j x_i} \qquad (6.6.2)$$

values of α_{ij} close to unity indicate that separation by methods relying on the formation of vapour and liquid phases will be difficult. For example, at 0·1 MPa α_{ij} for propane:butane is 5·5, whereas for butane:but-2-ene it is 1·05. Separation of the former mixture by simple distillation would be easy but for the latter a very large distillation column would be required.

6.6.2.2 *Ideal mixtures*. Ideal mixing is a good approximation for mixtures of molecules of similar types (e.g. alkanes) at low to moderate pressures.

For ideal mixtures the vapour-liquid equilibrium is described by *Raoult's law*

$$Py_i = P_i^0(T)x_i \qquad (6.6.3)$$

where P is the total pressure and $P_i^0(T)$ is the vapour pressure of component i at the system temperature T. It follows that the K-values in an ideal mixture are given by

$$K_i = P_i^0(T)/P \qquad (6.6.4)$$

and the relative volatility by

$$\alpha_{ij} = P_i^0(T)/P_j^0(T) \qquad (6.6.5)$$

The relative volatility is a function of temperature but not composition or pressure. Over moderate ranges of temperature α_{ij} can often be treated as a constant.

6.6.2.3 *Binary mixtures*. In binary mixtures there is only one independent composition variable in each phase. It is conventional to work in terms of the composition of the *more volatile component* (mvc). The relative volatility of the mvc to the *less volatile component* (lvc) is simply written as α. α is thus always greater than or equal to unity except for azeotropic systems.

The relationship between equilibrium compositions in a binary mixture may be conveniently expressed in terms of the relative volatility

$$\alpha = \frac{y(1-x)}{x(1-y)}$$

Rearranging this equation gives

$$y = \frac{\alpha x}{1 + x(\alpha - 1)} \tag{6.6.6}$$

Equation 6.6.6 is quite general for binary mixtures but becomes particularly useful if the assumption of constant relative volatility may be made.

6.6.3 Binary distillation

6.6.3.1 *Distillation columns.* Distillation is one of the most commonly used industrial separation processes. A high degree of separation can be achieved because distillation is a multistage process.

A schematic diagram of a distillation column is shown in Fig. 6.33. The column contains a series of *trays* (also called plates or stages) which are usually perforated metal plates. Vapour passes up the column countercurrent to the

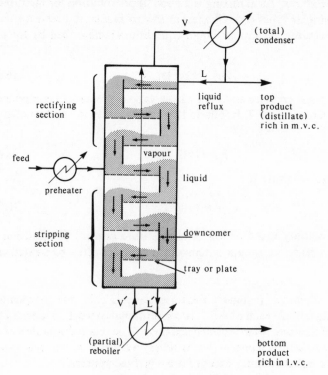

Figure 6.33 Simple distillation column.

liquid which flows across each tray and down the downcomers. As vapour bubbles through the liquid on a tray there is mass transfer between phases and their compositions approach equilibrium. Vapour passing up the column becomes enriched in the mvc whilst the liquid is enriched in lvc as it passes downwards. At the top of the column vapour is condensed and part is returned to provide a liquid *reflux*. Similarly, part of the liquid at the bottom is evaporated in the reboiler to provide a vapour phase. The separating agent is the heat supplied in the reboiler.

6.6.3.2 *Material balances.* The notation used to describe the column is defined in Fig. 6.34. Vapour leaving a typical plate n has a flowrate V_n and composition y_n, the corresponding liquid flowrate is L_n and the composition x_n:

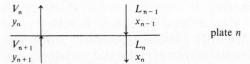

For a *theoretical plate* or stage the vapour and liquid streams leaving are in equilibrium, i.e. $y_n = K_n x_n$. This represents the best possible separation, never obtained in practice. The number of actual plates required to achieve a

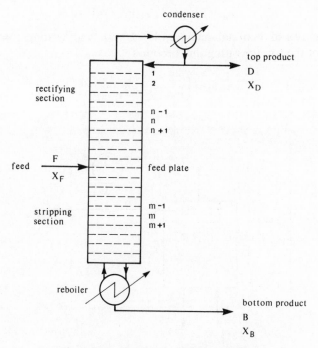

Figure 6.34 Notation for distillation column.

separation may be related to the number of theoretical plates using plate efficiencies.

In order to relate compositions of liquid and vapour streams on successive plates it is necessary to carry out material balances around the top and bottom sections of the column. Consider a balance around the top of the column down to plane n (above the feed plate) as shown in Fig. 6.35.

A balance on the mvc gives

$$y_{n+1}V_{n+1} = x_n L_n + x_D D$$

$$\Rightarrow y_{n+1} = \left(\frac{L_n}{V_{n+1}}\right)x_n + \left(\frac{D}{V_{n+1}}\right)x_D$$

This is the equation of the top *operating line* giving the desired relationship between compositions. In general an energy balance is required to determine V_{n+1} and L_n. A major simplification is possible for many problems by assuming *constant molal overflow*, i.e. constant vapour and liquid flowrates. This is approximately correct provided that enthalpy changes on evaporation are about equal for the two components. Constant molal overflow is a good assumption for systems with constant relative volatility. Making the assumption of constant molal overflow

$$L_n = L_{n+1} = L_{n+2} = \cdots = L$$

$$V_n = V_{n+1} = V_{n+2} = \cdots = V$$

Using the overall material balance, $V = L + D$, and writing $r = L/D$ the equation of the top operating line becomes

$$y_{n+1} = \left(\frac{r}{r+1}\right)x_n + \frac{x_D}{r+1} \qquad (6.6.7)$$

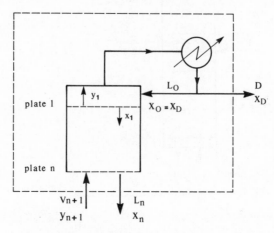

Figure 6.35 Material balance around top of column.

r is the ratio of liquid returned to the column as reflux, to the top product flowrate; usually called the *reflux ratio*.

An analysis similar to the above may be carried out for the bottom of the column below the feed plate. The equation of the bottom operating line, relating compositions on successive plates, is

$$y_m = \left(\frac{s+1}{s}\right) x_{m-1} - \frac{x_B}{s} \tag{6.6.8}$$

where $s = V'/B$. The vapour flowrate in the bottom of the column is V' and the liquid flowrate is L'. Because of the feed introduced between top and bottom sections the flowrates in the two sections are not equal.

6.6.3.3 *McCabe–Thiele graphical method.* The equations obtained in the preceding section can be represented graphically on a McCabe–Thiele diagram. The operating lines, equations 6.6.7 and 6.6.8, are plotted on x–y coordinates along with the *equilibrium line*, which is the locus of equilibrium (x, y) compositions. The latter are obtained from tabulations or equations like 6.6.6.

A schematic McCabe–Thiele diagram for the rectifying section of a column is shown in Fig. 6.36. We assume that the top product composition x_D and the recycle ratio r are known. The composition of the liquid reflux is $x_0 = x_D$. Substituting in equation 6.6.7 gives the composition of vapour leaving plate 1 as $y_1 = x_D$. The top operating line can now be plotted passing through the

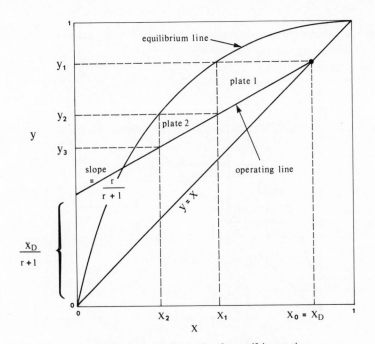

Figure 6.36   Operating line for rectifying section.

point (x_D, x_D) with a gradient $r/r + 1$. Liquid and vapour phase are in equilibrium leaving each plate. Thus, x_1 is the composition corresponding to $y = y_1$ on the equilibrium line. The vapour composition y_2 can now be obtained from the operating line and the process repeated down the column. Each theoretical plate is represented by a step on the x–y diagram between the operating line and equilibrium line.

The intersection point of top and bottom operating lines is determined by the composition and thermal condition of the feed. A material and energy balance around the feed plate gives the locus of intersection points as

$$y = \left(\frac{q}{q-1}\right)x - \frac{x_F}{q-1} \tag{6.6.9}$$

where $q = \dfrac{\text{enthalpy required to evaporate feed}}{\text{enthalpy required to evaporate liquid of feed composition}}$

The q line (equation 6.6.9) is a straight line passing through the point $(x_F, y = x_F)$ with a gradient $q/q - 1$. The q-lines obtained for different feed conditions are illustrated in Fig. 6.37.

A complete procedure for determining the number of theoretical plates in a column is summarized below:

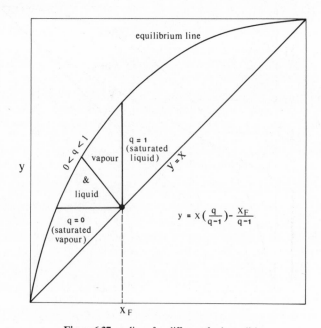

Figure 6.37 q-lines for different feed conditions.

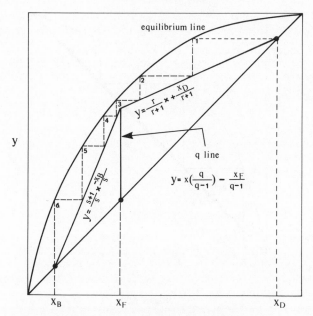

Figure 6.38 Complete McCabe–Thiele construction.

Given x_B, x_D, x_F, r, information on state of feed and equilibrium data;

  (i) Draw equilibrium line and $x = y$ line.

 (ii) Draw top operating line through point $(x_D, y = x_D)$ with gradient $r/r + 1$.

(iii) Determine q and draw q line through point $(x_F, y = x_F)$ with gradient $q/q - 1$.

 (iv) Draw bottom operating line from point $(x_B, y = x_B)$ to intersection of top operating line and q line.

  (v) Step off theoretical plates between operating lines and equilibrium line, transferring from top to bottom operating line after the intersection point.

Fig. 6.38 shows a complete McCabe–Thiele diagram. Feed is introduced as liquid at its boiling point ($q = 1$). The column requires 6 theoretical plates.

6.6.3.4 *Limiting operating conditions.* As the reflux ratio is increased the gradient of the top operating line, equation 6.6.7, approaches unity and the number of theoretical plates required for a given separation is reduced. The *minimum number of theoretical plates* corresponds to *infinite reflux* conditions. At infinite reflux the operating lines are coincident with the $y = x$ line as shown in Fig. 6.39. Columns are often tested under infinite reflux conditions by reducing feed and product streams to zero.

    The *minimum reflux ratio* is the opposite extreme in operating conditions. As r is reduced the operating lines move towards the equilibrium line and the

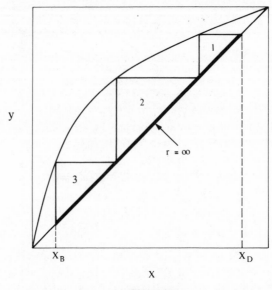

Figure 6.39

number of theoretical plates increases. When operating lines intersect on the equilibrium line the reflux ratio is at its minimum value but the separation requires an infinite number of plates because the driving force for interphase mass transfer becomes zero. Figure 6.40 shows a McCabe–Thiele diagram for this situation. The actual reflux ratio selected for a column must lie between the two extremes. A high ratio reduces the number of plates but increases the

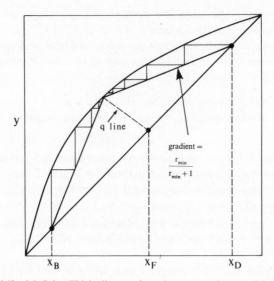

Figure 6.40   McCabe–Thiele diagram for column operating at minimum reflux.

column diameter and energy costs. The optimum reflux ratio depends on physical and cost data and hence varies from column to column. A value of $r = 1 \cdot 2r_{min}$ is typical.

The following examples illustrates the use of the McCabe–Thiele method.

Example 6.16

A continuous distillation column is required to separate a mixture containing 0·695 mole fraction heptane and 0·305 mole fraction octane to give products of 95% purity. The feed is liquid at its boiling point.
(a) What is the minimum reflux ratio?
(b) Estimate the number of theoretical plates if $r = 2r_{min}$.
Equilibrium data (mole fraction heptane):

x	0·1	0·2	0·3	0·4	0·5	0·6	0·7	0·8	0·9
y	0·188	0·343	0·475	0·588	0·686	0·767	0·840	0·901	0·955

Solution

The equilibrium data are plotted in Fig. 6.41.

(a) Minimum reflux corresponds to the top operating line and q-line intersecting on the equilibrium line. For saturated liquid feed the q-line is vertical. The intersection point is at $x = 0 \cdot 695$, $y \simeq 0 \cdot 835$.

The gradient of the top operating line is

$$\frac{r_{min}}{r_{min} + 1} = \frac{0 \cdot 95 - 0 \cdot 835}{0 \cdot 95 - 0 \cdot 695} = 0 \cdot 451$$

$$r_{min} = 0 \cdot 82$$

(b) For $r = 2r_{min}$ the top operating line has the equation $y = 0 \cdot 62x + 0 \cdot 36$. This line is plotted in Fig. 6.41 together with the bottom operating line which intersects it on the q-line. The q-line is vertical because the feed is saturated liquid.

From the construction in Fig. 6.41, 11 theoretical plates are required.

6.7 Process control

Automatic control is an essential feature of large-scale continuous processes. Many modern chemical plants are so complex that it would be impractical, unsafe and unprofitable to run them manually.

By its nature process control is concerned with the dynamic behaviour of systems. It is no longer sufficient to make the steady-state assumption. Material and energy balances for unsteady systems must include the accumulation terms so far omitted. Because of the extra mathematical complexity involved in a quantitatve treatment of control this section will,

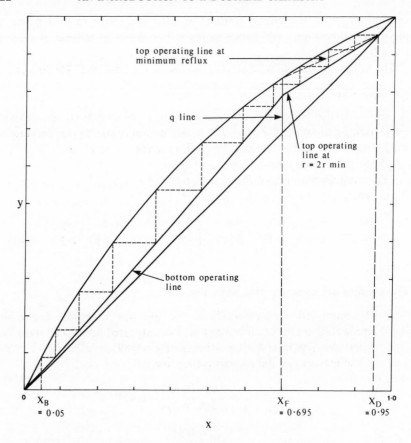

Figure 6.41

instead, concentrate on general concepts rather than detailed analysis of specific applications. For a more extensive treatment see the appropriate references[17,18,19,20].

6.7.1 Objectives of process control

Ensuring safe plant operation is one of the most important tasks of a control system. This is achieved by monitoring process conditions and maintaining variables within safe operating limits. Potentially dangerous situations are signalled by alarms and a plant may even be shut down automatically.

Safe operation is consistent with the economic objective of process control which is to operate the plant in such a way to minimize total costs. This involves maintaining product quality, meeting production targets and making efficient use of utilities such as steam, electricity, cooling water and compressed air.

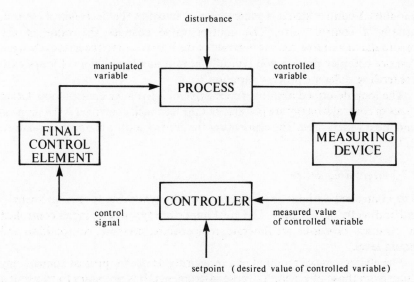

Figure 6.42

6.7.2 *The control loop*

The functional layout of a process control loop is shown in Fig. 6.42. A process operates subject to some disturbances e.g. the 'process' may be a preheater and the 'disturbance' a varying feed temperature (Fig. 6.43). We identify a variable which needs to be controlled, e.g. the temperature of the stream leaving the preheater. This quantity is measured by a suitable measuring device such as a thermocouple. The output from the measuring device is passed to a *controller* where it is compared with the *setpoint* for the controlled

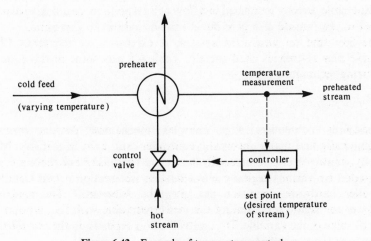

Figure 6.43   Example of temperature control.

variable. A control signal is generated and activates the *final control element*, usually a control valve. The control valve changes the value of the manipulated variable, e.g. the flowrate of the hot stream to the heat exchanger. A more complex process will typically have a number of control loops each controlling different process variables.

The loop described above is an example of a *feedback* control loop. Other types of control strategy are possible but the feedback controller is the simplest and most widely used. The elements of the control loop will now be examined in turn.

6.7.3 *Measuring devices*

The controlled variable in a process must be amenable to measurement— either directly or indirectly. The most important types of variables controlled in chemical processes are flowrate, temperature, pressure, composition and liquid level.

The characteristics required of a measuring device for process control may differ from those of general purpose instruments. It is necessary to transmit a signal representing the measurement from the measuring element to the controller—at one time this was done pneumatically but now an electrical or optical signal would be used. For example, a mercury-in-glass thermometer may provide a sufficiently accurate temperature measurement but it is difficult to convert it into an electrical signal. By contrast, the output of a thermocouple is naturally an electrical signal and hence this device is a more suitable choice for control applications. Another important characteristic is the dynamic response of the measuring system. For stable and effective control it is important to minimize the time lag between a variable moving off its setpoint and corrective action being taken. A measuring element must therefore respond quickly to changes—this is not always easy to achieve. For example, a thermocouple directly immersed in a flow will respond more quickly than one welded to the outside of a pipe but it may be subject to corrosion.

The literature on measuring systems is extensive, see references 17–20 inclusive plus references cited therein. Table 6.5 lists some of the common measuring techniques used.

6.7.4 *The controller*

Traditionally controllers were complex mechanical devices operated pneumatically and many pneumatic controllers still exist in industry. More recently, analogue electronic controllers have been used but these are now superseded by various digital control devices (see section 6.7.6). Details of controller hardware operation are given by Johnson.[20] The controller compares the signal representing the measured value with the setpoint (i.e. desired) value of the variable. The control action depends on the *control mode* selected for the controller.

Table 6.5 Measuring methods and devices

Variable	Method/principle
Flowrate	Differential pressure devices:
	pitot tube, orifice plate, venturi meter
	Turbine meter
	Rotameter
	Hot wire anemometer
Temperature	Thermistor
	Thermocouple
	Resistance thermometer
	Expansion thermometer
	Optical pyrometer
Pressure	Manometer
	Differential pressure cell (mechanical or
	semi-conductor)
	Bourdon gauge
Composition	Chromatography
	Refractive index
	Density
	Conductivity
	pH
	Infrared/ultraviolet absorption
Liquid Level	Differential pressure
	Float
	Weight of vessel

6.7.4.1 Proportional control. The simplest continuous control mode is proportional. The control signal produced by the controller is proportional to the error signal ε, defined as

$$\varepsilon = s - m \qquad (6.7.1)$$

where s is the setpoint and m is the measured value. The controller output is given by

$$c = K_c \varepsilon \qquad (6.7.2)$$

where K_c is the *gain* or *proportional sensitivity*.

The response of a system to a disturbance is shown schematically in Fig. 6.44. The behaviour without control is represented by the solid line. Proportional control (P) reduces the maximum deviation of the controlled variable but gives rise to long-lived oscillations. At the new steady state there is a finite deviation of the controlled variable, known as the *offset*. An offset is characteristic of proportional control which requires an error in order to generate a control signal, equation 6.7.2. If the error (offset) is reduced to zero there is no control action.

6.7.4.2 Proportional–integral control. The proportional–integral (or PI) mode is also often referred to as 'proportional plus reset' since this control mode eliminates the offset associated with proportional control alone. The controller output is described by the equation

$$c = K_c \varepsilon + \frac{K_c}{\tau_I} \int_0^t \varepsilon \, dt \qquad (6.7.3)$$

where τ_I is the *integral time*. The output signal is proportional to the error plus the time integral of the error. The values of K_c and τ_I may be set independently to give the best control action. With a PI controller the error itself can be reduced to zero whilst the control action is maintained because the integral of the error remains non-zero. Fig. 6.44 shows a response with PI control. The offset is eliminated but oscillatory behaviour still persists.

6.7.4.3 Proportional–derivative control. The addition of a derivative mode to a proportional controller results in faster action because the control signal is also proportional to the rate of change of the error:

$$c = K_c \varepsilon + K_c \tau_D \frac{d\varepsilon}{dt} \qquad (6.7.4)$$

where τ_D is the *derivative time*. Derivative action reduces oscillations, thus allowing a higher proportional gain setting and consequently smaller offset.

6.7.4.4 Proportional–integral–derivative control. This is a combination of the previous modes and is described by the equation

$$c = K_c \varepsilon + \frac{K_c}{\tau_I} \int_0^t \varepsilon \, dt + K_c \tau_D \frac{d\varepsilon}{dt} \qquad (6.7.5)$$

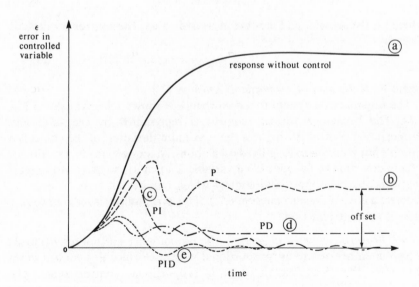

Figure 6.44 Typical response of system with different control modes.

The integral action eliminates offset and derivative action reduces oscillations but with three modes it is more difficult to select suitable controller settings.

The choice of control mode is influenced by the extent to which an offset or oscillatory behaviour may be tolerated. With pneumatic mechanisms the cost advantage of a simple proportional controller was significant—this is no longer the case.

6.7.5 Final control element

The signal from the controller is used to activate a final control element. In the chemical industry the control element is almost always a valve which controls a flowrate. Changes in flowrate can be used indirectly to change any of the other variables listed in Table 6.5.

Figure 6.45 shows an air-to-close pneumatic control valve. The valve stem is attached to a diaphragm which is held in the up position by a spring. Increasing the air pressure depresses the diaphragm and close the valve. Different spring arrangements can be used to give an air-to-open valve. The electrical signal from the controller must be converted to a pneumatic signal to activate the valve. Pneumatic valves are still almost universally used.

6.7.6 Computer control

Because of the low cost of computer hardware, computer control is now widely used. Computer control schemes can be powerful and flexible. They have the

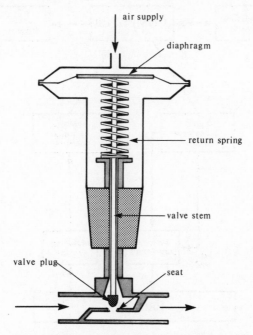

Figure 6.45 Pneumatic control valve.

potential to make a process easier to control and more profitable. The different types of computer control are described below.

6.7.6.1 *Supervisory control.* A large plant may have hundreds of PID controllers each operating on a separate control loop. Provided setpoints and controller settings have been well chosen the plant will operate satisfactorily. However, the tasks of supervising the plant, recording data and changing setpoints and controller settings to achieve optimum plant performance are rather complex. In a supervisory control system these jobs are carried out by the computer. Figure 6.46 shows the relationships between the main components in such a system.

As process variables change it may be advantageous to alter setpoints in certain loops. This is accomplished by the supervisory computer which monitors process variables and uses a mathematical model of the process to estimate optimal settings. Sophisticated control strategies are possible because settings of several loops can be altered simultaneously. The task of the plant operator is made easier because all important data relating to the process can be displayed in a compact form at a console. Any abnormal conditions can be quickly identified and acted upon.

6.7.6.2 *Direct digital control.* The conventional controller (as described in section 6.7.4) is a small-scale analogue computer operating on electrical or pneumatic principles. In direct digital control (DDC) measurements of process

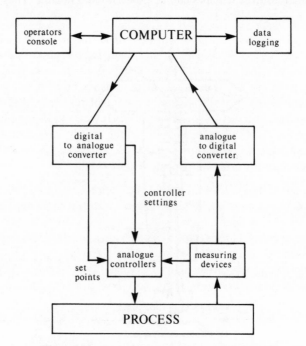

Figure 6.46 Supervisory computer control system.

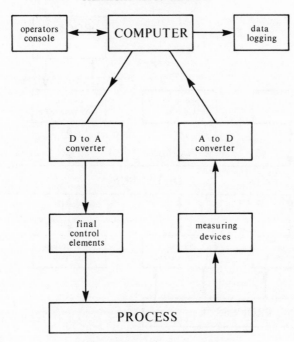

Figure 6.47 Direct digital control.

variables are transmitted directly to a digital computer which compares each signal with an internally stored setpoint. The mathematical operations associated with the desired control mode are carried out and a control signal transmitted to the final control element. Figure 6.47 shows a DDC system.

DDC has most of the advantages of supervisory control but with extra flexibility in choice of control strategies. *Any* control mode can be simulated — the control engineer is not restricted to PID. However, because none of the control loops can function independently of the central computer, elaborate measures may have to be taken to allow for computer failure. In critical applications a standby computer would be provided.

6.7.6.3 *Distributed digital control.* DDC became economically practical with the advent of mini-computers, and the concepts have been extended by the use of microprocessors. A microprocessor-based control computer can handle a small number of control loops (typically 1–16). Microprocessors are so cheap that, once again, it becomes cost effective to put the control logic in the part of the plant where it is used rather than in a central control room. The costs of cabling a large plant are now significant compared to the cost of the computer.

Distributed control refers to a number of control computers which are linked to each other and to peripherals such as operators' consoles by high speed data channels. A hierarchical structure may be imposed by having certain

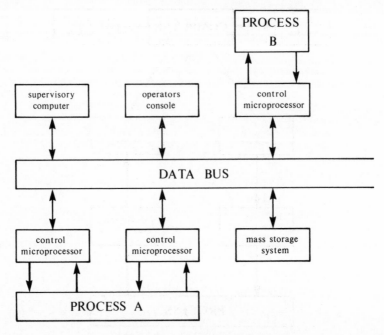

Figure 6.48   Distributed digital control.

processors operating in a supervisory mode. Figure 6.48 shows a distributed control system. Such a system is less liable to failure than the central computer based DDC system in Fig. 6.47 because each component can operate independently of the others.

Appendix

Answers are given in brackets at the end of each problem.

Problems

1. An evaporative crystallization process is used to produce Na_2SO_4 crystals from aqueous solution. Fresh feed, 22% Na_2SO_4, is mixed with recycle solution and fed to an evaporator where it is concentrated to 45%. This stream is cooled and crystals of Na_2SO_4 are removed in a filter unit as a wet filter cake containing 95% solids and 5% of a 38% Na_2SO_4 solution. The filtrate, also 38% solution, is recycled. For an evaporator with a maximum capacity of 5000 kg hr^{-1} calculate:

    (a) the production rate of solid Na_2SO_4;
    (b) the fresh feed rate;
    (c) the recycle ratio.

All compositions are on a mass basis.

((a) $1468\,kg\,hr^{-1}$, (b) $6468\,kg\,hr^{-1}$, (c) $1\cdot68$.)

2. Natural gas, (essentially pure methane at $1\cdot5$ MPa, $25\,^{\circ}C$) and steam ($1\cdot5$MPa, $450\,^{\circ}C$) are mixed in the ratio $4\,mol$ steam: 1 mole methane and contacted with a catalyst to produce synthesis gas for a methanol convertor. The following reactions occur:

$$CH_4 + H_2O \rightarrow CO + 3H_2 \tag{1}$$

$$CO + H_2O \rightarrow CO_2 + H_2 \tag{2}$$

90% of the methane is converted to CO by reaction (1) and $\frac{1}{3}$ of the CO produced goes to CO_2 by reaction (2). If 2×10^6 mol hr^{-1} of mixed gases leave the reactor/heat exchanger unit at $600\,^{\circ}C$, at what rate must heat be supplied?

($19\cdot9$ MW—using data from references 2 or 5).

3. 60% sulphuric acid is pumped at a flowrate of $12\cdot6$ kg s^{-1} through a smooth pipe to a storage tank. The liquid surface of the storage tank is 15.2 m above the pump outlet. The pipe is in two sections. The first section has an internal diameter of 78 mm and a length of 366 m. The second section has an internal diameter of 63 mm and is 122 m long. Estimate the absolute pressure at the pump outlet. Changes in kinetic energy may be neglected. (Density of acid $=$ $1\,530\,kg\,m^{-3}$, viscosity $= 6\cdot4 \times 10^{-3}$ Pa s.)

($0\cdot82$ MPa)

4. The temperature of oil leaving a cocurrent heat exchanger is to be reduced from 370 K to 350 K by lengthening the cooler. The oil and water flowrates, inlet temperatures and other dimensions of the cooler will remain constant. The water enters at 285 K and the oil at 420 K. The water leaves the original cooler at 310 K. If the original length is 1 m, what must be the new length? Assume that heat transfer coefficients and physical properties do not change.

($1\cdot23$ m)

5. A mixture of benzene and toluene containing 25% benzene is to be separated in a distillation column to give a top product containing 90% benzene and a bottom product containing 10% benzene. The feed enters the column as liquid at its boiling point. A reflux ratio of $8\,mol/mol$ distillate is to be used. Estimate the number of theoretical stages required for the separation. A constant relative volatility of $2\cdot25$ may be assumed.

(7)

References

1. *C&R*, vol. 6, ch. 4 (for full details of references denoted by *C&R* see first item in the bibliography).
2. D. M. Himmelblau, *Basic Principles and Calculations in Chemical Engineering*, 4th edn., Prentice-Hall, 1982.
3. K. E. Bett, J. S. Rowlinson and G. Saville, *Thermodynamics for Chemical Engineers*, Athlone, 1975.
4. R. C. Reid, J. M. Prausnitz and B. E. Poling, *The Properties of Gases and Liquids*, 4th edn., McGraw-Hill, 1987.
5. R. M. Felder and R. W. Rousseau, *Elementary Principles of Chemical Processes*, Wiley, 1978.
6. R. B. Bird, W. E. Stewart and E. N. Lightfoot, *Transport Phenomena*, Wiley, 1960.
7. B. S. Massey, *Mechanics of Fluids*, 5th edn., Van Nostrand Reinhold, 1983.
8. M. M. Denn, *Process Fluid Mechanics*, Prentice-Hall, 1980.
9. *C&R*, vol. 1, ch. 1.
10. *C&R*, vol. 1, ch. 7; vol. 3, chs. 1 & 2; vol. 6, ch. 12.
11. D. Q. Kern, *Process Heat Transfer*, McGraw-Hill, 1950.
12. W. L. McCabe and J. C. Smith, *Unit Operations of Chemical Engineering*, 3rd edn., McGraw-Hill, 1976.
13. *C&R*, vol. 6, ch. 12.
14. E. J. Henley and J. D. Seader, *Equilibrium-Stage Separation Operations in Chemical Engineering*, Wiley, 1981.
15. C. J. King, *Separation Processes*, 2nd edn., McGraw-Hill, 1980.
16. *C&R*, vol. 6, ch. 11.
17. *C&R*, vol. 3, ch. 3.
18. Instrument Society of America, *Fundamentals of Process Control Theory*, 1981.
19. D. R. Coughanowr and L. B. Koppel, *Process Systems Analysis and Control*, McGraw-Hill, 1965.
20. C. D. Johnson, *Process Control Instrumentation Technology*, Wiley, 1977.

Bibliography

J. M. Coulson and J. F. Richardson, *Chemical Engineering*, Pergamon.
Vol. 1: *Fluid Flow, Heat Transfer and Mass Transfer*, 3rd edn., 1980.
Vol. 2: *Unit Operations*, 3rd edn., 1980.
Vol 3: *Chemical Reactor Design, Biochemical Reaction Engineering including Computational Techniques and Control*, 2nd edn., 1979.
Vol. 6: *Introduction to Design*, 1982.
A. W. Westerberg, H. P. Hutchison, R. L. Motard and P. Winter, *Process Flowsheeting*, Cambridge University Press, 1979.
R. Aris, *Elementary Chemical Reactor Analysis*, Prentice-Hall, 1969.
The above, together with references 3, 5, 8, 14, and 19, provide a basic coverage of the subject.

CHAPTER SEVEN

ENERGY

J. McINTYRE

7.1 Introduction

Energy is one of the four principal resources required for the production of chemicals, and in the manufacture of a number of materials its cost contribution may well exceed that of the other resources, namely raw materials, personnel and capital.

7.1.1 *Energy required by the chemical industry*

The chemical industry needs an energy input for the following operations:

   (i) to drive endothermic reactions
  (ii) to provide optimum reaction conditions
 (iii) to drive separation processes
 (iv) to effect mass transfer operations
  (v) to power control and communication systems
 (vi) to provide a satisfactory work environment.

Different types of energy are required, e.g. mechanical energy to create pressure reaction conditions, gravitational energy to achieve mass transfer between vessels, electrical energy to illuminate the work-place. Different quantities of energy are needed e.g. the large thermal input in a naphtha-cracking furnace compared with the small amount required in a battery to power walkie-talkie sets used by process operators to communicate with control rooms.

7.1.2 *Sources of energy*

The energy used in the chemical industry is derived from the fossil fuels coal, oil, and natural gas augmented by the utilization of combustible by-products. Combustion of these organic fuels is a very suitable high-grade heat source, particularly for endothermic reactions, but is less convenient than electricity for driving machinery. Electricity is self-generated on some large sites or purchased from the Central Electricity Generating Board (CEGB). It is highly

233

efficient as an energy source at the user point, but overall efficiency from fossil fuel is not high because of losses incurred in the energy transformation chain: chemical energy (latent in the fuel) → thermal energy (hot combustion gases, high pressure super-heated steam) → mechanical energy (turbine rotation) → electrical energy (generated in the alternator). The change from disordered (thermal) to ordered (mechanical) energy is limited by Carnot Cycle efficiency. Heat in stack gases, frictional losses, etc. further restrict efficiency with a figure of 38% being quoted as achievable in the best modern fossil-fired plant[1]. Allowing for distribution losses, 1 J electricity consumed = 3 J primary fuel burnt.

Electricity is thus categorized as a secondary source of energy and statistics for industrial energy consumptions may quote consumptions on an energy-supplied basis or on a primary-energy-consumed basis to make comparisons between industries more meaningful. Figures for energy consumption in the chemical industry are also complicated by the fact that 'fuels', e.g. naphtha and natural gas, are used as major chemical feedstocks as well as energy sources. The importance of energy consumption in the chemical industry relative to total U.K. consumption can be seen in Table 7.1 which has been derived from information published by the U.K. Government Statistical Service[2].

The chemical industry consumes 21% of the energy supplied to all industrial

Table 7.1 U.K. consumption of energy in 1987

	$MTOE^a$	TJ^b	%
Total inland consumption	208·3	8 852 800	
Petroleum for non-energy use	9·4	399 500	
Inland consumption for energy use	198·9	8 453 300	100
of which: coal	68·4	2 907 000	35
petroleum	64·3	2 732 700	32
natural gas	50·5	2 146 300	25
primary electricity[c]	12·9	548 300	7
imported electricity	2·8	119 000	1
Consumed by energy conversion processes		2 585 600	
Supplied to final consumers		5 867 700	100
of which: industry		1 660 900	28
transport		1 694 000	29
domestic		1 725 300	29
others		787 400	14
Supplied to chemical industry	8·04	341 800	100
as coal, coke, etc.		35 400	11
petroleum		75 600	22
natural gas[d]		182 000	53
electricity		48 800	14

[a] MTOE = million tonnes oil equivalent
[b] TJ = terajoules = 10^{12} joules
[c] Generated in nuclear-fired stations and from hydro-power
[d] Includes consumption as process feedstock.

sectors and to this 8.0 MTOE must be added 5.8×10^6 tonnes petroleum products, principally naphtha[3], used as feedstock in the petrochemical industry. The remaining non-energy uses of petroleum products are in lubricants, waxes, industrial spirits and bitumens all of which are essentially products of the petroleum refining industry.

The consumption of natural gas as chemical feedstock is not recorded separately, but a minimum of 33 000 TJ ($865 \times 10^6 \, m^3$) will be required for its most important use—an annual manufacture of 1.5×10^6 tonnes of ammonia. There are also substantial uses for methanol and hydrogen production.

The chemical industry is a major consumer of energy resources. The pattern has changed with time and the current predominance of gas for energy (oil is still predominant for feedstock) has built up since North Sea gas became available. It was preceded by oil in the 1960s and coal before then. The two major step changes in the price of oil during the 1970s (see below) have reduced its usage both in the chemical industry and also in the U.K. as a whole (from a peak of 113×10^6 tonnes in 1973 to 75×10^6 tonnes in 1987[4]). Partial reversion to the use of coal-fired boilers was initiated in the early 1980s when coal to large industrial consumers was priced at around 60% of that of fuel oil on a calorific value basis[5]. However, the world oil supply situation has eased and heavy fuel oil prices have dropped from £126 per tonne in 1983 to £78 per tonne in 1987[5]. Thus, for the moment, there is no incentive for further reversion to coal.

7.1.3 Properties of fuels

The various fuels have different advantages in use and the properties of significance can be summarized as follows:

7.1.3.1 Calorific value.
The calorific value of coal varies widely due to the class of coal available in different locations and to the proportion of contaminating rock present in the seam or necessarily extracted with the seam. Coal treatment plants remove much of this gangue, but coal for power plants can be used directly as mined if the impurity level can be tolerated. An average calorific value for coal is 26 GJ per tonne.

Oil has a higher calorific value because of its higher hydrogen content and because it is effectively fully combustible. The range is narrower than for coal and a figure of 43 GJ per tonne can be assumed. The quoted figures refer to gross calorific value. After combustion stack gases are exhausted to atmosphere above 150°C to minimize corrosion due to sulphur oxides and to provide thermal punch since a cool exhaust has a floppy plume due to condensed water and is environmentally unacceptable.

The use of gross calorific values or net calorific values (which assume the water of combustion is present as vapour), plus the average coal quality assumed, explains the difference in coal:oil equivalence figures, usually in the range 1.5:1 to 1.7:1, quoted by various authorities[6,7].

The calorific value of natural gas is a function of the 'inerts' content. North Sea gas is of high purity, is essentially methane, and has a specification of $1000\,\text{Btu}\,\text{ft}^{-3} = 37\,250\,\text{kJ}\,\text{m}^{-3}$ giving a calorific value of 52 GJ per tonne.

7.1.3.2 *Transportability*. Gas is easily transported by pipe-line and there is even some intercontinental trade in cryogenic ships. Oil distribution is principally by ship and road tanker. On the other hand, unless distances are very short, when conveyor belt or suspension pumping can be used, coal is distributed in 20–30 tonne unit loads by lorry or rail-car. Clearly much more mechanical handling and abrasive wear is involved than for pumping oil or gas.

7.1.3.3 *Usability*. Combustion of gas, distributed at pressure, is easily controlled in a simple gas jet whereas a burner head with provision for atomization of pre-warmed fuel is necessary for fuel oil combustion. However, for coal pre-grinding to pulverized fuel and complex ignition systems are necessary or less efficient grate stokers. There are also problems with removal and disposal of clinker and ash from the stack gases. Automatic firing control is more easily achieved with fluids as ignition, flow control and metering are complex when handling solids.

Thus, because of the combination of high calorific value, ease of transportation, better controllability, greater reliability, and lower maintenance, fluid fuels have premium value. Other factors also play a part, e.g. gas is more useful when combusted gases from the fuel are in direct contact with product. Additionally the high power to weight ratio and vaporizing properties of hydrocarbon fractions makes gasoline cuts particularly useful as transport fuels. For combustion the preferred fuels in order of preference are gas, hydrocarbon liquids, heavy fuel oil, and coal. Environmental acceptability follows the same sequence. The only disadvantage of gas is the inability to have significant on-site storage.

The presence of carbon chains of appropriate length in certain hydrocarbon fractions make them very suitable as chemical feedstock for the petrochemical industry where their value is clearly greater than that as fuel, for which substitutes are available (cf. Chapter 2, p. 20). However, at under 10%, the proportion of the barrel used for chemical manufacture is small and thus the price of petroleum is determined by world energy demand.

7.1.4 *Cost of energy*

The effect of North Sea gas becoming available in the late 1960s/early 1970s and the effect of the OPEC crises of late 1973 and 1979 when the price of Kuwait Crude moved from $1·6 bl^{-1} in 1970[8] peaking to $40·0 bl^{-1} for spot delivery in the early 1980s are clearly seen in Table 7.2.

The relative price movements explain why the chemical industry effectively standardized on oil-fired furnaces in the 1960s and moved over to gas where

Table 7.2   Prices of fuels used by U.K. industry*

	1968	1973	1978	1983	1987
£GJ^{-1}					
Coal	20	32	84	181	179
Fuel oil	21	30	120	294	182
Gas	63	29	111	228	208
Electricity	179	206	528	807	803
Index[†]					
Coal	0·6	1·0	2·6	5·7	5·6
Fuel oil	0·7	1·0	4·0	9·8	6·1
Gas	2·2	1·0	3·8	7·9	7·2
Electricity	0·9	1·0	2·6	3·9	3·9

*Adapted from data published by Her Majesty's Stationery Office (HMSO).
[†]Index 1·0 at pre-oil crisis, 1973.

possible in the 1970s. Furthermore with the cost of hydrocarbons escalating and the price of electricity increasing more slowly there is a trend toward larger purchases from the CEGB rather than installation of more energy-efficient combined heat and power plants.[9]

With the cost of energy for energy purposes in the U.K. chemical industry exceeding £1 billion in 1987 there are major incentives to ensure efficient utilization.

7.2 Types of energy

Different types of energy are involved in chemical processing on the production scale. These include chemical energy, thermal energy, electrical energy, mechanical energy, radiation energy, gravitational energy, and kinetic energy. The most powerful form, nuclear energy, is utilized in the nuclear-industry, normally for the generation of energy for conversion to electricity, but also for the manufacture of plutonium.

Although all forms of energy have important parts to play, considerable variation in the order of magnitude of each type is found. It should be no surprise that *chemical energy* is of high magnitude because powerful bonds between atoms are being broken and reformed. Together with our need for volume-produced chemicals, this is the major reason why the chemical industry consumes a large proportion of the industrial energy used in the U.K. Two examples illustrate this. The manufacture of one tonne of primary aluminium requires 270 GJ, of which 228 GJ is consumed at the smelting stage[10]. In the petrochemical industry the manufacture of ammonia from hydrocarbon feedstock in an integrated single-stream plant requires 14 GJ per tonne to meet energy requirements[11]. Ammonia plants constructed in the 1980s use an improved process.

The energy input for endothermic reactions is normally supplied as heat,

Table 7.3 Heat properties of liquids

	Specific heat capacity $Cp/kJkg^{-1}K^{-1}$	Specific latent heat of vaporization at b.p. $LH/kJkg^{-1}$
Benzene	1·70	394
Cyclohexane	1·80	393
Ethanol	2·50	839
Nitrobenzene	1·40	330
n-Octane	2·20	364
Acetone (propanone)	2·20	522
Carbon tetrachloride (tetrachloromethane)	0·84	195
Water	4·19	2260

but much *thermal energy* in addition to that which is converted into chemical energy of the product is needed to drive preparation and separation processes. A major consumption is in the vaporization of liquids. Heat capacities of substances are normally numerically 2 orders of magnitude lower than their latent heat of vaporization (see Table 7.3).

Electrical energy is used in electrochemical processes, electrostatic precipitation, dielectric drying, area lighting, and control and communication systems. Note however the principal use of electricity is in machine drives. Electric motors are efficient and, most important, very reliable. Furthermore the use of flexible armoured cabling means there are no constraints on the positioning of motors. Large horse-power motors are normally associated with gas compression or grinding machines. In contrast pumping of fluids is not energy-intensive unless large hydrostatic heads are involved.

The transportation of liquids (and occasionally solids) between vessels in a chemical manufacturing unit is frequently achieved by *gravity*. It is very reliable (the motive power never failing!) but it is not free as liquids have to be pumped to the top of the plant initially. These fluids flowing through pipes possess *kinetic energy*. Accurate energy balance calculations in sections of large plants have to take account of this, but normally the quantity involved relative to the accuracy of other consumptions is so small that it can be ignored. This is best illustrated by examining the duty of a typical pumping system (see also section 6.4.4.2.).

7.2.1 *Power requirements for fluid flow*

In a liquid pumping duty, power is required to overcome gravity, supply work to overcome resistance to flow (dissipated as heat), and supply kinetic energy. Expressed mathematically:

$$W = mg\Delta h + Q\Delta P + \tfrac{1}{2}mv^2 \quad \text{(cf. also example 6.13)}$$

where

$$W = \text{power requirement (watts)}$$
$$m = \text{mass flow rate (kg s}^{-1})$$
$$g = \text{acceleration due to gravity (9.8 m s}^{-2})$$
$$\Delta h = \text{height differential (m)}$$
$$Q = \text{volume flow rate (m}^3\text{s}^{-1})$$
$$\Delta p = \text{pressure differential (N m}^{-2})$$
$$v = \text{flow velocity (m s}^{-1})$$

The formula for pressure drop in turbulent flow (turbulent flow is normal; the carrying capacity of a pipe where flow is laminar is relatively small) in a circular pipe is

$$\Delta p = \frac{2\rho f l v^2}{d}$$

where

$$\Delta P = \text{pressure differential (N m}^{-2})$$
$$\rho = \text{fluid density (kg m}^{-3})$$
$$f = \text{fanning friction factor (dimensionless)}$$
$$l = \text{pipe length (m)}$$
$$v = \text{flow velocity (m s}^{-1})$$
$$d = \text{pipe diameter (m)}$$

Taking as an example the pumping of water at a flow rate of $1200\,\text{m}^3\,\text{hr}^{-1}$ from a river through a 600 mm diameter pipeline 1 km long into a site cooling-water storage tank 60 m above the river, the component energy requirements can be calculated as follows.

Gravitational energy: $mg\Delta h = Q\rho g\Delta h$

For
$$Q = 1/3\,\text{m}^3\,\text{s}^{-1}$$
$$\rho = 1000\,\text{kg m}^{-3}$$
$$g = 9{\cdot}8\,\text{m s}^{-2}$$
$$h = 60\,\text{m}$$

Power requirement $= 1/3 \times 1000 \times 9{\cdot}8 \times 60 \times 10^{-3} = 196\,\text{kW}$

Frictional resistance: $Q\Delta P = \dfrac{2Q\rho f l v^2}{d}$

Now v (velocity) $= \dfrac{\text{flow rate}}{\text{cross-sectional area}}$

i.e.
$$v = \frac{Q}{\pi\left(\dfrac{d}{2}\right)^2} = \frac{4Q}{\pi d^2}$$

For
$$Q = 1/3\,\mathrm{m^3\,s^{-1}}$$
$$d = 0\cdot6\,\mathrm{m}$$

$$v = \frac{4}{3 \times \pi \times 0\cdot6^2} = 1\cdot18\,\mathrm{m\,s^{-1}}$$

For
$$Q = 1/3\,\mathrm{m^3\,s^{-1}}$$
$$\rho = 1000\,\mathrm{kg\,m^{-3}}$$
$$f = 0\cdot0045 \text{ for normal flow rate in steel pipe}$$
$$l = 1000\,\mathrm{m}$$
$$v = 1\cdot18\,\mathrm{m\,s^{-1}}$$
$$d = 0\cdot6\,\mathrm{m}$$

$$\text{Power requirement} = \frac{2 \times 1000 \times 0\cdot0045 \times 1000 \times 1\cdot18 \times 1\cdot18}{3 \times 0\cdot6 \times 1000}$$
$$= 7\,\mathrm{kW}$$

Kinetic energy: $\frac{1}{2}mv^2 = \frac{1}{2}Q\rho v^2$

For
$$Q = 1/3\,\mathrm{m^3\,s^{-1}}$$
$$\rho = 1000\,\mathrm{kg\,m^{-3}}$$
$$v = 1\cdot18\,\mathrm{m\,s^{-1}}$$

$$\text{Power requirement} = \frac{1000 \times 1\cdot18 \times 1\cdot18}{2 \times 3 \times 1000} = 0\cdot2\,\mathrm{kW}$$

In this instance the kinetic energy is three orders of magnitude smaller than the energy required to overcome gravity. This example also shows the relatively small amount of energy (7 kWh) needed to pump 1200 tonnes water through 1 kilometre horizontal distance.

7.2.2 Variation in energy content requirement

The different order of energy involved in different aspects of chemical processing operations is illustrated below:

Chemical energy:
    Burning 1 tonne octane releases   45 000 MJ
Thermal energy:
    Boiling 1 tonne octane requires   300 MJ
Gravitational energy:
    Raising 1 tonne octane through 100 m requires   1 MJ

Frictional resistance:
Pumping 1 tonne octane 1 km requires   0·01 MJ
Kinetic energy:
1 tonne octane flowing at $1\,m\,s^{-1}$ 0·001 MJ

That chemical energy is of a different order of magnitude than mechanical energy should be no surprise. Were it not so the industrial revolution would not have been possible based as it was on the conversion of the chemical energy in coal to mechanical energy in machines using very inefficient boilers.

Some energy values and consumptions are placed in perspective in Table 7.4. The figures quoted are rounded and should be used only for approximate equivalence calculations.

Table 7.4   Energy comparisons

	J
Joule	1×10^0
Calorie	$4·2 \times 10^0$
British thermal unit	1×10^3
Latent heat of water (g^{-1})	$2·2 \times 10^3$
Kilowatt hour	$3·6 \times 10^6$
Butter (kg^{-1})	30×10^6
Petrol (l^{-1})	35×10^6
Coal $(tonne^{-1})$	26×10^9
Natural gas $(1000\,m^{-3})$	37×10^9
Oil $(tonne^{-1})$	43×10^9
House usage (y^{-1})	100×10^9
747 transatlantic flight	3×10^{12}
2000 MW station output (y^{-1})	50×10^{15}
U.K. energy consumption (y^{-1})	10×10^{18}
World energy consumption (y^{-1})	300×10^{18}

7.3 Use of energy in the chemical industry

As mentioned in section 7.1.1, energy is required for a multitude of purposes in the chemical industry. For some, e.g. illumination, electricity is the only sensible energy source (apart from very special circumstances), but for other purposes substitute energy sources are possible, e.g. use of high-pressure steam or direct firing as heat source for a distillation operation. Many applications, e.g. communication and control, use small quantities of energy. In contrast reaction energy, and a limited number of preparation and separation processes, account for a high proportion of the energy consumed in the chemical industry.

7.3.1 Reaction energy

In reactions involving electrolysis energy input is added directly to the reactants. In operations similar to the production of aluminium there is direct electricity input to provide the energy for the endothermic reaction. For most reactions however indirect heat is used. Thus fuel is burnt and the heat in the combusted gases is transferred through the wall of a vessel or tube into the space where reaction is taking place. The high-grade heat available by direct firing may be unnecessary. In some situations direct heating may be unsafe because of fire hazard; and a multitude of small furnaces will consume fuel inefficiently. Low-grade reaction heat is supplied conventionally by steam raised in a central boiler plant.

7.3.2 *Preparation and separation energy*

Energy demands are heavy in gas compression, comminution, and in those separation operations involving vaporization, evaporative concentration, drying and distillation. Distillation involves revaporization of material condensed for reflux.

7.3.2.1 *Gas compression.* Not only must energy be supplied to force the gas molecules into a smaller volume but the input energy which appears as heat must be dissipated to prevent the compression machinery from seizing up due to gross differential expansion or to lubrication failure if the temperature gets too high. High-pressure machines are therefore necessarily multi-stage in action with a 4:1 compression ratio at each stage being fairly average.

7.3.2.2 *Comminution.* Particle size reduction of large lumps makes use of flaws within crystal lattices, rupture along these flaws occurring on compression, e.g. crushing. But reduction to finer particles involves abrasion and this is very energy-intensive, partially because of the energy absorbed by the machines themselves.

7.3.2.3 *Distillation.* The process of distillation involves the vaporization of a liquid mixture and the interchange between this vapour rising up a column and condensed liquid descending the column, with contact being assured by the presence of plates or packing in the column. The degree of separation of the components in the mixture can be improved by using a taller column with more trays or by increasing the reflux ratio, i.e. the ratio of condensed vapour from the top of the column returned to the column to that removed as distillate. For some difficult separations reflux ratios higher than 20:1 have been employed. As every kilogram of reflux has to be revaporized, and latent heats of vaporization are much higher than specific heats, it is clear that distillation is an energy-intensive process. Latent heat has also to be supplied in drying operations; maximum removal of fluid mechanically, e.g. in pressure filters, is always undertaken before drying providing the dryer can cope with the physical form of the feed.

7.3.3 *Heat transfer media*

A few operations such as induction heating make use of an energy source directly but most heating (and cooling) processes involve conductive and/or convective transmission via a heating medium. Steam and water are most commonly used for heating and cooling respectively, though steam generation is used as a coolant in high-temperature applications and warm water is used for mild heating. However, there are occasions when the temperature range $10°C \rightarrow 250°C$ normally available with these two media is inadequate.

Cooling water requirements in the chemical industry relative to availability in the environment are such that on large sites the demand cannot be met by once-through cooling. Recirculation through cooling towers is essential. These cooling towers are designed to promote droplet formation in a counter-flow draught of air with cooling being achieved principally by increasing the water-vapour content of the air. The latent heat required to evaporate 1–2% of the water lowers the temperature of the remainder. The required air-flow can be achieved by the use of fans or by the chimney-effect exemplified by the cooling towers associated with inland electricity generating stations. Because of the importance of water recovery, direct air-cooling using fans and vaned heat-exchangers is becoming more popular.

Steam is useful as a heat transfer-medium because:

(i) it possesses a high heat content
(ii) it can be distributed easily
(iii) flow is easily controllable
(iv) it is non-combustible and will not support combustion
(v) it is non-toxic and relatively non-corrosive, and
(vi) it is produced from water which is relatively cheap and abundant.

Steam gives up its heat by condensing. Its latent heat is 1000 times greater than the heat capacity of gaseous steam. Thus dry saturated steam at a temperature differential for good controlled heat flow (about 30°C) is ideal, but steam temperature in the presence of any condensate is a function of steam pressure. It is not possible to have steam of optimum pressure available at each point of use. Consequently steam is normally distributed at 2 atmospheres and perhaps at two other pressures on each site with a maximum of 40 atmospheres.

One further aspect is that steam can be used as a source of power as well as heat: power can be extracted in machine drives and the residual heat in the steam exhausted at lower pressure can be utilized allowing high overall efficiency. If high-pressure steam from a boiler plant is passed through a turbine/alternator set to generate electricity, then exhausted and distributed at a lower pressure to supply process heat, most of the energy in the original fuel is usefully used rather than rejected to atmosphere. This contrasts with conventional power stations where maximum conversion to mechanical then electrical energy is the target. Self-generation of electricity—the system now known as CHP (combined heat and power)—is only feasible where there is a steady continuous demand for thermal energy, such as is required for processing operations in a chemical complex.

The desirable properties of a high-level heat transfer medium include low cost, non-flammability, nil toxicity, compatibility with common metals, remaining liquid at ambient temperature and, above all, thermal stability. This is a tall order and materials in use do not meet all the criteria. Examples include:

Petroleum oils. Low cost, non-toxic, non-corrosive, usable at operating temperatures up to 315°C, but flammable and subject to oxidative degeneration. The latter are countered by operating under a nitrogen blanket, but thermal cracking will still occur.

Dowtherm. 'Dowtherm' is a proprietary generic name applied to a range of heat transfer fluids, but in the U.K. normally refers to a mixture of diphenyl (73·5%) and diphenyl oxide (26·5%). The mixture is also marketed as 'Thermex'. It is fluid down to 12°C, leading to a solidification problem in cold weather. High maintenance standards are necessary because of its searching characteristics. For prolonged operation, temperature restriction to 370°C should apply, but it is usable up to 400°C.

Hygrotherm. Hygrotherm is a trade-name for tetra-aryl silicate fluids which, being silicon derivatives, are expensive. However they are non-toxic, non-corrosive, will burn only at high temperatures, and have good heat transfer coefficients. They can be used at temperatures up to 355°C.

Heat transfer salt. This is a mixture of sodium nitrite (40%), sodium nitrate (7%) and potassium nitrate (53%) and is liquid above 142°C. It is stable up to 455°C and usable up to 535°C under nitrogen blanketing. Thus it has a higher temperature range than organic-based materials. However the mixture is a powerful oxidizing agent and leaks must not be allowed to come into contact with flammable material.

Sodium metal. This is used as a coolant in fast reactors because of its superior efficiency, but overcoming problems associated with its reactivity with air and water makes this system costly, and restricts its use to nuclear applications.

Carbon dioxide. Carbon dioxide at pressure is used in U.K. nuclear power plants to transfer heat from the reactor core to the steam boilers. The low heat capacity and high circulation costs restrict the application of gases for heat transfer purposes to special circumstances such as this. The use of combusted gases for heating and air for cooling is regarded as direct use.

Glycol. Ethane-1,2-diol/water mixtures are used in intermittent cooling applications where freedom from frost worries when off-line is desired.

Low level heat transfer media. Although ammonia and chorofluoromethanes are used in refrigeration plants, it is not unusual to locate a refrigeration plant to serve a number of chemical plants and use *ice* for direct addition to reaction media where water dilution is acceptable. Alternatively 'cold' may be transmitted to its points of use by circulating *sodium chloride brine* (23%) for temperatures down to $-21°C$ or the more popular *calcium chloride brine* (29%) which can be taken down to $-40°C$. For even lower temperatures liquid propene or liquid ethene are used if available.

7.4 Efficient utilization of energy

Because of the high energy input required for many major chemical processes, the chemical industry has been in the forefront of the development and application of efficient energy utilization techniques, and the recent escalation in fuel prices has made it even more energy-conscious. With this change the capital cost of installing additional heat-recovery equipment or thicker layers of insulation becomes more acceptable. All such 'add-on' features help, but the major economies will come from the development of new energy-saving processes (e.g. the recently commissioned ICI LCA ammonia plants on Severnside) and less energy-intensive separation techniques.

7.4.1 *Exothermic reactions*

Reaction exotherms (and endotherms) are usually substantial, so why are there very few energy-exporting chemical processes? Firstly, the temperature rise associated with many exothermic reactions must be limited—when the temperature rises the reaction rate increases and a reaction runaway, with or without undesirable by-product formation, must be avoided. Additionally of course Le Chatelier's Principle states that favourable equilibria for exothermic reactions are associated with lower temperatures. Thus if the temperature must be kept low ($< 370\,K$) then reaction heat must be dissipated into the reaction diluent (many reactions are undertaken in suspension or solution) or into cooling coils. In a few cases there may be heat interchange to pre-heat another flow, but mostly this low-grade heat is necessarily dissipated to the environment, as the energy content in warm water is of little use.

This aspect of energy grade is of prime importance. Irrespective of the quantity of heat available, heat will not flow from a cooler to a warmer body. High-grade heat from continuous exothermic chemical processes is always recovered in heat exchangers, either by direct interchange (reactor outlet flow pre-heating reactor feed) or by raising steam which is subsequently used to drive separation processes or turbine machinery. The success of many modern petrochemical processes is based on the integration of the various chemical processing and separation stages with the heat recovery equipment in order to make the best overall use of the available energy from reaction exotherms and imported fuel. An example is the modern process for the manufacture of ammonia (NH_3) from natural gas (CH_4). The ammonia synthesis stage is exothermic, but to achieve a commercially feasible equilibrium (15–17% conversion per pass) and rate of reaction the temperature must be maintained between 400°C and 500°C. The quality of the recoverable heat is insufficient to meet the requirements of the front end of the plant where the endothermic steam reforming reaction

$$CH_4 + H_2O \rightleftharpoons CO + 3H_2$$

requires temperatures between 700°C and 750°C. Thus primary heat must be added. The heat in the furnace flue gases (and in the process gas stream downstream of the secondary reformer where air is injected to supply the nitrogen) is used to generate steam for the turbine-driven process air compressor, synthesis gas compressor, and gas circulator in the synthesis loop. Exhaust steam from the turbines is used in the regeneration column boiler to strip CO_2 from the liquor used in the absorption system and for other duties. Some plants burn less fuel, recover less heat, and need to import electricity to drive the big machines. Theoretically there is a net evolution of 34 kJ mol^{-1} of heat in the manufacture of ammonia from natural gas, steam, and air, but achievement of the necessary reaction conditions and the preparation and separation processes involved requires an energy input of approx. 240 kJ mol^{-1}[1,12].

Hydrocarbon partial oxidation reactions are highly exothermic but reaction temperatures are normally kept down to prevent by-product formation. Any energy recovered in the form of steam is usually fully utilized in the distillation train and/or other processes which are used to separate the desired product in sufficient purity from the accompanying spectrum of by-products. One process which is a net exporter of energy is the naphtha oxidation process for the manufacture of acetic (ethanoic) acid (section 11.5). The only two other processes which are significant exporters of energy are the manufacture of sulphuric acid from sulphur and the oxidation of ammonia to nitric acid.

7.4.2 Separation processes

In *distillation* operations the heat supplied to vaporize the feed and column reflux is dissipated into non-recoverable heat in condenser cooling water. Thermal economy can be achieved in re-designed plants by using longer columns fitted with more plates (or additional packing) thus allowing a reduced reflux ratio, and hence lower reboiler energy demand. Alternatively the product specification may be relaxed, since for many purposes very high purity may be unnecessary.

The efficiency of *evaporators* is improved by multi-effect operation. In this the evolved vapour from the first stage is condensed in a heat exchanger and gives up its latent heat to the second stage, etc. The necessary temperature differential across each heat exchanger means a progressive reduction in boiling point. This is achieved by having a pressure (vacuum) gradient. Theoretically the heat evolved on condensing 1 kg of vapour equals the heat required to evaporate 1 kg of the same liquid. Efficiencies of 0·85 for each stage are achievable. Increasing the number of stages improves the thermal economy but at increased capital expenditure, and therefore each system must be optimized. However, not all evaporators are multi-effect. For example evaporative concentration of thermally unstable materials requires short residence time and hence once-through operation in film evaporators.

Drying operations also involve the removal of liquid by evaporation. Pre-concentration of slurries by mechanical means, e.g. press filtration, centrifuging, etc., is normal practice in order to reduce the evaporative load in the dryer. The degree of diluent removal is determined by the particle size of the solid and the feed characteristics (pumpable slurry, preformable paste, hard lumps, etc.) of the selected dryer which, in turn, is determined by the properties and the wanted physical form of the product.

Electrostatic precipitation is another energy-intensive separation process. But it is the most efficient method of restricting particulate emission in off-gases. Consequently, its use for environmental control purposes in large installations is fairly widespread.

7.4.3 *Restriction of losses*

Provided the economics are favourable any energy which can be re-used is recovered. Effectively this means when it is available at a sufficient temperature differential and in quantity and continuity to justify the capital expenditure on a heat exchanger. The recovery of heat which is only available intermittently is seldom worthwhile. Some rejection to the environment is inevitable, but more heat can be kept within the system—and therefore less heat input is required—by minimizing convection and radiation losses from hot surfaces, and by restricting leaks of heat transfer fluids.

Many plant items and interconnecting pipework are *lagged*, principally to limit heat (or cold) loss but also for personnel protection. Apart from very special applications, asbestos is now out of favour. Lagging of hot pipes is principally with magnesia, calcium silicate or mineral wool whereas expanded polystyrene is frequently used on cold pipework. Lagging thickness varies from 30 mm to 200 mm dependent upon duty[13]. Thicker layers are becoming more economical as energy costs continue to rise. As an example the heat loss of 8400 W m^{-2} from a bare surface at 325°C in still air can be reduced to 200 W m^{-2} by fitting 100 mm-thick lagging[14].

In some instances, *leaks* in heat transfer systems cannot be stopped completely. The number of apertures for air supply, burner guns, feed and withdrawal pipes, peep-holes, etc. in a normal furnace, coupled with the differential expansion of the materials of construction, means that tight sealing is impossible. Furnaces are therefore operated at fractional negative pressure to avoid outward escape of hot gases, but this means an inward leak of cold air. Although they cannot be eliminated, such leaks should be minimized. On the other hand, determined efforts should be made to repair leaks from flanges and valve packings on steam distribution systems. Initiated by thermal cycling (and ageing), the pressure differential to atmosphere ensures flow through any weakness and due to erosion this flow increases with time. Damage may then extend from the gasket material to guttering of flange faces and thus lead to a more complex repair job. Clearly wherever possible leaks should be repaired as they arise. On key areas of continuous plant and factory distribution systems immediate shut-down may not be practicable. In these cases temporary online repairs can be effected by fitting a jacket round the leaking flange and filling the space with a special heat-setting material. Steam leaks can lose a lot of energy, e.g. there will be a wastage of $60\,kg\,hr^{-1}$ from a 3 mm dia. hole in a 20 atmosphere steam pipe[9]. Note also that leaks are continuous and this can represent a loss of 10 tonnes of high-quality steam, i.e. 28 GJ, per week.

7.5 Conclusions

The energy balance of a chemical process is of prime importance. Continuous processes have the advantage over batch processes in that there is no heat wasted discharging and recharging batch reactors; additionally, recoverable heat is available continuously, making it easier to justify the capital cost of recovery equipment. Reaction endotherms/exotherms involve the absorbtion/release of relatively large quantities of energy. Therefore in large continuous plants the absolute quantity of energy is very large and every effort is made to extract energy from hot flows possessing a satisfactory temperature differential. This, or imported, energy is used to drive preparation and separation processes, some of which are very energy-intensive. Successful process designs require the integration of reaction and separation stages, and heat recovery equipment to make best use of the energy input. Sound fabrication to minimize potential leaks and good insulation for heat retention are also necessary.

For the future, the development of less energy-intensive processes is a priority. When fundamental changes are not possible (e.g. on an existing plant) improved instrumentation and control techniques will yield better reaction mixture products and reduce demands on separation equipment. Further minor improvements, e.g. avoidance of overcooling reflux and better lagging, will effect cumulative economies. Waste combustible material is being or will be collected and fed to boilers. There will be a slow changeover to coal for large boiler plants, but smaller furnaces on plant will continue to use natural gas or liquid hdrocarbons because of interruptibility and superior controllability. Most large chemical companies now have personnel who have been assigned specifically to analyse, and devise, economic methods of reducing energy requirements of their plants. That the progressive introduction of energy efficiency measures is having success across the country as a whole can be demonstrated by comparing the annual temperature-corrected primary inland energy consumption with the deflated gross domestic product

(GDP). Dividing the energy consumption by GDP yields the *energy ratio*[15] which can be regarded as an approximate overall indication of the comparative efficiency of energy utilization. From 1950 to 1973 there was a steady rise in U.K. energy consumption associated with a doubling of GDP. In this 23 year period the energy ratio decreased by 0·9% per annum. Since the first OPEC oil crisis in 1973 the GDP has been erractic, but by 1987 had risen 27%. However, energy consumption fell and over this 14-year period there was a fall of 2% per annum in the energy ratio. Thus the rate of increase of efficiency of energy utilization has doubled. This was the response to the steep change in energy costs.

Appendix

Calorific values (CVs)

1 tonne coal	26 GJ
1 tonne oil	43 GJ
1000 m^3 natural gas	37 GJ

These are average gross calorific values. Net calorific values, which assume water produced by combustion remains in the vapour phase, are a better indication of equivalence of fuels (but remember the dirt content in coal).

	Gross CV $MJ\,kg^{-1}$	Net CV $MJ\,kg^{-1}$	Gross less Net $MJ\,kg^{-1}$	%
C (coal)	32·8	32·8	Nil	Nil
C$_{20}$H$_{42}$ (crude oil)	47·2	44·2	3·0	6·4
CH$_4$ (natural gas)	55·5	50·1	5·4	9·7

Quoted average equivalences of fuels vary according to source, but usually between 1·5 tonne coal = 1 tonne oil[7] and 1·7 tonne coal = 1 tonne oil[6].

References

1. *CEGB Statistical Yearbook*, supplement to Central Electricity Generating Board Annual Report and Accounts 1987/88, pp. 9–11.
2. *Digest of United Kingdom Energy Statistics* 1988, HMSO, 1988, pp. 11, 13 and 28.
3. *Digest of United Kingdom Energy Statistics* 1988, HMSO, 1988, p. 17.
4. *BP Statistical Review of World Energy* 1988, British Petroleum 1988, pp. 7, 8.
5. *Digest of United Kingdom Energy Statistics* 1988, HMSO, 1988, p. 97.
6. *Digest of United Kingdom Energy Statistics* 1988, HMSO, 1988, p. 111.
7. *BP Statistical Review of World Energy* 1988, British Petroleum 1988, p. 36.
8. *Energy Statistics: Prices of Fuel Oil 1960–1974*, Eurostat, 1974, p. 168.
9. C. D. Grant, *Energy Conservation in the Chemical and Process Industries*, Institution of Chemical Engineers, 1979, p. 39.
10. P. F. Chapman and F. Roberts, *Metal Resources and Energy*, Butterworth, 1983, p. 199.
11. C. D. Grant, *Energy Conservation in the Chemical and Process Industries*, Institution of Chemical Engineers, 1979, p. 31.
12. *The Chemical Engineer*, March 1983, 30–31.

13. P. M. Goodall, *The Efficient Use of Steam*, Westbury House, 1980, p. 101.
14. O. Lyle, *The Efficient Use of Steam*, HMSO, 1968, p. 124.
15. *Digest of United Kingdom Energy Statistics 1988*, HMSO, 1988, p. 4.

Bibliography

Principles of Industrial Chemistry, C. A. Clausen and G. Mattson, Wiley, 1978.
The Energy Question, G. Foley, Penguin, 1976.
Energy and the Environment, J. M. Fowler, McGraw-Hill, 1975.
Energy Conservation in the Chemical and Process Industries, C. D. Grant, Institution of Chemical Engineers, 1979.
Kirk-Othmer's Encyclopedia of Chemical Technology, 3rd ed., Wiley, 1978–1984.
The Efficient Use of Steam, O. Lyle, HMSO, 1968.
Energy Resources and Supply, 2nd ed., J. T. McMullen *et al.*, Arnold, 1983.
Energy Around the World, J. C. McVeigh, Pergamon, 1984.
Chemical Engineering in Practice, G. Nonhebel, Wykeham, 1973.
Perry's Chemical Engineers' Handbook, 6th ed., eds R. H. Perry and D. Green, McGraw-Hill, 1984.
Energy and Introduction to Physics, R. H. Romer, Freeman, 1976.
Ullmann's Encyclopedia of Industrial Chemistry, 5th ed. W. Gerhartz, VCLI Verlagsgesellschaft, 1985.
Our Industry Petroleum, British Petroleum Co. Ltd., BP Publishing Co., 1977.
Annual Abstract of Statistics, Central Statistical Office, HMSO, annually.
Digest of U.K. Energy Statistics, Central Statistical Office, HMSO, annually.
Coal and the Environment, Commission on Energy and the Environment, HMSO, 1981.
Combined Heat and Power Generation in the U.K., Department of Energy, HMSO, 1979.
Development of the Oil and Gas Resources of the U.K., Department of Energy, HMSO, annually.
A Guide to North Sea Oil and Gas Technology, Institute of Petroleum, 1978.

CHAPTER EIGHT

ENVIRONMENTAL POLLUTION CONTROL

K. V. SCOTT

8.1 Technology and pollution

Pollution is any process whether natural or man-made which leads to the harmful or objectionable increase in the amount of any factor in the environment. Frequently encountered factors are scrap, refuse, sewage, noise, and biological, radioactive or chemical substances. Pollution may arise from the accidental escape of materials or from the disposal of waste products.

Pollution problems have long existed, especially in volcanic regions. Virtually all the inhabitants of Pompeii were killed during an eruption of Mount Vesuvius in A.D. 79 by the effects of dust, fumes and heat. From earliest times there have also been the adverse effects of smoke and combustion gases from forest fires. One of the earliest recorded reports on the adverse effects of coal burning was in 1257 when Queen Eleanor left Nottingham for Tutbury on account of the smoke nuisance[1]. By the reign of Queen Elizabeth I the problem had become so serious that coal burning was banned in London while Parliament was sitting.

It was during the Industrial Revolution that the problems from pollution dramatically increased. In addition to the massive increase in smoke and combustion products from coal burning, there was a wide variety of gaseous and aqueous effluents from the developing industrial processes which had a marked effect on the environment. With the accompanying growth of towns, sewage and domestic waste disposal became major problems.

In factories and similar industrial installations such as water treatment plants, an accidental release may occur from a process or storage, or from a pipeline or container when material is being transferred or is awaiting shipment or use.

In recent years much work has been carried out to identify sources of accidental release, to establish the size and duration of such releases, and with the aid of sophisticated computer programs their impact on the general public. Some companies have set targets which they feel can be defended, and where the estimated risks have exceeded the targets considerable

251

expenditure has been made in the form of plant modifications and changes to operational methods to attain the target hazard frequencies.

In dealing with waste material man has three basic options. The first option is to dump the waste into the nearest convenient environment such as the air, a river or lake, on the soil, or in a well or in the ocean. The second option is to contain the waste and treat it within a delimited environmental waste park where engineered systems along with natural ecosystems such as ponds, spray irrigated forests and landfill do most of the work of decomposition and recycling. The third option is to treat the waste in artificial chemo-mechanical regenerative systems.

With the first option the basic philosphy is that 'the solution to pollution is dilution'. Until recently this was the principal waste disposal system and this is still the case in many parts of the world. Because of the potential for using this option, industry and cities have tended to concentrate along waterways which provided free sewers. This option is no longer acceptable in developed countries and is being phased out as quickly as possible, though slowed somewhat by the costs involved.

The second option proved the most economical method of avoiding general environmental pollution by the fairly dilute, but large-volume, wastes that now reduce so badly the quality of man's living space and endanger his health. Setting aside large areas for environmental waste treatment would also preserve valuable open space that not only protects the environmental quality in general, but provides other uses as well, such as food and fibre production, recreational areas etc. To exercise this option means setting aside large areas for the purpose, and while this is practicable in the U.S.A. it is unlikely to be adopted in the U.K. However, if this second option is not adopted, it leaves only the much more expensive third option.

The third option involves treatment to remove or almost completely remove the pollutants. Abiotic treatment and recycling is necessary for some types of wastes, especially in densely populated industrial areas. With air pollution mechanical treatment is the only option for some components which have to be stopped or reduced at source. Where this is not possible a substitute energy source or a different industrial process must be found as the cost of air pollution is no longer being tolerated. This may involve expensive artificial treatment of biodegradable waste as well as the poisons, but the cost is then borne by the product or enterprise and not by the population in the area of the production unit.

Partial treatment, particularly in relation to sewage (secondary treatment of sewage), gives an effluent which if discharged to closed waters such as lakes results in progressive eutrophication (nutrient enrichment) and the total loss of animal life in the water. The growth of algal blooms and oxygen depletion are early signs of this. Where it has occurred, especially in the U.S.A., intense political pressure has resulted in considerable capital expenditure on tertiary treatment plant for sewage in particular. Once the contamination of the water has stopped, either as a result of the effluent being diverted elsewhere, or due to tertiary treatment, the body of water gradually returns to normal life as the quantities of phosphates, nitrates and other nutrients decrease towards their level before pollution.

8.1.1 *Air pollution*

Pollution of the atmosphere arises from many sources and in a variety of ways:

(a) continuous release, in a dilute form, of toxic or corrosive gases (gases which give corrosive solutions on contact with water);
(b) intermittent and normally infrequent release, in a more concentrated form, of toxic or corrosive gas;
(c) foul-smelling gases which may be harmful or only a nuisance;
(d) gases which are not harmful to man in the concentrations encountered but which are harmful to vegetation;
(e) dust which may be toxic, or an irritant, or just a nuisance;
(f) smuts, including contaminated ones;
(g) sprays including aerosols.

Some examples of the above are as follows.

8.1.1.1 *Pollution by sulphur dioxide.* Of all the gases subject to the control of the authorities, the one emitted in the greatest volumes is sulphur dioxide, mainly from the combustion of fossil fuels. The low concentration in stack gases, before discharge, makes its removal both difficult and costly. Until recently, reliance has been placed on the use of chimneys of adequate height and design, to ensure sufficient dilution of the gases before they reach the ground.

The introduction of air quality standards for sulphur dioxide, and concern about transfrontier pollution (see also 8.6.1), combined with the development of commercial processes to greatly reduce the emission of sulphur dioxide, has forced the review of present and future positions. While the urban ground level concentrations of sulphur dioxide have been greatly reduced in the U.K. in recent years[2], the total amount released has fallen slowly, with an increase from industry coinciding with the mid 1980s recovery in activity.

Flue gas desulphurization (FGD), as agreed for some existing power stations, will reduce their sulphur dioxide emission by at least 90%.[43]

8.1.1.2 *Toxic gas emissions.* One form of intermittent pollution that has come to the fore in recent years is the occasional release of toxic gas from a site which can have acute effects on the outside population as well as on the employees on site.

With a rapid increase in chlorine production capacity at one of ICI Mond Divisions factories in the late 1960s there was an increase in the emissions of chlorine which caused a nuisance to the population near to the factory.

A method was developed[3] for examining the position by comparing it with previously agreed standards, and of assessing the value in terms of risk reduction of proposed changes. The aim was to reduce the risk to the public to a rate that was considered socially acceptable. The methods developed were applied to chlorine and other toxics across Mond Division and can be applied elsewhere in the chemical industry provided care is taken in deciding the criteria for what is socially acceptable.

The impact on the public is assessed for each potential release—based on the emission rate and duration—using a sophisticated computer program which takes account of wind and weather conditions and their probability. By combining the impact and the frequency of each release an overall assessment of risk can be made. The method covers a wide range of releases from those which had no effect, through those which were a nuisance, those which caused some distress, to those which could lead to injury or risk to life.

8.1.1.3 *Foul-smelling gases.* In some ways these pose a significant problem because it is difficult for the public to distinguish between what is harmful and what is merely a nuisance. Hydrogen sulphide is probably the most frequently encountered and is particularly dangerous, not only because of the low concentration at which it is toxic but because the ability to smell it is lost after a fairly short exposure.

Organic compounds containing sulphur, mainly the mercaptans (thiols), often have very persistent smells. Some can be quite unpleasant in concentrations below 0·01 parts per million, and though not toxic at this level they are a definite nuisance and give rise to frequent complaints. Some are added as stenching agents to odourless but hazardous gases (e.g. Calor gas for domestic use and North Sea gas).

8.1.1.4 *Gases not harmful to humans.* Although it can lead to oxygen deficiency or give rise to an explosive mixture, ethylene (ethene) in air is not harmful to humans, but it does have an adverse effect on vegetation. Ethylene systems are monitored for leaks to avoid fires and explosions. In addition it is a fairly expensive material. Small amounts may be discharged to atmosphere through stacks, but if a large discharge becomes necessary it is normally flared.

8.1.1.5 *Dusts.* Minerals processing covers a wide variety of products from blocks to powder, broken stone to special cements, and sand to diamond dust. Wherever minerals are extracted, transported, reduced in size or heated to induce chemical change, dust and fume may be released. The limitation, containment and collection of these pollutants have led to a wide variety of processes and equipment.

Following the Clean Air Acts of 1956 and 1968, control was imposed initially in 1971 by a Works Order under the Alkali Act requiring the registration of such works. In 1974 the Alkali and Clean Air Inspectorate issued a document entitled 'Notes on best practicable means for roadstone plants'. This document has been used subsequently as a basis of negotiations on similar processes in other industries, including cement works and lime kilns. The standard laid down in 1974[4] restricted the discharge of particulate matter to atmosphere as follows:

0·2 grains per cu ft ($460 \, mg/m^3$) for volumes up to 25 000 cu ft/min (700 m^3/min)

and 0·1 grains per cu ft ($230/mg/m^3$) for volumes over 50 000 cu ft/min (1400 m^3/min).

with a sliding scale for intermediate flows.

Particulate matter was also defined as follows:

grit—particles larger than 75 microns.
dust—particles smaller than 75 microns but larger than 1 micron.
fume—particles smaller than 1 micron.

There are four basic types of dust collectors; cyclones, electrostatic precipitators, bag filters and scrubbers. Many of the factors relating to the control of dust are dealt with in the following case study.

Some years ago, Imperial Chemical Industries PLC, when examining a project for the use of new Maerz two-shaft kilns, fired by natural gas and producing 300 tons per day burnt lime, required a dust-removal system. Water supplies were limited, and as recovered dust could be sold or reused, a dry collection system was specified. Bag filters and cyclones were considered and ruled out because of their size, the requirement for the units to be on the ground, and the long lengths of expensive ducting which could allow the temperature of the gases to fall below its dewpoint.

Tests carried out on a similar installation showed the dust burden to be relatively low. A collection efficiency of 96·8% was necessary to meet the 0·1 grains per cu. ft. emission limit. A Holmes–Rothmulhe Multi-Cell cyclone was chosen, as the pressure drop was small enough to avoid having to install additional fans, and it could be installed on top of the kiln because of its compact size, thus avoiding ducting and associated heat losses. To provide further protection, casing insulation was used and the unit enclosed within the kiln penthouse. Also incorporated was a facility to recycle exhaust gases at start-up until the temperature reaches about 65°C. This shows how a simple collector can prove to be suitable even though more sophisticated systems were initially considered.

8.1.1.6 *Smuts*. With the introduction of the Clean Air Acts the amounts of smoke and smuts discharged to atmosphere have been drastically reduced. Smuts from power station chimneys are a relatively rare occurrence, but when discharged as a result of some upset or equipment failure they may be contaminated with acid and damage items, such as cars, on which they land.

At some power stations ammonia gas is fed into the flue gases to ensure that any sulphurous (sulphuric (IV)) or sulphuric acid forming is neutralized.

8.1.1.7 *Sprays, including aerosols.* These can arise either from process upsets or by being blown from a work area by a high wind, although in the latter case the problems are usually confined to the factory concerned.

8.1.2 Water pollution

Pollution of waterways (including lakes, streams, rivers, canals and tidal waters) in the past has been mainly due to the deliberate discharge of effluents. With the introduction of legislation and its gradual enforcement this type of pollution has decreased greatly, and many waterways that were dead (containing no fish) are now clean and alive.

Although water pollutants can be classified generally as physical, chemical or biological, several categories are of particular concern[5]:

(a) excreta and organic industrial wastes
(b) infectious agents
(c) plant nutrients
(d) organic pesticides
(e) waste minerals and chemicals
(f) sediments
(g) radioactive substances
(h) heat.

8.1.2.1 *Excreta and organic industrial wastes.* If the organic constituents are excessive even after the treated or untreated waste water is diluted by discharge into a body of water, the biochemical oxygen demand of the stabilizing organisms may deplete the available oxygen content of the water, producing pollution. The effluents from sewage plants, food processing, paper making, malting and distillation processes, all contribute to the biochemical oxygen demand made on natural waters. The biochemical oxygen demand (B.O.D.) should be reduced to 20 mg/1, and suspended solids to 30 mg/l before discharge to a body of water. (B.O.D. is a measure of the amount of oxygen which will be used in the biological decomposition of the organic matter present in the effluent.)

8.1.2.2 *Infectious agents.* Arising mainly from excreta, pathogenic bacteria must be prevented from contaminating potable waters. Coliform organisms which can be detected and estimated may be used to monitor water systems. It should be noted that a negative test does not mean that the water is virus-free.

8.1.2.3 *Plant nutrients.* These arise from two main sources: sewage not subjected to tertiary treatment, and agricultural land drainage. The build-up of these nutrients in water systems is called eutrophication. The nutrients lead to a rapid increase in algae, normally seen as blue-green algae on the surface, and because of their high consumption of oxygen from the water, the body of water ceases to be able to support animal life.

8.1.2.4 *Organic pesticides.* In the early part of the century, with the aid of simple chemicals, mainly inorganic salts, it was possible to keep people well fed on small diversified farms where labour was plentiful and cheap, and cultural practices blocked the build-up of pests.

There followed a range of organo-chloro compounds, organo-phosphates and other broad-spectrum poisons—(referred to as the DDT generation). These were supposed to usher in a new era of industrialized agriculture and to solve pest problems for ever. The environment was saturated with these persistent poisons, pests developed resistance to them, and many entered food chains with serious results. The use of this type of pesticide has virtually ceased. The most recent types of pesticide are the hormones (narrow-spectrum biochemicals) and biological controls (parasites) which aim to pinpoint control without poisoning the ecosystem.

8.1.2.5 *Waste minerals and chemicals.* Heaps of mineral waste or chemical waste can be leached by rain or oxidized by air to give rise to harmful solutions and airborne releases. Many different chemicals have been discharged into waterways and done immense damage. Toxics may alter the biological activity of streams. Changes in pH due to releases of acids or alkalis may kill off animal life in the water. To combat the effects, upper and lower pH limits may be set for discharges, and upper limits have been set for many toxics including heavy metals, cyanide and phenols.

8.1.2.6 *Sediments.* These can cause turbidity in lakes and streams, with reduction of air and light thereby interfering with photosynthesis. Sludge can form, particularly on the beds of lakes, and as a result of anaerobic decomposition of the sludge noxious gases (such as hydrogen sulphide) are released.

8.1.2.7 *Radioactive substances.* These can be very dangerous, having the capability of producing cancer, disrupting biological processes, and causing genetic changes to organisms including man.

Two procedures are adopted: 'D.D.' and 'C.C.' D.D. (dilute and discharge) can lead to a situation where the radioactive decay cannot keep up with the discharge of new contamination. C.C. (concentrate and confine) is much more costly but must be the method used if the nuclear power industry is to grow.

8.1.2.8 *Heat.* In many processes energy is produced which cannot be used directly or indirectly, and therefore has to be transferred to some heat sink outside the system, either in water to the local river, or through air coolers or cooling towers to the atmosphere.

Normally water has been used to carry away the heat along with other forms of effluent. In some cases the heat can be satisfactorily dissipated in the water body, but in other cases a process similar to eutrophication occurs with rapid growth of plant life, reducing the oxygen content of the water body.

8.2 Methods of pollution control

Why permit any pollutants to contaminate our work and living space? This is a very simple question but the answer is complex. There is no absolute zero, but pollution should be controlled as far as is reasonably practicable. The disposal of waste is the cause of environmental pollution. The economy of highly industrialized nations is to a large extent based on a philosophy of waste. Such waste products create problems in the home, the work place, outdoors, in rivers, lakes and oceans, in the atmosphere and even in nearby outer space.

Control requires efficient manufacturing, treatment and energy conservation processes, conscientious effort to eliminate waste at the source, measurement of the effects on human health, plants, animals and structures, study of the ability of natural processes to cope with the waste discharged, economic evaluations, political and legislative actions, and the education of policy makers to obtain realistic support for abatement measures.

Many methods are used for pollution control but can be covered by the following groupings:

(a) elimination of effluent at the source
(b) reduction of effluent volume
(c) water re-use and recovery of materials
(d) physical, chemical and biological methods of treatment.

8.2.1 Elimination of effluent at source

Earlier it was shown that sulphur dioxide is a major pollutant even after the reduction achieved over the past 20 years. It arises from sulphur present in fossil fuels. One way to prevent sulphur dioxide formation is the Holmes–Stretford process for the removal of hydrogen sulphide from fuel gases.

8.2.1.1 *The removal of hydrogen sulphide from fuel gases*[6,7,8]. The problem of smoke and sulphur dioxide from the combustion of fossil fuel is recorded as far back as the Middle Ages. It was the development of the manufacture of fuel gases for domestic and industrial processes which brought about the necessity to remove, before combustion, the sulphur (which originated in the plant life from which the fuels were formed). The basic chemistry used is simple:

$$2H_2S + O_2 = 2H_2O + 2S.$$

About 1800 the first alkaline process was used to remove hydrogen sulphide and carbon dioxide from coal gas, but the resultant sulphide when dumped reacted with dissolved carbon dioxide in rain water and released hydrogen sulphide to atmosphere. About 1850 the ferric (iron (III)) oxide process, in which the hydrogen sulphide was oxidized to elemental sulphur, was first used and continued to be used for about a century.

The changing conditions after World War II stimulated the need for a washing process, and from this the Stretford Process was developed. The initial system was essentially a simulation of the solid process, except that the gases were washed with a suspension of hydrated ferric oxide. The process, known as the Manchester or Ferrox Process, worked well when there was no hydrogen cyanide in the gases entering the process. Sulphur was removed as a filter cake mixed with iron oxides, and there was less than 1 ppm hydrogen sulphide in the cleaned gas. When hydrogen cyanide was present iron/cyanogen/sulphur complexes were formed, giving rise to foul smells when the filter cake was autoclaved, and the sulphur produced was not clean. Plants using this process were erected between 1944 and 1956.

The need to use a true aqueous solution for the removal of hydrogen sulphide led to the development of the anthraquinone system, using its redox cycle in the oxidation of hydrogen sulphide. The anthraquinone was partially sulphonated to improve its water solubility. The ADA (anthraquinone disulphonic acid) process was effective in reducing the hydrogen sulphide level to less than 1 ppm, but had a number of disadvantages. Only 50% of the sulphide was converted to sulphur, while about 30% was converted to thiosulphate, and the balance to sulphate. The formation of oxysulphur acids was the real problem and in that sense the process was a failure.

By 1960 it was clear that the processes based on a suspended metal oxide or on a quinone alone were unsatisfactory. It was opportune to develop a mixed process in which a water-soluble polyvalent metal could be used to oxidize the hydrogen sulphide instantaneously to free sulphur, but in which the quinone would act as the oxygen carrier in the oxidation stage. Dr T. Nicklin of the North West Gas Board found the almost perfect solution using the pentavalent and quadravalent forms of vanadium in conjunction with anthraquinone. Some problems arose from the deposition of vanadium salts, but this was overcome by the use of a sequestering agent (citric acid seems to be the best). A diagram of the process is given in Fig. 8.1.

Exactly how A.D.A. operates is not fully understood, but it acts as an oxygen carrier in a similar way to quinone, has a molecular arrangement that is not reactive to hydrogen peroxide, and therefore has an indefinite life.

In the absence of multivalent metals, the following two reactions occur in nearly equal proportions.

$$2H_2O_2 + 2SH^- = 2H_2O + 2S + 2OH^-$$
$$4H_2O_2 + 2SH^- = 5H_2O + S_2O_3^{--}$$

However, in the presence of a multivalent metal, reaction is almost exclusively in accordance with the first equation and only small proportions of

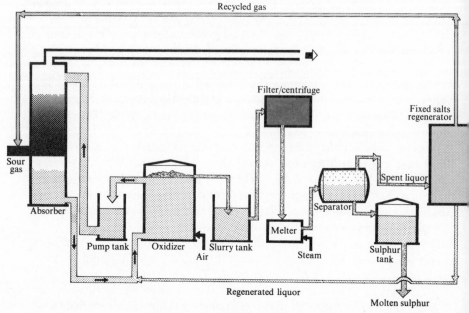

Figure 8.1   The Holmes–Stretford process.

thiosulphate and sulphate are formed. Their formation cannot be suppressed economically.

If the effluent contains only sodium sulphate and sodium thiosulphate, they can be crystallized out by spray-drying and dumped, but where sodium thiocyanate is present biodegradation is used to convert the thiocyanate and thiosulphate ions to inoffensive forms (carbon dioxide, nitrogen and sulphate). Complete detoxification occurs at pH 6–7·5 and a temperature of 20–25°C. The process will only accept thiocyanate at less than 1000 ppm and thiosulphate at less than 500 ppm. Control of the process is very difficult.

Oxidative combustion to yield only sodium sulphate requires the expensive spraying of alkaline solution into the flame to neutralize the acid gases as they are formed.

Reduction of the solids from spray evaporation or oxidative combustion can be achieved at about 800°C, using hydrogen and carbon monoxide to yield a mixture of sodium carbonate, sodium sulphide, and hydrogen sulphide, which can be returned to the process and result in savings and a nil effluent.

8.2.2 *Reduction of effluent volume*

An example of this technique is when dealing with suspended matter which is not biodegradable. Settling is often aided by the use of flocculating agents, sludge concentration, filtration and incineration, all used to obtain a purer water for discharge, the waste after treatment being used for landfill.

8.2.3 Water re-use and recovery of materials

There are many processes which achieve one or the other and some which achieve both. The recovery of materials in many instances is doubly important in that pollution is prevented, and a scarce and expensive material is recovered. In the other cases, the recovery of water is of paramount importance because of limited supplies of adequate-quality water in the area. A typical example of the latter is at the vegetable freezing unit, used as a case history later in this chapter—(in section 8.3.1.5). The most widespread examples of water re-use are in cooling-tower systems, and the conversion of water from a sewage treatment plant to a suitable quality, even to potable water.

8.2.3.1 *Recovery of materials.* With the dwindling of the world's resources of many materials in everyday use, particularly some metals, there is an increasing need to recover these scarce materials. The main problem lies in the way these materials are dispersed. Examples include chromium plating on steel, and small stainless steel parts. In other cases there are established recovery systems for substances such as scrap iron, waste paper, glass in the form of bottles and aluminium in a variety of forms—scrap, milk bottle tops, used drinks cans etc. However, the recession has affected the economics of many of the recovery systems: scrap iron is being exported, waste-paper collection is giving very poor rewards, the establishment of bottle banks is not being pursued enthusiastically, and the value of scrap aluminium has fallen.

In laboratories for many years it has been common practice to recover precipitates containing silver, where, because of its scarcity and value, recovery systems were economical and therefore well established. More recently, because of limitation of what could be included in refuse, laboratory samples (i.e. samples of product taken to the laboratory for analysis/testing) are being collected and returned to the process stream. Also off-specification material which was formerly dumped is being sold at reduced prices to manufacturers who can convert it into useful products.

In some cases recovery processes have to be developed as the material concerned is no longer tolerated in the environment. When removal of a material from the effluent is obligatory, it is often possible with little additional cost to recover a material which is scarce or expensive or both.

8.2.4 Physical, chemical, and biological methods of treatment

8.2.4.1 *Physical.* Foremost amongst the physical methods is the separation of solid waste and its burial. Usually there is some attempt to segregate some of the components of domestic and industrial waste. Large items can be removed by hand, and iron by magnetic separation. Some authorities encourage households to separate materials such as waste paper and glass bottles. In some cases domestic refuse is reduced in size before combustion to reduce the amount which has to be buried, but this may add to atmospheric pollution unless appropriate plant is used.

8.2.4.2 *Chemical.* The main way in which toxic, irritant and noxious gases are prevented from contaminating the atmosphere is by careful containment and the scrubbing of the contaminated air before release to atmosphere. Normally the scrubbing is carried out using water or an aqueous solution. The scrubbing liquor in some cases is converted to a saleable product, but frequently has to be treated chemically before discharge to the environment. Contaminants trapped by solution in organic solvents or adsorbed (e.g. on activated charcoal) can often be recovered for reuse.

Many acid gases are removed from air as strong solutions in water, or are absorbed in alkaline solutions to give a saleable product, e.g. hydrogen sulphide to give sodium bisulphide or sodium sulphide, chlorine to give sodium hypochlorite (bleach), and hydrogen chloride to give hydrochloric acid.

Gases such as ammonia which produce alkaline solutions in some ways present a greater problem. Contact with caustic alkalis can lead to the liberation of ammonia, and if ammonium salts are discharged to the environment they can lead to eutrophication.

Solutions containing heavy-metal salts have to be chemically treated to give insoluble compounds which can be separated using physical means (filtration, settling, centrifuging etc.) and then further treated to recover the metal.

Catalytic converters and traps are being used to clean up exhaust gases from automobiles. The catalyst ensures complete cumbustion of the (lead-free) gasoline and the trap helps to remove undesirable gases such as oxides of nitrogen.

8.2.4.3 *Biological.* Methods using biological agents have in the past been used only on organic waste (e.g. food processing effluent, sewage etc.), but as can be seen later in this section, applications to problematic inorganic effluents are now emerging.

8.2.4.4 *Examples of treatment.* One industry where the problems of effluents are dealt with mainly by chemical treatment, resulting in the recovery of valuable materials, is the electroplating industry[9]. Typical effluents, which can amount to 3 to 4 million gallons per week total, are alkaline rinse liquors containing zinc and copper as well as up to 500 parts per million of sodium cyanide, and acid waste liquors containing nickel and chromium. Great care is required to keep the cyanide containing the liquors separate from the acid liquors, where mixing would result in the release of highly toxic hydrogen cyanide.

$$NaCN + HCl = NaCl + HCN.$$

The alkaline solution is treated with sodium hypochlorite to give harmless products.

$$NaCN + NaOCl + H_2O = CNCl + 2NaOH.$$

$$2CNCl + 3NaOCl + 2NaOH = N_2 + 2CO_2 + 5NaCl + H_2O.$$

Any surplus hypochlorite must be treated or decomposed before the excess alkalinity can be neutralized, otherwise chlorine is liberated.

$$NaOCl + 2HCl = NaCl + Cl_2 + H_2O.$$

Treatment is often with sodium thiosulphate or sodium bisulphide.

A recent development in the processing of cyanide waste using a biological process is described below. It avoids the use of hypochlorite and the necessity to treat any excess used.

ICI Agricultural Division[10] in 1982 put on the market this enzyme based process. The enzyme in pellet form was used in vertical columns with downward flow of the effluent. The liquid discharged from the columns contained not greater than 20 parts per million of cyanide.

Typical treatment of the acidic chromium-containing effluent is with sulphur dioxide or sodium bisulphite[11].

$$2H_2CrO_4 + 3SO_2 = Cr_2(SO_4)_3 + 2H_2O.$$

The effluents are then mixed with slaked lime (particularly if available as a cheap by-product of acetylene (ethyne) generation from carbide). Nickel, chromium, zinc and copper are all precipitated as hydroxides which may be removed for recovery of the metals. Because of the low solubility of lime in water the process is slow and is usually carried out in large lagoons. The optimum pH for precipitation is different for different metals. Zinc is very problematic because with excess alkalinity it readily forms soluble zincates. In any case, the precipitate is gelatinous, containing up to 99% water after settling, and requires concentration before recovery or dumping.

The supernatant liquor from such lagoons is very hard, due to the calcium and chloride present. It has at least to be diluted before discharge to waterways, and has been the subject of many disputes with the river authorities.

A more-up-to-date and compact system for treatment and recovery of heavy metals is the method that Sulzer[12] has developed, called the fast mixing pipe process, where neutralization and detoxification takes place in an effluent line, thus avoiding costly reaction vessels. The process is used for example to treat dilute chromic acid (rinse waters and concentrates from plating operations).

From a collecting tank the liquid is pumped into the effluent line. 20% sulphuric acid is added through a control system to reduce the pH to about 2.5. Sodium bisulphite solution is then added, being controlled by the redox potential, and finally 40% caustic soda is added to bring the pH back to about 8.5. The precipitated heavy metal hydroxides are filtered off leaving a neutral solution for disposal. Capacities from 2 to $60\,m^3$ per hr are available. The process takes up less space than conventional systems, is cheaper to install and easier to modify.

The process which utilizes physical, chemical and biological methods is sewage disposal which is dealt with in some detail below. All students should be able to gain access to a treatment plant either near to their college or near to their home, and see the process operating.

8.2.4.5 *Sewage disposal.* Some 26 centuries ago[13] the Chinese had learned to use human and animal waste as fertilizer, and horse-drawn carts of 'night soil' were taken out from the cities to the nearby countryside. The Romans also used sewage waste as fertilizer, and even in the nineteenth century American farmers came into the cities to collect waste from cesspools, privy valves, stables, restaurants and markets. By the end of the century the practice had almost died out. Indoor plumbing had resulted in a big increase in water used, flooding cesspools which by then were recognized as breeding grounds for disease. As running water was recognized as a great purifier, sewage systems were led to the nearest stream or river. This resulted in massive pollution of waterways.

Nowadays most sewage in industrialized countries is at least partially treated before release to rivers, although much raw sewage is discharged into estuaries and into the sea.

Modern treatment plant produces an effluent that is relatively clean. The major problem is what to do with the sludge. There are three options: (a) incineration, which is costly and leads to air pollution; (b) ploughing under and compaction, or using for landfill—this is also expensive, and problems arise from methane gas seeping from the landfill and occasionally giving rise to explosions; (c) converting to fertilizer. *Raw sludge*, the solid matter which settles out of raw sewage by gravity, is dangerous and spreads disease. *Digested sludge* (anaerobically fermented until it decomposes) can be air-dried, heat dried or centrifuged, or can be kept liquid to aid spreading. *Activated sludge* (digested sludge put through a secondary (aerobic) treatment) is air-dried to be so inoffensive that it can be bagged and sold as fertilizer. There can be some danger from metal residues so analysis is necessary before sale.

Treatment of raw sewage is usually carried out in up to three stages[14,15]:

(i) Mechanical screening and sedimentation of solids which are burned and buried.
(ii) Biological reduction of organic matter.
(iii) Chemical treatment to remove phosphates, nitrates, organic residues and other materials.

Stage 1. Primary treatment. Raw sewage is passed through a series of screens to remove mainly inorganic matter. Special provisions are required to handle storm water. Grease, scum and froth are then skimmed off before the liquor, containing fine solids, emulsions, colloids, and soluble matter, passes to sedimentation.

Sedimentation, carried out in large tanks with low crossflow rate, is encouraged by the use of flocculating agents, but some, such as alum, may give problems later. The clear effluent passes to stage 2, while the sludge (95% water) is treated and used as a fertilizer or is thickened, incinerated and buried.

Stage 2. Secondary treatment. In the U.K., if adequate land is available, so-called trickling filters may be used. These are not filters, but beds 3–6 ft deep, filled with broken rock or slag on which the liquor is sprayed. Biological growths on the surfaces live on the sewage constituents and oxidize them. Little maintenance or power is required for operation. Where space is limited the activated sludge system is used. This comprises either a tank where a spinning cone throws out the sewage to aerate it, or a tank where compressed air is bubbled through the sewage. Micro-organisms flourish and oxidize the sludge. With this latter system power costs are high and more maintenance is required but better purification is obtained. Sedimentation is then required, with the clear effluent going to stage 3 or a river, but the sludge is nasty to handle due to a foul smell. It may be carted away for use by farmers as fertilizer. Closed tank digestion, giving off methane and carbon dioxide mixture which will burn, yields a sludge without odour and which dries easily but has little value as a fertilizer.

Stage 3. Tertiary treatment. The effluent from secondary treatment is highly polluting in terms of eutrophication, and is unfit for human use, but can be used for irrigation at limited rates. Otherwise suitable plant is required to remove nitrates, nitrites, phosphates, and residual organic matter.

A tertiary system is illustrated in Odum's *Fundamentals of Ecology*[16]. The secondary effluent is treated with lime, and after flocculation is settled with the sludge passing to a lime thickener. The effluent, almost freed of phosphate, passes forward to an ammonia-stripping tower, where the ammonia is later converted to nitrogen and discharged to atmosphere. The liquor from the tower is carbonated to prevent calcium hydroxide passing forward, and the settled sludge passes to the lime sludge thickener. Polyelectrolyte or alum is added and the solids are separated in a bed. The solids are back-washed to a decanting tank and fed to the lime thickener. The effluent may be subjected to a further clarification before passing through an activated carbon bed to remove remaining organic compounds. This water is suitable for recycle as drinking water after normal chlorination. Lime mud from the thickener passes through a furnace before recycle to the inlet of the system.

This degree of treatment is only required if the water is for re-use locally. If the water is being fed to a river, only phosphate removal is necessary as 80% of phosphate in rivers comes from sewage, whereas only 20% of the nitrogen and other contaminants (e.g. silicon, potassium, chloride, sulphate) do and the balance is introduced as run-off from agricultural land[17]. In this situation the nitrogen is best dealt with at water abstraction points.

Some plant effluents, such as that from a casein plant[18], may require treatment based on a biological method.

8.3 Economics of pollution control

While it is more difficult to quantify the benefits of pollution control, rough estimates[19] in the U.S.A. by the Council on Environmental Quality, released in April 1980, showed the annual benefit to the U.S.A., as a result of air pollution control, was $21·4 billion in 1978, in the form of improved human health and reduced damage to crops and vegetation, among other things. The same study showed that benefits of $12.3 billion were obtained in the same year (1978) as a result of water pollution control in the form of reduction of some waterborne diseases, reduced water treatment costs (at water intakes) and increased use of lakes and waterways for recreation.

The American Lung Association estimated that as much as $10 billion per year was being spent to combat the effects of air pollution on health.

In 1977[20] according to some estimates it was expected that the expenditure on pollution control equipment in the U.K. during that year would be of the order of £250 million (about 25% of the predicted figure for new plant investment).

Waste treatment and disposal was only one of the pressures being applied to the process industries in the name of environmental protection. Other environmental topics, such as safety of plant and safety in health terms of the product, were eating into company budgets, both in capital and research expenditure, allegedly limiting growth and profitability. The concern for the environment was also alleged to be inhibiting innovation in the process industries, but it was accepted that industry did have the responsibility to see that its operations did not spoil the environment.

The high estimated cost of pollution control measures and the resultant impact on the price of goods and services, delayed the enforcement of the Control of Pollution Act 1974, particularly in relation to aqueous systems. The expenditure being discussed in 1989 to reduce sulphur dioxide emissions from power stations, to improve the quality of drinking water, and to raise the standards of seawater at beaches in order to meet EEC Regulations or Directives, will have a significant effect on the cost of electricity and on water rates.

8.3.1 Treatment plant/processes

Treatment in many cases requires either the use of large tracts of land or the use of compact, expensive plant and processes.

In the case of sewage treatment for the secondary stage there is a choice between:

(a) Waste stabilization ponds—very high land usage but with little other capital cost and very low running costs.
(b) Trickling filters—moderate in land usage and fairly low in capital and running costs.
(c) An activated sludge system—very compact but relatively high in capital and running costs.

In selecting a treatment plant and process apart from the consideration of the effectiveness of the plant, the factors which must be assessed are:

(a) The cost of the land and the ongoing costs of rates from use of the land.
(b) The capital cost of the plant and the rate at which it should be depreciated.
(c) The running cost of the plant in terms of labour, services, materials and maintenance.
(d) The risk from the effects of faults that can arise on the system.

8.3.1.1 *Heat removal.* Because of ever-increasing water tariffs there has been a move towards closed cooling systems with cooling towers, but here capital costs are high and running costs are considerable. Air-cooled heat exchangers[21] where the heat is dissipated harmlessly have also been used. Capital costs and runing costs are moderate on such systems, but the systems are fairly difficult to control and undercooling is a significant problem. The use of variable-speed fans increases the capital costs, and only gives marginal savings in power, but results in lower noise levels at lower fan speeds.

8.3.1.2 *Treating organic effluents, mainly sewage.* At various locations, due to gradual increase in the loading, traditional stone-packed biofilters have become overloaded. Faced with major expenditure for replacement by larger units or the installation of additional units, a much cheaper solution was found by replacing the stone with Cascade Filter packs supplied by Mass Transfer International[22]. These supplies resulted in increased throughput and higher BOD removal efficiency.

Data by Stephens and Weinberger (1968) quoted by Odum[23] give some comparison of the capital and running costs of the three stages of sewage plants.

Stage of treatment	Capital cost ($ million)	Operating cost (cents per 1000 gallons*)
Primary	10	3–5
Secondary	20	8–11
Tertiary—to remove nutrients		17–23
to produce drinking water	25	30–50

*Lower figures are for units processing 100 million gallons per day.

Desalination to produce drinking water would have cost at least $1 per 1000 gallons.

Greater London Council costs for sewage treatment in 1970/71 are given by Reuben and Burstall[9] as follows:

	Cost (pence per 1000 gallons)
Reception of sewage into sewer and conveyance to plant	1·48
Primary treatment (screening and sedimentation)	0·92
Biological treatment (mainly activated sludge)	1·23
Sludge treatment (mainly anaerobic digestion) and disposal	1·54
	5·17

8.3.1.3 *Dust control.* The costs involved with a dust control system are (a) the initial cost of the equipment and its installation, and (b) the cost of running and maintaining the equipment. Highly efficient dust collectors (e.g. bag filters up to 99·99% efficiency) are more costly than the less efficient collectors (e.g. cyclones, wet deduster, about 96% efficiency), although these latter may be combined to obtain high-efficiency units.

The financial assessment regarding what to install has until recently been mainly based on the marginal savings in material recovered using the more efficient system, though more recently what can be discharged to atmosphere has become the overriding factor.

The main running cost is for the energy required to operate the dust control system. The satisfactory operation of a system depends on the correct evaluation of the mass flow and attention to design to reduce the resistance of the system.

Three different dust removal plants are considered and costed (at 1971 prices) for the treatment of flue gases from a limestone dryer[24].

In the *fabric positive filter*, dust-laden vent gases from the limestone dryer are drawn through a fabric filter medium. The dust is retained by the fabric and the cleaned gas passes forward through a fan to atmosphere. The fabric is in the form of a number of tubes operated in parallel. Periodically the gas flow is interrupted and the tubes are partly deflated, are shaken by a vibrator, and cleared of dust by a scavenging flow of air. The main advantage of this collector is that it collects particles down to 0·01 micron.

The *inertial separator* exploits the tendency of the comparatively large dust particle to continue to travel in a straight line when the direction of flow of the gas in changed, by the use of a cyclone or bank of cyclones. The second stage exploits the fact that collection efficiency of particles rises with the mass of individual particles. This is achieved by wetting the dust in the air from the cyclone in a venturi scrubber. The venturi is followed by a spray separator from which the gases pass through the fan to atmosphere. The wet scrubber makes the system efficient at dust removal, but there is the problem of disposal of the slurry and there may be corrosion problems.

In *electrofiltration*, the gases from the dryer pass into an electrostatic precipitator where a cloud of electrons passes from negative electrodes held at high voltage (25–50 kV) to earthed 'positive' electrodes. The dust particles acquire a negative charge and are attracted to and adhere to the positive

Table 8.1 Treatment of flue gases from a limestone dryer

Type of collector	Fabric filter	Inertial separator	Electrostatic precipitator
Construction	Tubular	Venturi	1-Field
Pressure losses			
Ducting	3	3	3
Pre-collector	—	—	2
Collector	6	15	1
Total (in. w.g.)	9	18	6
Capital excluding Ducts & civil wk.	£18 300	£9 100	£19 000
Running cost 5 years at 5760 hours/year	£12 000	£18 800	£7 600
Maintenance Cost for 5 years	£5 200	£1 200	£1 300
Total cost	£35 500	£29 100	£27 900

electrodes. The particles are removed by vibration or by flushing with water. This technique allows even very small (0·001 micron) particles to be collected at high efficiency and low cost. To maintain the high collection efficiency regular maintenance of the system is essential.

Table 8.1 gives some typical costs (at 1971 prices) for a system to handle 22 000 cfm at 20°C with an inlet dust burden of 14 grains per cu ft and an outlet specification of less than 0·05 grains per cu ft.

8.3.1.4 *Water saving scheme*[25]. At the works of a high-quality metal-plating specialist, the requirements were to treat 18 m³ per hour of metal finishing swill water for recovery and re-use. A Permutit-Boby effluent treatment plant with manually operated controls was recommended. The intake of water was reduced from 32 000 to 15 000 m³ per annum—a saving of 17 000 m³—and the effluent discharge was reduced from 21 000 to 6 500 m³ per annum—a reduction of 14 500 m³. The cost savings are of the order of £3 000 per annum based on water costs of 10·3 pence per m³ and effluent discharge costs of 8·2 pence per m³. An additional benefit is that the consistently high quality of demineralized swill water ensures a superior standard of product finishing. There are also savings in time and labour over the previous treatment.

It should be noted that this case history only covers the recovery of water aspect. It is probable in practice that it would be associated with a system for the recovery of the metals.

8.3.1.5 *A treatment scheme for organic waste*[25]. At a new vegetable freezing unit in the frozen food processing industry, effluent treatment was required because the discharged water had to be of suitable standard for discharge to a watercourse. A plant was installed to handle a flow of 1670 m^3 per day with an input BOD of 2 800 ppm. It comprised flow balancing, primary settlement, high-rate plastic media filtration, activated sludge, sludge thickening and dewatering. Tertiary filtration and chlorination were also included.

The plant provided an effluent suitable for discharge to the water course, and additionally, the tertiary treatment allows about 50% of the water to be re-used within the process. The overall cost of the scheme was £300 000 of which only £40 000 was for the tertiary treatment. The cost of fresh water was 9·7 pence per m^3, but the cost of recovered water (power plus chemicals) was only 0·6 pence per m^3 showing a saving of 9·1 pence per m^3. The total daily savings work out at £76; thus a complete write-off of the capital can be achieved in 540 days. Add to this that the supply of fresh water was limited and the case becomes even stronger.

The above costs and savings are questionable, in that the cost of recovered water is falsely low (total labour costs are assumed to be unchanged and while this may be true for process labour there will be extra costs for the maintenance of the system) and therefore the savings are falsely high. There is, however, a very strong financial case for the tertiary treatment, even if legislation did not necessitate it.

8.3.2 *Disposal into sewers*

In some areas local authorities will agree to the discharge of certain industrial effluents to sewers. The local authorities will have to be assured that the effluent will be controlled within specified limits, that there is no danger to public health from the effluent, or from any interactions that may result in the sewers, and that the local authorities treatment plant can handle the material discharged to sewer. Sampling facilities will also have to be agreed. Charges are made by the local authority, and are based on the volume discharged, its solids content and on its B.O.D.

8.3.3 *Recovery of materials*

In costing the recovery of materials it is frequently found that once the equipment has been provided to meet current or impending legislation, the additional cost to recover the material in a re-usable or saleable form is low in relation to the return, and a strong case for the additional expenditure can be made.

8.4 Industrial health and hygiene

8.4.1 *Introduction*

Health has been defined in the Preamble to the Constitution of the World Health Organization as 'a state of complete physical, mental and social well-

Table 8.2 Deaths in specified years for England and Wales

Cause of death	1900	1930	1950	1960
Tuberculosis	55 326	33 750	15 969	3 435
Diphtheria	9 345	3 497	49	5
Cancer	26 721	50 704	85 270	98 788
Cardiovascular	55 712	112 477	186 560	174 660

being and not merely the absence of disease or infirmity'[26]. This ideal is far from being attained in the world in general and in industry in particular at the present time.

While health and the conquest of disease are not subjects which lend themselves readily to measurement, by keeping accurate records over the years it has been possible to compare the health of communities in different locations and to identify trends (Table 8.2)[27].

In industry while statistics are readily available on serious injuries and deaths from mechanical factors, it is much more difficult to obtain data on deaths, injuries and disease arising from exposure to chemicals and irritant dusts. In many spheres it is difficult to obtain reliable data on factors such as carcinogens where the number of known deaths is small but the issues are very emotive and receive much publicity, or where there is a long delay between exposure and death as in the case recently reported[28] where there was 40 years' delay between the last exposure to asbestos and death.

Hygiene is defined by the *Encyclopaedia Britannica*[35] as the science of preserving health, embracing all agencies affecting the physical and mental well-being. It includes, in its personal aspects, consideration of food, water and other beverages; clothing; work; exercise and sleep; personal cleanliness; specific habits, such as the use of tobacco, narcotics, etc; and mental health. In its public aspect it deals with climate; soil; character; materials and arrangement of dwellings; heating and ventilation; removal of waste materials; medical knowledge on the incidence and prevention of diseases; and the disposal of the dead.

While most of the above relates to an industrial environment emphasis is different to that in the non-industrial environment. Emphasis has changed over the years due to public opinion, trade union action, legislation and the actions of enlightened management.

8.4.2 Health hazards

The risks to health encountered in industry fall into the following categories: chemical hazards, disease, noise, radiation, as well as mechanical and electrical hazards. All should be controlled as far as possible by containment, but the effects of loss of containment must be assessed and appropriate measures taken to reduce to an acceptable level the exposure of personnel to the hazards. Containment is achieved by the use of equipment (pipework, vessels, etc.) of adequate strength and constructed of materials suitable for all envisaged conditions, supplemented by adequate control and relief systems.

8.4.2.1 *Chemicals.* Chemical hazards may be divided into three main types.

(a) Poisons—usually referred to as toxic substances. In this section those giving rise to acute symptoms are dealt with; those giving rise to chronic symptoms are covered under diseases. The toxic substance may enter the body by ingestion, absorption through the skin and by inhalation. Poisons result in a malfunction of part of the body.

(b) Corrosives—e.g. acids and alkalis, which damage body tissue with which they make contact—eyes are particularly sensitive.

(c) Flammables, giving rise to fires and explosions.

The elements of industrial hygiene include[29]:

(i) the establishment of criteria
(ii) the reduction of emissions
(iii) ventilation of the working atmosphere
(iv) monitoring the working atmosphere
(v) monitoring of personnel
(vi) skin care.

These are some of the steps taken to ensure that workers are not exposed to unacceptable risks from industrial chemicals.

In establishing criteria for many of the frequently used toxic chemicals, maximum permissible levels in the atmosphere (Threshold Limit Values—TLVs) have been established. A list of these TLVs is given in *Guidance Note EH* 40 published annually by the Health and Safety Executive. In the note definitions are given, along with the system for dealing with mixtures of toxics, and attention is drawn to toxics which can be absorbed through the skin and carcinogens.

Reductions of emissions can be achieved in various aspects (frequency, quantity, rate and duration). These data when processed with a dispersion model can give an estimate of the effects of such emissions on personnel away from the source of emission, e.g. on people outside a works. Ventilation is one method frequently used to keep the level of toxics in the working atmosphere down to acceptable levels, to dispose of smells and to give a comfortable air temperature at the work place. The atmosphere is monitored where it may be contaminated. This is especially important where the threshold of smell is in excess of the TLV. Monitoring systems are frequently coupled to alarm systems and sometimes to trip systems.

It is also practice to monitor personnel who may be exposed to hazards. Personal monitors may be worn by personnel to assess the total exposure to a specific hazard over a period. Personnel are given regular medical checks (lung function, urine and blood tests and X-ray examinations).

Skin care is usually in the form of protection against contact with toxics which can be absorbed through the skin, corrosives, solvents which remove oil from the skin giving a risk of dermatitis, irritant chemicals and protection of any break in the skin (e.g. cut) against the ingress of bacteria.

8.4.2.2 *Diseases.* Diseases can be divided into various classes:[30] cancer, skin diseases, lung diseases, effects of chronic poisons and diseases caused by germs.

The basic cause of *cancers* (a very emotive topic) is not known, but it is now accepted that there are many factors which include long exposure to small amounts of certain substances (carcinogens), or a short small exposure to others with a long dormant period, e.g. very limited exposure to blue asbestos has resulted in death from cancer some 30 years later. Exposure to chemicals or irritants can trigger cancer in many different parts of the body.

The most common *skin disease* is dermatitis—an inflammation of the skin—which can be caused by primary irritants, skin oil removers, substances to which the worker has become allergic, and excess heat and radiation. Other skin diseases include chronic ulceration, chloracne, tar wart and folliculitis[31].

Lung diseases caused by dust are given the group name of pneumonicosis, and include silicosis (silica), coal miner's pneumonicosis (coal dust), asbestosis (asbestos fibres), byssinosis (cotton dust), bagassosis (sugar cane), and farmer's lung (mould spores)[32]. Other lung diseases include asthma (due to various substances such as toluene di-isocynate, maleic (cis-butenedioic) anhydride and some enzymes used in making biological detergents), and emphysema, developing from other lung diseases and chronic bronchitis.

Probably the best known *chronic poison* is lead, which tends to accumulate in the blood because the body gets rid of it so slowly. Others include mercury, arsenic, cadmium and nickel, although exposure to nickel carbonyl may result in acute symptoms. In many cases clear correlation has been established between degree of exposure and content in the blood.

The most common *disease caused by germs* is septicaemia, generally referred to as blood poisoning—it usually arises from germs entering the body at a wound. Deaths have been greatly reduced by improvement in hygiene and the use of antibiotics. Other diseases such as brucellosis, Weil's diseases and anthrax are usually caught from infected animals or their excreta.

8.4.3 Industrial medicine

This branch of medicine is concerned with the maintenance of health and the prevention and treatment of disease and accidental injuries in the working population, and more recently the population adjacent to industrial establishments. Modern occupational medicine is considered by many to have started with Ramazzini, an Italian 17th century physician who drew the attention of other physicians to the need to determine patients' occupations in order to learn causes of their complaints. With the Industrial Revolution there was a rapid increase in numbers exposed to work hazards, and serious injuries became frequent as did diseases due to exposure to dusts, fumes and toxic gases. The development of industrial medicine in the U.K. and U.S.A. is covered in more detail in *Encyclopaedia Britannica* which gives an extensive bibliography[36].

The aims of an industrial health service are summarized briefly by Banks and Hislop[33]. More extensive summaries can be found elsewhere[34].

8.5 Legislation[37,38]

8.5.1 *U.K. legislation*

In England the law relating to pollution control is based on two separate sources: common law, a set of principles based on the decisions of judges in cases tried by them, and statute law, a set of rules prescribed by Parliament in the form of Acts and Statutory Instruments.

The aspect of common law which applies to pollution is the law of tort (a loss which can be compensated). It deals with civil wrongs which are not solely a breach of contract, where a person who has been wronged by the action of another has the right to compensation for the loss that has been sustained. The wrongdoer must have breached his legal duty or the injured parties' legal rights must have been infringed, and secondly the law limits the damages which the injured party can claim. To prevent a repetition of a harmful action the victim can take out an injunction.

A public nuisance is a criminal offence consisting of an act or omission which materially affects the comfort or convenience of the public. Private nuisance is where the conflicting interests of individuals lead to an inconvenience or annoyance.

In broad terms, the responsibility for ensuring a satisfactory local working environment is clarified and emphasized in the Control of Substances Hazardous to Health (COSHH) Regulation 1988, which is administered by the Health and Safety Executive. The general environment is administered by H.M. Inspectorate of Pollution (part of the Department of the Environment), aided by local authorities and regional water authorities.

8.5.1.1 *Water pollution.* Common law on the protection of rights in rivers and waterways is fairly well developed in that the owner of land forming the bank of a river has the right to receive unpolluted water into his property and may sue for trespass or nuisance if a person upstream pollutes the river.

Statutory intervention began early in the 19th century when the public health movement began to exert its influence in the improvement of water supplies and tackled the problem of sewage and its disposal. The first major piece of legislation was the Rivers Pollution Prevention Act, 1876. For many years, however, water pollution was only considered as one aspect of public health, and it was not until after the Second World War that its importance was recognized. There followed the development of an administration machine as well as the development of legislation.

On the administration side, in 1948 a system of River Boards was set up, but were replaced in 1963 by the River Authorities with Central Water Resources Board. In 1963 nine English Regional Water Authorities were established along with the Welsh Water Authority, and a National Water Council replaced the Resources Board.

Legislation developed at the same time, with the Rivers (Prevention of Pollution) Act 1951 replacing the 1876 Act, and controls were extended in the 1961 Act. The Clean Rivers (Estuaries and Tidal Waters) Act 1960 extended similar controls to specified coastal waters. The Water Resources Act 1963 and the Water Act 1973 were concerned with the problems of water. The 1963 Act covered the control of the discharge of effluent to underground strata including emergency powers to remedy pollution. The 1973 Act allowed Water Authorities to fix charges for facilities provided and services performed, but they must disclose methods of calculating their charges.

Most of the above legislation was replaced by the Control of Pollution Act 1974, but the parts dealing with water pollution took a very long time to be brought into effect, superseding the Rivers (Prevention of Pollution) Acts 1951 and 1961.

Information on application data and samples may not be disclosed by the Authority, without the consent of the person who supplied the information, except in the connection with proceedings.

The Water Resources Act, 1963, gave the Authorities power to control the discharge of trade and sewage effluent to underground strata by means of borehole pipe or well, consent by the Authority being required for discharge. The Act also gave the Authorities power to remedy pollution, the power to enter premises to do so and means of recovering the cost.

The Control of Pollution Act, 1974, though similar to previous legislation, is more embracing covering virtually all waters:

(a) controlled waters—sea within 3 miles of the low-water mark
(b) tidal waters—tidal waters and other areas where vessels are moored
(c) specified underground water—any underground water in an area under an Authority which might be used for any purpose
(d) stream—any river, waterway or inland water, whether natural or manmade (including lakes, sewers, docks etc.)

Relevant waters include streams, controlled waters and specified underground waters.

Offences under the 1974 Act are very similar to those under earlier legislation.

It is an offence to cause or permit

(a) any poisonous, noxious or polluting matter to enter any relevant waters
(b) any matter to enter a stream so as to impede its flow or be likely to aggravate pollution from other causes
(c) any solid waste matter to enter a stream or restricted water.

Trade effluent and sewage are covered by the same legislation. Application must be made for consent to discharge effluent into rivers or coastal waters, and the data required are as with previous legislation and the forms of consent are as previously.

Sewage to a large extent had previously been covered by the Public Health Act 1936 and 1961, and The Public Health (Drainage of Trade Premises) Act 1937. The provisions of these Acts still hold but are reinforced by those of the Control of Pollution Act 1974. The responsibility for administering the legislation rests with the various Regional Water Authorities.

Some trade effluents may be discharged to sewers after prior consent. Charges are made based on the amount of treatment required by the local Authority which depends on the volume discharged, the solids content, and the biochemical oxygen demand (B.O.D.).

8.5.1.2 *Control of atmospheric pollution.* The first statutory control was the Alkali Act, 1863, which arose from a widespread reaction to atmospheric discharges of large quantities of gases from new chemical processes to atmosphere. The most notorious was the first stage of the Le Blanc process

$$2NaCl + H_2SO_4 = Na_2SO_4 + 2HCl$$

where all the hydrogen chloride was released into the atmosphere, giving rise to acid rain and killing off almost all vegetation round towns such as Widnes.

Subsequent amendments were consolidated into the Alkali, etc., Works Regulations Act, 1906. This Act laid down the maximum permitted release of chemical pollutants to atmosphere from industrial processes. These are now covered by the Health and Safety at Work, etc., Act, 1974, with additional coverage being given by the Clean Air Acts and the Control of Pollution Act, 1974.

The 1906 Act was enforced by the Alkali Inspectorate, and the current legislation is enforced by H.M. Inspectorate of Pollution.

The original Act specified the maximum release of hydrogen chloride to atmosphere, but has been extended over the years to cover a wide variety of gaseous, spray and dust emission, the 1906 Act being amended when required to take under control new processes and gaseous effluents.

The new legislation—The Health and Safety at Work etc., Act, 1974— covers a much wider area, allows for the use of prohibition and improvement notices in this field, and the introduction of codes of practice for the handling of specific materials. It also places the responsibility for safe working on both employers and employees. Although far reaching (it applies, for example, also to teaching establishments), shortage of space precludes further discussion about the Act here.

The Alkali Acts only covered certain emissions from specific classes of premises, but the Clean Air Acts, 1956 and 1968, extended the coverage to smoke. Dust, to some extent has been covered by the Public Health Act, 1936. The Clean Air Act, 1956, in addition to dealing with smoke also dealt with grit and dust from furnaces.

The responsibility for enforcement lies with H.M. Inspectorate of Pollution, but local Authorities may obtain authority to enforce by application to the Secretary of State.

The Control of Pollution Act, 1974, only changes the Clean Air Acts marginally, e.g. concerning the contents of oils (including motor fuel), as well as research and publicity, and the powers of local Authorities to obtain information and to give publicity to problems.

8.5.2 U.S. legislation

Both state and Federal government are sources of legislation relating to health, safety and the environment. In both cases use is made of national standards and codes of practice. Standards produced by the American National Standards Institute (ANSI) and the National Fire Codes by the National Fire Protection Association (NFPA) are widely referred to in legislation.

Some of the areas of legislation which relate to the topic of pollution control include occupational health and safety, toxic substances, pollution, noise, transport and siting[39].

8.5.2.1 *Occupational health and safety.* This topic is principally covered by the Occupational Safety and Health Act, 1970 (OSHA), which directs that the Secretary of Labor should lay down safety standards, inspect work places and assess penalties. A separate commission deals with violations of the Act. The Act is similar in many ways to the Health and Safety at Work Act, 1974, especially in that it is an enabling Act imposed on existing legislation, provides a framework for applying standards, lays down duties of employers and employees, and has provision for inspection and enforcement.

The Act establishes the National Institute for Occupational Safety and Health (NIOSH) to develop and establish standards.

8.5.2.2 *Toxic substances.* The main item of legislation is the Toxic Substances Control Act, 1976. As well as requiring an inventory of existing chemicals it requires notification of new chemicals or of new uses and can require testing of chemicals for toxicity and can delay or ban manufacture.

The Act is administered by the Environmental Protection Agency (EPA) which contains an Office of Toxic Substances.

8.5.2.3 *Pollution.* The Clean Air Act 1977 and subsequent amendments gave the EPA power to adopt and enforce air pollution standards. These have led to significant improvements in urban air quality, but the stringent National Ambient Air Quality Standards have not been met in several regions, and are unlikely to be met in the foreseeable future in some others. Threshold Limit Values (TLVs), and intended changes, are published annually by the American Conference of Government Industrial Hygienists (ACGIH). The Federal Water Pollution Control Act 1972 gave the EPA power to adopt and enforce water pollution standards.

8.5.3 EEC legislation

Most of the legislation relating to the environment is produced in the form of EEC Directives, which have to be implemented by the member states by prescribed dates. Basically, the individual states have to achieve the required results but do have some flexibility in the methods of achieving the results.

Examples of items covered include the reduction of use of CFCs, the reduction of sulphur dioxide from large fossil fuel burning installations such as power stations, the quality of drinking water, and the quality of seawater at and near beaches.

8.5.4 Responsibility

Responsibility for the prevention and control of pollution is universal. This is aptly stated by Cousteau in *The Rights of Future Generations* which covers the general responsibility. There are, however, well-defined responsibilities on specific groups of people.

8.5.4.1 *Government.* Governments have the responsibility of providing a framework of legislation under which industry can operate in a responsible manner and with inspectors having the power to enforce the legislation where the requirements are not met. Within the framework there should be facilities to introduce standards (preferably agreed beforehand with industry) to ensure safe and non-polluting operation.

8.5.4.2 *Industry.* Industry has a variety of responsibilities, a legal responsibility to operate in such a way as to ensure the safety and health of its employees and neighbours (this may involve the development of new technology), a social responsibility not to pollute the environment (e.g. with unpleasant odours, noise, dust, etc.), an additional responsibility to employees and the area to ensure continued provision of employment and a responsibility to shareholders to provide an adequate return on their investments. These responsibilities may pull in different directions. For example, expenditure made in meeting legal and social requirements may reduce profits and thereby reduce the money available for reinvestment in new plant and equipment to ensure the continuation of the business and for distribution to the shareholders.

8.5.5 Effects of legislation and standards

The standards imposed and the legislation enacted in different countries can have a significant effect on the costs of production and thereby influence international competition. It is not the standards set that are important but the standards that are enforced. For many years in the USSR many TLVs have been much lower than in the U.K. or U.S.A., but the actual working levels have been much higher in practice in the USSR because the standards have not been enforced as in the U.K. or U.S.A.

8.6 Currently headlined environmental topics

A number of major concerns are dealt with in this section. In recent years (1985–1988) they have been given increasing media space and increasing attention by politicians, ensuring that resources are made available to obtain good data and aid in decision taking. The data being released do not seem to decrease any of the problems but rather emphasize the need for urgent action.

All the topics included have had to be dealt with rather briefly and each topic is much more complex than indicated here.

8.6.1 Acid rain

The term 'acid rain' was first used by Smith in 1872 to describe the polluted air in Manchester which damaged vegetation, bleached coloured fabrics and corroded metal surfaces. Current concern arose when Odén analysed the 1956–1966 data from 160 European sites, and showed precipitation to be increasing in acidity.

The acidity is commonly expressed in terms of pH values. Unpolluted precipitation is assumed to have a pH value higher than 5.6[44], the slight acidity being due to carbonic acid formed from dissolved carbon dioxide. Natural emissions, such as volcanic sulphur dioxide and hydrogen sulphide, lower the pH value but the human factor results in values between 4 and 5 at some locations[45], mainly due to combustion products from burning fossil fuels and industrial effluents being converted to strong acids (e.g. sulphuric and nitric). Pre-industrial data, from glacier analysis in Europe and North America, indicate precipitation pH values of 5–6, whereas in recent years values regularly have been below 4.5 and occasionally much lower during rainstorms (e.g. pH 2.4 at Pitlochry in Scotland on 10 April 1974).

Two major difficulties in dealing with the problem of acid rain are (i) that the acidity often comes to earth in a different country to where it was generated and (ii) the cost of removal of the acid gases from the effluent is expensive, particularly at power stations.

The effects on aquatic ecosystems[46] indicate that even at pH 6.5 the water is harmful to fish eggs and fry, and between 6 and 5 there is the potential high toxicity due to dissolved aluminium.

The effects on terrestrial ecosystems include increased soil acidity, decreased nutrient availability, release of toxic metals, leaching of important soil chemicals, and change of species present and decomposer organisms. There has been pronounced damage to trees in North America and in Europe[47].

Hazes, caused by the increase in sulphates present in the atmosphere, are becoming more widespread and may ultimately affect climate.

8.6.2 Ozone depletion

The ozone layer is a layer with relatively high ozone concentration in the stratosphere at a height of about 16–18 km in polar latitudes and about 25 km over the equator. The ozone layer absorbs solar ultraviolet radiation (UV-B, 280–320 nm wavelengths) warming the stratosphere and producing a steep temperature inversion between 15 and 50 km. The warm stratosphere acts as a lid to vertical motion in the troposphere (i.e. the conductive process which produces cloud and rain). Weakening of the inversion would upset atmospheric circulation and thereby weather and climate. Reduction of the ozone layer (i.e. ozone concentration) would allow more UV and visible radiation to reach the ground, warming the lower atmosphere and the earth's surface. Opposing this the reduced temperature of the stratosphere would promote surface cooling. The net effect is very difficult to predict.

The adverse effects of increased UV-B radiation reaching the earth would include ageing of the skin and increased skin cancer. Also affected would be growth, composition and function (including photosynthesis) of a wide variety of plants, including important crops such as soya bean.

UV-C radiation (wavelengths less than 290 nm) is completely absorbed by the ozone layer. Any failure of the ozone layer resulting in UV-C radiation reaching the earth's surface, would have a much more marked effect, as UV-C is far more damaging than UV-B because it can kill cells or cause them to mutate (e.g. to become cancerous) as a result of attack on DNA.

Ozone is produced at stratospheric levels by the photodissociation of oxygen by radiation (wavelengths less than 242 nm), and removed by the classical reaction of $O + O_3 = 2O_2$ at levels below the stratosphere. Ozone also participates in many complex reactions involving trace substances in the stratosphere. Human activities are increasing the concentrations of substances such as oxides of nitrogen, oxides of hydrogen, chlorine and bromine to such an extent that these substances (through catalytic cycles) may be removing ozone faster than it is being produced. The number and complexity of the reactions in the stratosphere make it uncertain that all the potential threats have yet been identified. Factors known include combustion gases from supersonic aircraft, oxides of nitrogen from nuclear weapon testing and chlorofluorocarbons (CFCs) used in aerosol sprays, refrigeration systems, and industrial processes. The increased use and thereby release to atmosphere of CFCs especially $CFCl_3$ (Arcton 11), CF_2Cl_2 (Arcton 12) and CHF_2Cl (Arcton

22), and their stability has resulted in increased quantities building up in the atmosphere where, as a result of a long residence time (65–90 years)[48], they are removed slowly. Thus, even after the use and release of CFCs has stopped, the problem will remain for a long time.

Increased carbon dioxide concentrations are believed to marginally increase ozone formation, and recently it has been intimated that upper atmosphere aircraft operations may be resulting in increasing ozone formation rather than reducing it as first supposed.

8.6.3 Carbon dioxide and the greenhouse effect

Carbon dioxide occurs naturally in the atmosphere and it is an essential plant nutrient as well as being an important factor in the earth's atmospheric thermal balance thus controlling weather and climate. Though for a considerable time it was felt than man's activities were leading to an increase in carbon dioxide content of the atmosphere, it is only over the last 30 years or so that accurate measurements have been made. These measurements indicate that carbon dioxide will have increased from a pre-industrial concentration of about 260 ppm to about 600 ppm during the next century.

As carbon dioxide is almost transparent to incoming short wave radiation and is a strong absorber of outgoing terrestrial radiation (7–14 μm wavelengths) it traps energy resulting in the warming of the earth's surface and lower atmosphere. The doubling of the carbon dioxide content is estimated to result in an average temperature increase of $3 \pm 1.5°C$ (U.S. EPA 1983) which would result in the earth being warmer than it has been for 125 000 years, and possibly warmer than for 2 million years. The position is becoming progressively worse due to the combustion of fossil fuels and the clearing of large forests (double effect—carbon dioxide from burning the trees and the loss of trees which would convert carbon dioxide to oxygen). In addition, as the upper layers of the oceans become saturated with carbon dioxide, the fraction of atmospheric carbon dioxide absorbed by the oceans will decrease, adding to the problem.

While there is much speculation regarding the exact temperature increases occurring, it is thought that polar and subpolar temperatures may increase disproportionally (more than 4°C and up to 10°C).

Historical research indicates that 4500–8000 years ago, when average temperatures were 1.5–2.5°C higher at middle latitudes, the present corn belts of North America and the Soviet Union were drier but increased precipitation occurred in tropical and subtropical, currently arid, zones. The benefits of higher temperatures on growth are likely to be more than offset by the loss of precipitation. Any reduction in North American grain yield would result in higher prices and food shortages which would hit the poorer Third World nations hardest.

Action on a world basis is required to deal with the carbon dioxide problem and the resultant greenhouse effect and will require the co-operation of all nations. As the CFCs also contribute to the greenhouse effect, it is essential that the substitutes used to overcome the ozone depletion problem also reduce the greenhouse effect.

References

1. Information Leaflet No 1. *The History of Air Pollution and its Control in Great Britain*, N.S.C.A., 1.
2. *Industrial Air Pollution, Health and Safety*, 1981 H.M.S.O., 20.
3. N. C. Harris, *Hazard Assessment in the Chlor-Alkali Industry*, ICI plc. Mond Division.
4. M. Swift, Reprint of paper—'Dust and Fume Collection and Control for the Minerals Processing Industry', Peabody–Holmes p.l.c.
5. *Encyclopaedia Britannica* 1969, **18**, 183.
6. A. J. Moyes and J. S. Wilkinson, Reprint of paper—'Development of the Holmes–Stretford Process', W. C. Holmes & Co.
7. *Fuel Gas Processing*, Publication 90, Peabody-Holmes Ltd.
8. *Holmes-Stretford Process*, Publication 74, Peabody-Holmes Ltd.
9. B. G. Reuben and M. L. Burstall, *The Chemical Economy*, Longmans, 1973, 475.
10. *ICI plc. Mond Mail*, 1982, (17 Dec.), 2.
11. As ref. 9, 476.
12. Environmental Protection Survey in *Process Engineering*, 1979, 9.
13. Jacques Yves Cousteau, *The Cousteau Almanac*, Doubleday, 1981, 356.
14. As ref. 9, 468–475.
15. E. P. Odum, *Fundamentals of Ecology*, Saunders, 1971, 435–8.
16. *Ibid*, 436.
17. H. A. Tribe, *et al.*, *Ecological Principles*, Cambridge University Press, 1975, 149.
18. Environmental Protection Survey in *Process Engineering*, 1977, 5.
19. As ref. 13, 251.
20. As ref. 18, 3.
21. Environmental Protection Survey in *Process Engineering*, 1978, 36.
22. Plant Profiles Nos. 9791 and 9792, Mass Transfer International.
23. As ref. 15, 437.
24. F. G. Minifie and A. J. Moyes, Reprint of paper—Low Cost Electrostatic Precipitation, W. C. Holmes Ltd.
25. As ref. 18, 25.
26. A. L. Banks and J. A. Hislop, *Health and Hygiene*, University Tutorial Press, 1966, 70.
27. *Ibid*, 77.
28. *Daily Telegraph* 1983, (22 June), 32.
29. F. P. Lees, *Loss Prevention in the Process Industries*, Butterworths, 1980, 823.
30. P. Kinserley, *The Hazards of Work: How to Fight Them*, Pluto Press, 1973, 121.
31. *Talking About Health and Safety—Lubricants and Allied Products*, Burmah Castrol, 18. 19.
32. As ref. 30, 137.
33. As ref. 26, 297.
34. W. Hobson (Ed.), *The Theory and Practice of Public Health*, Oxford University Press, 1969, 367.
35. *Encyclopaedia Britannica*, 1969, **11**, 982.
36. *Encyclopaedia Britannica*, 1969, **12**, 206 et seq.
37. Control of Pollution Act 1974, H.M.S.O.
38. A. Walker, *Law of Industrial Pollution Control*, George Goodwin, 1979.
39. As ref. 29, 41.
40. *Ibid*, 42.
41. *Ibid*, 779–803.
42. As ref. 13, xx.
43. F.G.D. Report, *Engineering*, 1988, 620.
44. D. Elsom, *Atmospheric Pollution*, Blackwell, 1987, 75.

45. W. R. Hibbard, *Acid Precipitation Materials and Society*, 1982, **6**, 357.
46. As ref. 44, 87.
47. *Ibid*, 90.
48. K. L. Brice *et al.*, *Atmospheric Environment*, **16**, 2543.

Bibliography

Solid Waste Disposal, B. Baum and C. H. Parker, Ann Arbor, 1973.
Industrial Water Pollution Control, W. E. Eckenfelder, McGraw Hill, 1966.
Industrial Waste Water Control, F. C. Gurham (Ed.), Academic Press, 1965.
Air Pollution and Human Health, L. B. Lane and E. R. Siskin, John Hopkins Press, 1977.
Aspects of Pollution Control Associated with the Process Industries, D. H. Napier, Keith Shipton
    Developments Ltd., 1974.
Waste Water Systems Engineering, H. W. Parker, Prentice Hall, 1975.
Industrial Hygiene and Toxicology, F. Patty (Ed.), Wiley Interscience, 1962.
Industrial Pollution, N. I. Sax (Ed.), Van Nostrand Reinhold, 1974.
Fundamentals of Air Pollution, A. C. Stern, Academic Press, 1973.
Air Pollution Control, W. A. Strauss, Wiley Interscience, 1971.
Encyclopaedia of Occupational Health and Safety (2 .vols.), International Labour Office, 1972.
Guide lines on the Responsible Disposal of Wastes, C.B.I., 1982.
Atmospheric Pollution, D. Elsom, Basil Blackwell, 1987.
EEC Environmental Policy and Britain 2nd ed., N. Haigh, Longman, 1987.
SCOPE 29 The Greenhouse Effect, Climatic Change, and Ecosystems, eds. Bolin, Döös, Warwick
    and Jäger, Wiley, Chichester, 1986.

CHAPTER NINE

CHLOR-ALKALI PRODUCTS

S. F. KELHAM

9.1 Introduction

The production of chlorine and caustic soda is the story of the bulk inorganic commodities chemicals business. It is an example of a large scale manufacturing industry which uses high technology processes to create the basic raw materials used worldwide as the building blocks for industry.

The discovery of chlorine is usually attributed to Scheele in 1774, who referred to it as 'dephlogisticated marine or muriatic acid air' and who described its greenish yellow colour, its bleaching properties and suffocating effects on insects. However, it is certain to have been known earlier through the work of alchemists from the 12th century. The material was assigned the property of an element by Davey in 1810, and he also proposed the name 'chlorine', from the Greek 'chloros', meaning yellow-green.

Use of chlorine developed slowly. Bleaching of textiles with chlorine gas and later with chlorine water was not very efficient. More success was obtained using chlorine as a disinfectant. In 1790, Tennant was granted a patent on the production of solid bleaching powder. Manganese dioxide and hydrochloric acid routes were expensive and inefficient despite the introduction in about 1870 of the Weldon process for the recovery of manganese dioxide. The Deacon process, proposed in 1870, used air and a copper catalyst to oxidize waste HCl from the LeBlanc process to chlorine and gave a cheaper production route. An increasing demand as a bleaching agent for both textiles and paper followed. The first electrochemical production of chlorine was observed by Cruikshank in 1800, and the first patent, on a cell with a porous diaphragm, was granted to Watt in 1851. Both the lack of commercial electricity supplies and of suitable diaphragms slowed development. It was not until 1888 that the Griesheim cell was developed, making potassium hydroxide and chlorine on a batch basis using a diaphragm of porous cement. The forerunners to the modern diaphragm cells were the Hargreaves–Bird cell of 1890, which was first operated by the United Alkali company in Widnes, and the Le Sueur cell, developed in the U.S.A., which was the first to use a

284

percolating diaphragm. In the 40 years that followed, diaphragm cells of all shapes and sizes were developed. Now only a limited number of designs are in service, some of which are described in more detail later in this chapter.

The use of a mercury cathode in the production of chlorine was simultaneously discovered by Castner and by Kellner, in 1892. The inventors joined forces and the first rocking cell was built at Oldbury, with production moving to Runcorn in 1897. Rocking cells have been operated as recently as 1960 but, as with diaphragm cells, development of different types of mercury cell was rapid, with the evolution of one or two basic designs which are now standard within the industry. While diaphragm cells developed mainly in the U.S.A., the mercury cell development was concentrated in the U.K. and Europe.

Soda was known long before chlorine, as naturally occurring sodium carbonate, and was widely used in the growing industrial world. Artificial soda production started in France, as a result of the Napoleonic wars, when lack of natural soda led to a prize of 2400 livres being offered for a practical manufacturing process. In 1783, the prize was promised to LeBlanc, but never paid, and LeBlanc was forced to make his patents public, without payment, in 1793. Leblanc committed suicide in the workhouse in 1806, but his inventions laid the foundations of the chemical industry (and many of its problems) for more than a century afterwards. The LeBlanc Process burns coal, sodium sulphate (saltcake, made from salt and sulphuric acid) and limestone together in a ratio 35:100:100, leaches the product and treats the black liquor produced with lime to form sodium hydroxide solution. As the process developed, it became more integrated and the source of many other bulk inorganic materials. Muspratt introduced the process in Liverpool in 1823, and Widnes soon became the centre of the U.K. industry. Even after the introduction of the Solvay process in 1861 for making sodium carbonate directly, the LeBlanc process was still viable for caustic soda production until after world War I, when causticization of carbonate with lime became dominant. This process was superseded by the electrolytic routes which at that time were rapidly becoming established. Crossover of the electrolytic and chemical routes occurred in the mid to late 1930s.

The dominant production routes of mercury and diaphragm cells developed in parallel in the first half of the 20th century, with technological developments leading to lower energy consumptions and capital costs. Both systems used carbon based anodes, and this led to major cost inefficiencies. The use of titanium as a basic electrode material was first proposed in 1956, and platinum based coatings for chlor-alkali electrodes were patented by Beer in 1962. In 1967 titanium-ruthenium oxide catalytic materials were introduced, and these proved highly successful on the large scale. After the early 1970s, all new plants contained metal anodes, and conversion programmes were initiated throughout the industry to change from the inefficient and labour intensive carbon based anodes to the long lasting dimensionally stable metal anodes. This major invention of the 20th century once more revolutionized the chlor-alkali industry.

The principle of the chlor-alkali membrane cell has been known for a considerable time, and patents for this approach to chlor-alkali manufacture were granted in the early 1960s. Membrane cells combine the purity of mercury cell caustic with the power efficiency of diaphragm cells, while avoiding many of the operational and environmental concerns of both these systems. However, the membranes available at that time were unsuitable, and it was not until the development of new types of membrane by DuPont in 1962 and their use in a chlor-alkali cell in 1964 that the potential of this technology became realized. Membrane cell development followed rapidly, with the first commercial plants on line in the early 1970s, linking membrane, metal anode and cell developments together. Concerns around mercury, particularly in Japan in the late 1960s, were recognized, and the move away from mercury cells gave added impetus to membrane cell development. Today, almost all new plants utilize membrane cell technology, and, as with the introduction of metal anodes in the 1970s, conversion of many older diaphragm and mercury plants to membrane cell technology is starting to take place.

Production of bleaching powder in the U.K. had reached 150 000 tonnes/year by 1900, and the demands for chlorine and caustic soda grew continually at a compound rate of nearly 7·5% until the first oil crisis of

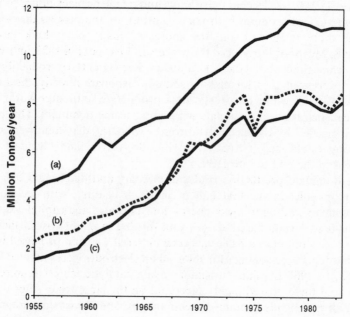

Figure 9.1   Western   European   chlorine   consumption   1955–1983.   (a) IIP   1975 = 100; (b) chlorine; (c) NaOH.

1973. Increased energy prices led to a major world recession and growth paused, being hit by the second energy crisis in 1979. Since then the position has recovered, and there is still a small growth of 1–2% per year. However, as environmental concerns around refrigerants and chlorinated compounds (CFCs) increase, the growth rate may slow even further. Since the 1970s, most of the growth has been in the Third World countries, whereas in the developed nations, growth has been very low, or actually decreased as old, inefficient or environmentally unacceptable plants are shut down. World capacity in 1987 was 40 m tonnes/year (excluding the Communist Bloc), and the growth rate in European production of chlorine since 1955 is shown in Fig. 9.1. Occupacity is currently at around 90% of installed capacity.

9.2 Uses of chlorine

Chlorine consumption is closely linked to standard of living in a country. Initial production of chlorine is for use as a disinfectant, for water treatment and the exploitation of natural resources, such as pulp and paper. Inorganic chemistry applications follow and organic chemicals begin to take large quantities as PVC becomes important in the building industry and for the production of consumer goods. As standards of living increase, so does the

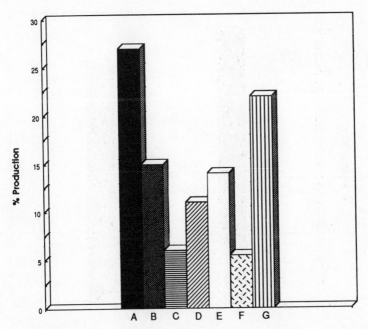

Figure 9.2 Market outlets for chlorine. A vinyl chloride; B, solvents; C, propylene oxide; D, chloromethanes; E, inorganic chemicals; F, pulp paper; G, other users.

usage of solvents, aerosols, dry cleaning materials and refrigerants. Typical market outlets for chlorine are shown in Fig. 9.2.

9.3 Uses of caustic soda (sodium hydroxide)

Caustic soda has similar outlets to chlorine and these are shown in Fig. 9.3. Large amounts are used in the organic and inorganic chemicals industries, for soaps and detergents, and for aluminium extraction.

9.4 Uses of hydrogen

Hydrogen, the third product of the electrolysis of brine route, is ideally used as a chemical feedstock, either separately, or integrated with other hydrogen plants. Electrolytic hydrogen is very pure, and can be compressed to 200 atm for transportation in cylinders and a wide variety of uses. However, hydrogen disposal is not normally an important consideration and it is often burned as a fuel, or in some cases, even vented to atmosphere, although as a raw material, it has considerable value.

Electrolytic production of chlorine and caustic soda results in fixed ratios of

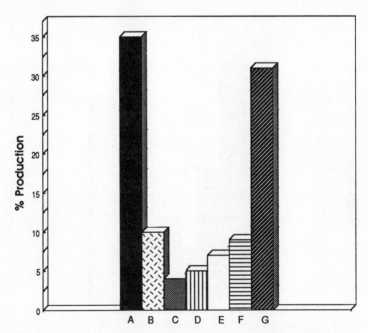

Figure 9.3 Market outlets for caustic soda. A, manufacture of chemicals; B, rayon, C, soap; D, alumina; E, neutralization; F, pulp paper; G, miscellaneous.

chlorine, caustic and hydrogen being produced. As uses of chlorine and caustic have developed, the balance of production has been critical. Historically, caustic soda demand has tended to be greater than that of chlorine, so processes such as the LeBlanc route and the production of caustic from sodium carbonate have enabled a balance to be maintained. Chlorine and caustic demands are currently essentially in balance, and although it has long been expected that chlorine demand will eventually exceed caustic demand this has not proved to be the case to date, and markets can be balanced by stocking and pricing policies. As caustic soda prices rise, alternative sources of alkali are found, and the market restabilizes. While caustic is a world scale commodity chemical, and is bought and sold freely, chlorine is not readily transported or stored in quantity. Large chlorine users are generally tied closely to the chlor-alkali plant. If chlorine has to be transported it is usually converted into ethylene dichloride which contains over 70% w/w chlorine and is an important intermediate in the manufacture of PVC. The high power consumptions and opportunities for process integration tend to move the chlor-alkali industry to areas of cheap power, natural gas or hydroelectricity, and in many cases, chlor-alkali production forms part of an integrated petrochemicals and plastics complex.

9.5 Types of cell

9.5.1 *Mercury cell process*

The mercury cell process has developed extensively since its inception in 1892. A number of distinct phases can be identified, from the rocking cell to the 400 000 A modern mercury cell unit. The cell chemistry is identical in all cases

Anode reaction (titanium or graphite)

$Cl^- - e = Cl^\bullet$ chloride ion loses an electron to form a chlorine atom

$2Cl^\bullet = Cl_2$ chlorine atoms combine to escape as chlorine molecules

Cathode reaction (thin film of mercury)

$Na^+ + e = Na$ sodium ion gains an electron to form a sodium atom

$Na + Hg = Na/Hg$ sodium atom immediately dissolves in the mercury film electrode to form sodium amalgam which passes into the decomposer where it reacts to form sodium hydroxide solution and hydrogen, leaving mercury which is recycled.

Decomposer reaction (essentially a short circuited cell)

$2Na/Hg + 2H_2O = 2NaOH + H_2 + 2Hg$ (iron/graphite electrodes)

The caustic soda is produced at up to 50% concentration. The reversible voltage of the cell overall is 3.1 V, to which has to be added the operating electrode overvoltages and resistance effects of the cell components and liquids within the electrolysis compartment. The design of the cell has centred on reducing these resistances, and increasing cell efficiencies. The major inefficiency of the mercury cell is the premature decomposition of the amalgam and the evolution of hydrogen within the brine cell rather than in the decomposer.

The first operating membrane cell was the rocking cell, largely developed by Baker, who was Castner's chief chemist, at Oldbury, and later at Runcorn. The history of the Castner Kellner plant at Runcorn gives the history of the development of the mercury cell, and indicates the way in which a technology has developed in the drive to increase production quantities and efficiencies while minimizing capital costs.

The 1896 rocking cell plant at Runcorn was of 1000 HP and operated at 575 A. The cells consisted of two outer brine compartments of $0.9\,m^2$ with graphite anodes and a central amalgam decomposing chamber of approximately $0.6\,m^2$ containing iron grids. The cell was rocked periodically from side to side so that the mercury flowed from one compartment to the other. Only one anode compartment was in use at a given time. While rocking cells continued in service until 1938, the slow movements and mass transfer limited performance and current densities, and the long cell concept was developed.

The long cell was first introduced by Baker in 1902 and has been the basis of all subsequent mercury cells. The cell was 50 ft long and consisted of a brine channel, with mercury flowing along the base, an amalgam channel where decomposition of amalgam occurred, returning the mercury to the front end of the cell, and a pump to return the mercury to its starting point. The brine cells were of concrete, with steel mushrooms in the base to provide electrical contact with the mercury. The slope was 1 in 300 to give a high mercury flow rate. The denuder was half the width of the cell and initially made of concrete, although this was later replaced with steel. The denuder contained graphite grids. An Archimedean screw pump was used, and the current rating eventually achieved 11 000 A with a base area of $7.3\,m^2$. Anodes were of graphite rods impregnated with wax which were sealed into the covers, and as these suffered appreciable wear, anode–mercury gaps increased and electrical efficiency deteriorated, leading to the need for frequent replacement. Concrete cell covers placed on wooden laths, which were progressively removed to lower the anodes, were used as a primitive adjustment system.

Ebonite (hard rubber) lined steel based cells were introduced in 1937, although concrete cells continued to run until 1967. Ebonite lasted several times as long as concrete and introduced fewer impurities into the cell. The retention of the steel mushrooms still limited current loadings, but these cells still continued to operate until 1971 at Runcorn. The current limitations were overcome by the use of steel baseplates, which became widespread after 1943, with the base area increased to $12.5\,m^2$. The denuder was mounted under the cell instead of alongside to conserve space and more efficient mercury pumps developed, leading to centrifugal rather than screw designs. Anode design and anode adjustment developed, with more emphasis on the optimum gaps and the ability to release gas bubbles. Individual anode adjustment was introduced. Vertical denuders, packed with iron impregnated graphite balls, were introduced in 1966 and by 1975 all the remaining cells at Runcorn had been converted to vertical denuders. Introduction of metal anodes in 1971 led to further improvements, and with the limitations on current density removed, currents increased, reaching a maximum of 90 000 A on the final versions.

The modern generation of mercury cells began in 1966 with the introduction of the 25 m^3 cell, still 48 ft long but double the width and with a slope of 1 in 100. A sketch of such a cell is shown in Fig. 9.4. Water cooled vertical denuders, centrifugal pumps and later, metal anodes were installed. Currents were raised to 225 000 A. The major design change came from the cellroom layout, with cells on a single level instead of several floors, and a much greater attention given to working conditions and mercury hygiene. Adiabatic denuders were installed in later designs, giving more efficient and compact units operating at higher temperatures, and aluminium started to replace copper for electrical connections. In 1976, the 33 m^2 cell was introduced. This was the first unit designed specifically to take advantage of metal anodes, with automatic anode adjustment and currents of up to 400 000 A.

As currents increase and power consumptions and efficiencies become more

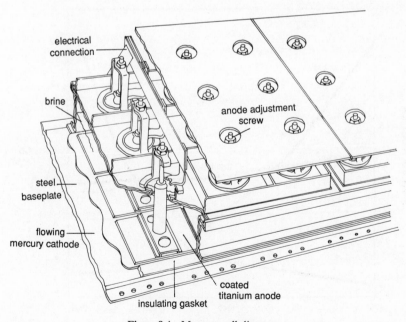

Figure 9.4 Mercury cell diagram.

critical, the problems of brine quality become more apparent. Brine purity is controlled at 20–40 ppm of calcium and less than 1 ppm of magnesium. Brine can be supplied on a resaturation circuit or on a once through basis. Poor brine quality can lead to high hydrogen levels in the chlorine and the formation of 'thick mercury', a buttery compound which has to be removed from the cell from time to time, and to denuder deactivation. As the mercury cell technology has reached its peak, chemistry, rather than engineering problems have started to become dominant, and the fundamentals of mercury cell operation have been increasingly investigated.

Of particular interest has been the relationship of electrical power consumption and current density to development activities. Current densities have increased continuously, giving greater production capability per unit area and hence for a given capital cost. Cell efficiencies, in the form of voltage loss reduction have increased, but the relationship between the two has led to a virtually constant power consumption, of about 3400 kW/tonne chlorine as shown in Fig. 9.5.

With the introduction of membrane cell technology, mercury cell development has virtually ceased, and it is highly unlikely that any significant new mercury cell capacity will ever be installed. However the modern mercury cell is still a highly effective way of producing high quality caustic soda, and provided mercury discharge levels can be maintained at sufficiently low levels mercury cells will remain viable until well into the 21st century.

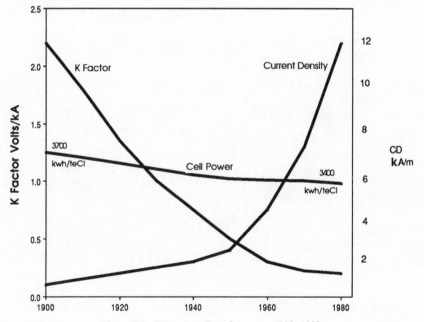

Figure 9.5   Mercury cell performance 1900–1980.

9.5.2 *Diaphragm cell process*

The diaphragm cell principles have been in use commercially since 1888, when the Greisheim cell was introduced in Germany. The principle is to reduce reverse flow and reaction of hydroxyl ions with chlorine by use of a porous diaphragm, across which a hydraulic gradient is set up, and the efficiency of the cell depends on the design of this separator.

The principal reactions are

Anode reaction (titanium or graphite)

$NaCl = Na^+ + Cl^-$ equilibrium of sodium and hydroxyl ions

$Cl^- - e = Cl^{\cdot}$ chloride ion loses an electron to form a chlorine atom

$2Cl^{\cdot} = Cl_2$ chlorine atoms combine to escape as chlorine molecules

Liquor containing ionic mixture flows through the diaphragm

Cathode reaction (steel)

$2H_2O + 2e = H_2 + 2OH^-$ water decomposition to form hydrogen and hydroxyl ions

$OH^- + Na^+ = NaOH$ hydroxyl ion combines with sodium ion to form caustic soda

The reverse reactions are

$$Cl_2 + OH^- = Cl^- + HOCl$$
$$HOCl = H^+ + OCl^-$$
$$2HOCl + OCl^- = ClO_3^- + 2Cl^- + 2H^+$$
$$4OH^- = O_2 + 2H_2O + 4e$$

With graphite electrodes, there is in addition

$$4OH^- + C = CO_2 + 2H_2O + 4e$$

It can be seen that in a commercial cell, the major byproducts are sodium hypochlorite, sodium chlorate, oxygen and carbon dioxide, which lead to overall current efficiencies in the range 90–96%. The electrolysis current only converts about half the 25% w/w sodium chloride feed to caustic soda, which leads to a maximum of about 12% w/w caustic in the product sodium chloride/sodium hydroxide liquor.

The early cells used a porous diaphragm made from concrete which included soluble salts leached out after setting. The anodes were carbon or magnetite and a steel pot was used as the cathode. The system was operated on a batch basis, initially with KCl. The first continuously percolating diaphragm cell was the Le Sueur cell in 1897 which used a concrete trough with a porous base to separate the liquids, and cells using this principle were still in operation in the U.S.A. until the mid 1960s. Cell design development rapidly moved towards more effective diaphragms and improved anodes in the form of graphite rather than carbon. Most of these had vertical diaphragms. Major breakthroughs occurred with the Marsh cell, with finger cathodes and side entering anodes, with an asbestos paper diaphragm, and the Hooker deposited diaphragm cell, the forerunner of all modern diaphragm cells, which was first introduced in about 1928. Current ratings were raised from 5000 to 30 000 amp as the technology developed. Filter press bipolar designs were developed by Dow and later by PPG. Some of these units are still in operation.

Hooker S type cells and the Diamond Alkali D type cells became the standard units for the industry, and had many similar features. The anodes consisted of vertical plates of graphite set in a shallow steel or concrete pan and attached to copper connection bars by pouring in molten lead which was then covered with pitch to protect it from the corrosive conditions in the cell. A steel gauze finger cathode with a deposited asbestos diaphragm was placed over the anode blades and insulated from the anode base by a rubber gasket. A concrete cover was then located over the cathode box. The asbestos diaphragm was deposited by sucking a slurry of asbestos fibres onto the gauze in a depositing bath. In operation, brine is admitted to the top of the cell, and percolates through the diaphragm and cathode gauze into the catholyte compartment. Chlorine is evolved from the anodes and is collected from inside the concrete cover. Hydrogen is collected from inside the cathode box, from where the caustic/brine mixture is also removed. As the percolation rates through the diaphragm vary, so does the level of brine in the cover, and the art is to produce a diaphragm which gives a long lifetime before the brine level in the cell becomes too high. Not surprisingly, as most of the components have limited lifetimes, the cells are relatively expensive to rebuild, and handling asbestos and lead can lead to health and environmental concerns.

As with the mercury cell, the introduction of metal anode technology led to a considerable review of cell design and the opportunity to optimize the anode–cathode gap, which as with the mercury cell, increased with time as the anode deteriorated, but unlike the mercury cell, could not be adjusted with the cell in service. Diaphragm technology improved, with the replacement of wet asbestos diaphragms with heat treated and polymer modified diaphragms. Lead bases became unnecessary, and the ICI DMT cell is typical of a modern diaphragm cell unit and shown in Fig. 9.6. This cell has a current capability of 120 000 A and a cathode area of 43 m^3, giving a current density of 2.8 kA/m^2. The titanium anodes are permanently welded onto a titanium sheet which is explosion bonded onto a massive steel base to minimize electrical resistance losses. The cathode box and base has explosion bonded copper plates to provide low resistance intercell connections and the anode–cathode gaps can be minimized by expanding the electrode structures into the cathode fingers once the box is in place. The asbestos diaphragm is heat treated to improve

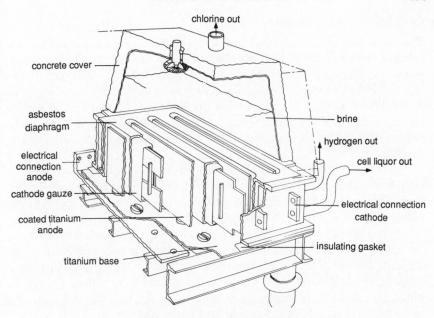

Figure 9.6 Modern diaphragm cell.

stability and lifetimes, and the cell cover is of GRP lined concrete to maintain hydraulic heads. Current leakage is minimized through flow interrupter devices, and the cell can operate without maintenance for in excess of a year, depending mainly on brine qualities. Asbestos is a magnesium rich material and, in operation, the pH changes in a diaphragm lead to deposition and leaching of magnesium. For long lifetimes and good permeability, the magnesium balance must be maintained by keeping the feed concentration of magnesium at 1–2 ppm in the brine.

Reversible voltages are of order 2.3 V, which leads to intrinsically significantly less electrical energy than for a mercury cell where the decomposition voltage is 3.1 V. However, the actual resistance losses (known as the k factor) and therefore added voltage, are significantly higher—typically 0.43 V per kA/m^2 compared to 0.1 V per kA/m^2 for a mercury cell. This is offset by the lower current density, typically 2.8 kA/m^2 compared to 12 kA/m^2. Total energy requirements must, however, include the need to purify the product caustic soda to remove the salt and increase the strength to the typically 50% sales specification. This is usually carried out in three or four effect evaporators. As the caustic concentration increases, the sodium chloride crystallizes out and is removed by filters and/or centrifuges. However, solubility levels are such that most diaphragm cell products still contain up to 1% w/w salt, and this can restrict its use, e.g. in the rayon industry. While processes have been devised for further purification of the caustic, they have never been a commercial success.

9.5.3 Membrance cell process

The membrane cell concept has been known for many years, but early work failed as a result of non availability of suitable ion exchange membranes which would resist the very demanding conditions within the chlor-alkali cell. It was also realized that the graphite anodes, which wore away in use, would present major difficulties from the engineering side, so it was not until metal anodes became available that interest revived. However, much of the early work and patents of the 1950s set the scene for the way in which this technology was to develop.

The membrane cell operates in a similar way to the diaphragm cell, with the same basic reactions. However, instead of the hydraulic gradient preventing reverse flow of hydroxyl ions, the cation exchange membrane will only allow sodium ions to pass through in the direction anode–cathode, and will inhibit the reverse flow of hydroxyl ions. The effectiveness of the system depends on the selectivity of the membrane, coupled with the electrical resistance of the cell itself. Instead of the hydrogen-producing reaction taking place in a brine solution, the catholyte liquor consists of a recirculating caustic system, in which the strength is increased on each pass through the cell, and the concentrations maintained by purge and water addition.

The first commercial membranes were fabricated by DuPont under the Nafion tradename. These had a backbone of carbon and fluorine atoms, to which were appended sidechains containing active sulphonic acid ion exchange groups (see Fig. 9.7). These end groups gave high conductivities and therefore low voltage drops, but had relatively poor current efficiencies. The major breakthrough came in 1978 when Asahi Glass introduced a carboxylic acid grouping to the sidechains. This gave high current efficiency but relatively poor voltage, and was attacked by the anolyte side conditions. The compromise was to produce a laminated membrane with a thick anolyte side of sulphonic acid structure to give low voltage and physical integrity, and a thin catholyte face containing carboxylic structures which provided a good

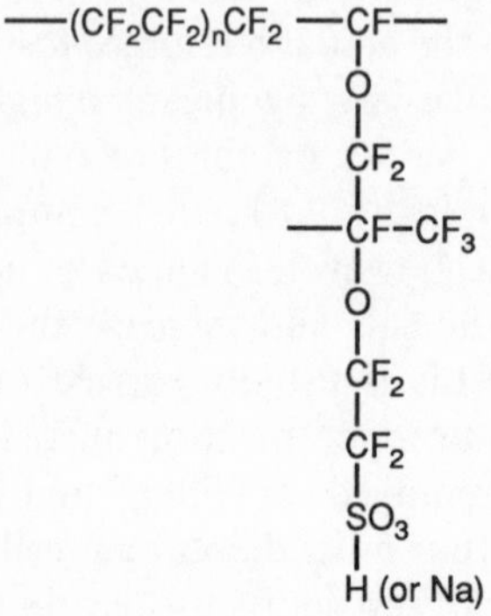

$$—(CF_2CF_2)_nCF_2 —CF—$$
$$|$$
$$O$$
$$|$$
$$CF_2$$
$$|$$
$$—CF–CF_3$$
$$|$$
$$O$$
$$|$$
$$CF_2$$
$$|$$
$$CF_2$$
$$|$$
$$SO_3$$
$$|$$
$$H \ (or \ Na)$$

Figure 9.7 Nafion structure.

current efficiency. These membranes are now manufactured by both DuPont and Asahi Glass. From time to time other manufacturers have produced membranes, but these are largely based on Nafion membranes with minor chemistry modifications. Further enhancement of membrane performance has been by addition of surface coatings to promote bubble release, reducing thicknesses, and incorporating reinforcing meshes. Membranes have typically a 96% initial current efficiency and voltage drop of about 250 mV. Lifetimes are of the order of 3 years. Caustic concentration is typically 30–35% so a simple evaporation system, to bring concentrations up to 50%, is still required. However, the product material contains only a few ppm of salt.

Early membrane cell development was dominated by the Japanese drive to eliminate mercury cell activities. Initial conversion was to diaphragm cells, but it was soon realized that the inferior grade of caustic would not fulfil all needs, and the early membrane systems were rapidly brought up to full commercial scale. Early cells were of the bipolar rather than the monopolar design. In a bipolar cell, the voltage is applied across a stack of membranes and electrodes such that all the individual components are running in series. The current passes through each compartment in turn, going in sequence anode, membrane, cathode; anode, membrane, cathode etc., and the design of the cell means that the anodes and cathodes are fabricated as a single unit. Monopolar cells, which are more akin to the conventional mercury and diaphragm cell layouts have the electrodes arranged in the cell in parallel. Each cell has the same voltage across it, and the current is spread between the cells in the unit. The sequence would be anode, membrane, cathode, membrane, etc., with the electrode structure being either wholly anode or wholly cathode based. Bipolar cells show voltage savings over monopolar equivalents as the voltage drop within the composite electrodes can be reduced to very low levels, but there are potential problems with materials, due to hydrogen diffusion and embrittlement and current leakage. For this reason, special precautions have to be taken in feeding liquors into the cell and removing them from it, using long plastic tubes to maximize electrical resistances and hence reduce current bypassing which, with a 300–400 V stack of plates, can be considerable. Asahi Chemical, in Japan, developed a successful bipolar cell design, using welded titanium anodes and steel cathodes approximately $2.5 \, m^2$ area and incorporating their own variety of membrane to make 22% caustic soda. This type of cell is closed with hydraulic rams, and a typical 50 000 tonnes/year plant would contain 5 of these cell units, operating at 350 V, 10 000 A and about $4 \, kA/m^2$. An advantage of this design of cell is that the rectifiers required are high voltage, low current, and are therefore relatively cheap. A disadvantage is that the loss of a single membrane can lead to shutdown of the whole cell, whereas for the monopolar cell, this is not a major hindrance.

Monopolar cells have been developed more extensively, and are more appropriate for larger installations. The ICI FM21 cell provides a radical departure from the traditional heavy engineering of electrochemical cells as

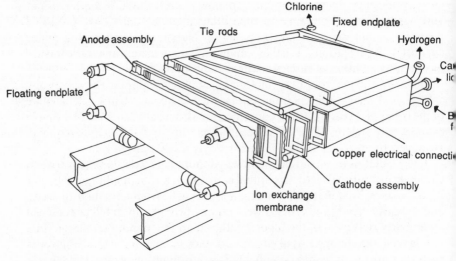

Figure 9.8 Membrane cell diagram.

shown in Fig. 9.8. The electrodes are pressed from sheets of nickel for the cathode and titanium for the anode, each with an active area of $0.21 \, m^2$. The electrodes are coated with electroactive materials, and the cell is assembled, sandwiching the membrane between alternate anode and cathode plates. Sealing is by rubber gaskets and compression is maintained by disc springs. As the voltage drop across the unit is only of order 3.5 V, current leakage is not a problem and liquor inlets and liquor-gas outlets are through internal porting systems. A typical cell will operate at up to 100 000 A and $4 \, kA/m^2$, using an option of membrane types. A 50 000 tonnes/year plant would contain 50 such cells.

Electroactive coatings for membrane cell anodes are essentially the same as for mercury or diaphragm cells. The cathodes, unlike those on diaphragm or mercury cell systems, can also be coated to reduce the hydrogen evolution overvoltage. These coatings are usually platinum or nickel based, and although capable of reducing overvoltages to as low as 50 mV, tend either to be expensive or have short lives, and so the economic justification for these has to be critically examined. Recoating operations have to be carried out, and the style of fabrication of the cell can significantly change the cost and complexity of this operation. Generally, the lifetime of an anode coating can be defined as 24/current density, i.e. typically 2 years for a modern mercury cell at $12 \, kA/m^2$, 8 years for a diaphragm cell at $3 \, kA/m^2$ and 6 years for a membrane cell at $4 \, kA/m^2$. The cost of electrode coatings is therefore virtually the same per tonne of chlorine whatever technology is utilized. The actual cost of this operation can vary considerably, depending on the cell design, the cost of dismantling and the loss of production incurred through the shutdowns involved.

Other manufacturers such as Asahi Glass, Tokuyama Soda, De Nora,

Oxytech, Chlorine Engineers and Uhde offer a variety of either monopolar or bipolar cells and most of these technologies now have representative plants. Membranes are almost all Nafion based.

The emphasis on improving power performance of membrane cells has led to a continual drive to reduce voltages and resistance losses in the system. Power consumptions are now less than 2400 kW/tonne of chlorine, and a comparison of costs and energy consumptions for mercury, diaphragm and membrane cell systems is shown in Fig. 9.9. Reversible voltages are fixed although there are areas where some changes could potentially still be made in the long term, for instance by the use of air or oxygen depolarized cathodes which produce water instead of hydrogen gas at the cathode. However, most of the other achievable improvements are based on the fine tuning of the cell design itself. The voltage breakdown components of the monopolar cell would be typically as shown in Table 9.1.

It can be seen that there are finite limits as to how far voltage performance can be decreased as internal resistances tend to smaller and smaller values. Further improvements can be made by reducing resistance losses within liquid films by bringing electrodes closer to the membrane and most modern membrane cells can accommodate essentially zero gap configurations so that the electrodes virtually touch across the membrane. However, this has been

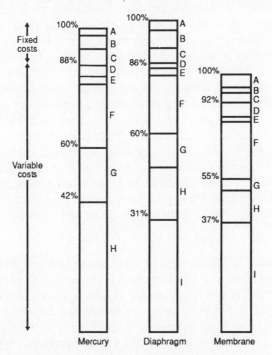

Figure 9.9 Production costs. A, operation; B, maintenance; C, others; D, cell renewal; E, others; F, salt; G, steam; H, resistance power; I, reaction power.

Table 9.1   FM21 cell voltages at $3\,kA/m^2$

Anode assembly and over voltages	210 mV
Cathode assembly and over voltages	140 mV
Anolyte	130 mV
Catholyte	100 mV
Membrane	250 mV
Reversible anode voltage	1220 mV
Reversible cathode voltage	1000 mV
Total	3050 mV

shown to create secondary problems within the cell, mainly resulting from high local brine depletion rates and deterioration of membranes, and the optimum is now viewed as being 1–2 mm. Brine circulation and degassing is important and again there is an optimum to be achieved between a tall cell for good liquor circulation through gas lift effects and a short cell to minimize bubble voidages and electrode voltage losses. Uniformity of operating conditions between compartments is essential. Overall the whole concept of cell design is one of optimization to ensure that the minimum operating costs in terms of membrane, power consumption and anode lives are achieved, together with the minimum installation costs for the cell itself. Membranes are the highest cost item, and membrane lives are currently predicted to be of the order of 2–3 years in service. It is therefore vital that there should be no requirement to take the cell off-line to recoat electrodes or change gaskets while the membrane is still usable. The drive in membrane cell design has therefore been to extend component lives to ensure that a membrane cell rebuild is governed by membrane life and not for other reasons.

The actual time when a membrane is changed depends very much on local power costs, the type of membrane and cell systems used. In the economic assessments for membrane installation high priority must be placed on obtaining the right system overall, including the costs of all the replacement operations and other maintenance activities and also the correct choice of current density. As current densities increase so do power consumptions, while capital costs reduce. However, higher current densities will tend to lead to shorter operating lives, so optimization studies are essential in any project assessment.

9.5.3.1 *Installation of membrane cells*   The electrolytic cell itself comprises only a small part of the total chlorine plant, perhaps 15–20% of the capital for a new installation. A typical schematic layout is shown in Fig. 9.10. The system of providing the brine as a salt source and handling the product chlorine, hydrogen and caustic soda adds significantly to the cost of the total installation, and rectification and power distribution systems for handling the tens or hundreds of megawatts involved lead to considerable capital cost. Ion exchange membrane cells require high purity brine with less than 20 ppb of

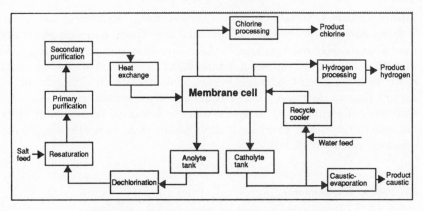

Figure 9.10 Chlorine production from membrane cells.

total hardness and this is achieved by using ion exchange systems. The brine is circulated around the anolyte side of the cell and is normally resaturated to return the feed concentration to near saturation. The catholyte circuit usually consists of 30–32% caustic which is concentrated within the cell to about 34–35% liquor strength. Externally to the cell this liquor is further evaporated to give a 50% sales specification. Hydrogen treatment is relatively straightforward with cooling, possibly filtration and compression before the product is used either as a fuel or for further chemicals manufacture. The chlorine product is cooled and filtered and then dried, usually with concentrated sulphuric acid before being compressed for use either as a chlorine gas stream or subsequent liquefaction. The major by-product from a membrane cell is oxygen in the chlorine from the back-migration and reaction of hydroxyl ions. Achieving a high current efficiency is important, not just from the electrical energy point of view but also to minimize this level of oxygen in the cell gas product. Energy conservation within the membrane cell circuit is important and various systems are devised to minimize heating and cooling duties so that a well-designed circuit should be essentially energy neutral. However, as the operation of membrane cells becomes more efficient the heat input due to resistance losses decreases and at low loads membrane cell systems often require additional heat input to maintain thermal equilibrium.

Conversion of an existing diaphragm or mercury cell plant to membrane technology is usually cheaper but not necessarily less straightforward than in a greenfield site plant. Much of the existing brine system for both mercury and diaphragm cell plants can be re-used for membrane cells but the purity levels have to be increased by addition of appropriate ion exchange systems. A diaphragm cell plant has a once through brine system in that all the brine is consumed by the cell and there is no liquid recycle. Salt recycle is through the evaporator stage and so extensive modifications may have to be done in this area. Chlorine and hydrogen systems are usually very little changed but the caustic system is usually entirely different for a membrane cell plant. Rectification systems in existing mercury and diaphragm cell plants are normally related to monopolar cells and so the conversion of an existing plant to membrane cells is much easier if monopolar membrane cells are to be used rather than the bipolar variety. However, in many cases rectifiers currently in use are old and there is a good financial case for replacement of these units with more efficient and up-to-date systems which could be used with either monopolar or bipolar geometries, although for large plants the monopolar system is usually preferred.

9.6 Future developments

Membrane cells present many advantages over the mercury and diaphragm alternatives. However, many of the existing mercury and diaphragm cell plants are relatively new. After metal anodes were introduced there was a major rebuilding programme in the 1970s and early 80s and nearly all producers upgraded their old graphite anode plants or built new plants to use the new technology. As a result, the majority of chlor-alkali plants which were built at that time are still in relatively good condition and unless there are major environmental or economic reasons for conversion then there is a high likelihood of these plants continuing to operate for some time to come. What has happened is that over the last few years there has been a significant increase in the number of small plants, particularly in the developing part of the world where on-site production of chlorine and caustic soda is essential to local industries. In addition the transportation problems associated with liquid chlorine are leading to more on-site generation in the developed countries and again a proliferation of relatively small plants away from the traditional centres of production. All these new plants are almost without exception ion exchange membrane cell systems. The drive to convert large plants is starting to take off and it requires a trigger mechanism to set the ball rolling. In Japan this was the Minamata mercury scare in the early 1970s which led to a Government requirement that all mercury cells were phased out. This initially led to conversion to diaphragm cells, but diaphragm cells were not able to produce the quality of caustic required for consuming industries and so membrane cell technology really became the only possible alternative to shutting down and as referred to earlier, this led to many of the major technological advances.

The current situation around the world is shown in Fig. 9.11. Mercury cells account for about 15 million tonnes/year of chlorine or 36% of capacity, diaphragm cells for 20 million tonnes/year or 49%, and membrane cells for 4 million tonnes/year or 10%. However, the rate of growth of membrane cell capacity has far outstripped the growth rate in the industry, suggesting that many of the other technologies have shut down plants to give way for the new

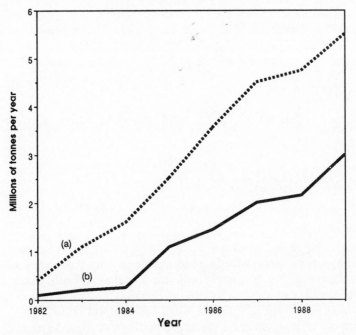

Figure 9.11 World membrane cell capacity. (a) Including Japanese licensees; (b) Excluding Japanese licensees.

technology introduction. If the Japanese indigenous capacity, which is now virtually all ion exchange membrane, is excluded it can still be seen that the growth rate in the rest of the world is significant and is likely to escalate rapidly during the next decade. As energy costs and environmental pressures continue to increase there is no doubt that membrane cells will become dominant within the industry and the major producers will be adopting this technology.

Bibliography

History of the Chemical Industry in Widnes, D. W. F. Hardie, ICI, 1950.
Modern Chlor-Alkali Technology, Vol. 2, C. Jackson (ed.), Ellis Horwood, 1983.
Modern Chlor-Alkali Technology, Vol. 1, Coulter (ed.), Ellis Horwood, 1980.
Diaphragm Cells for Chlorine Production, Soc. Chem. Ind., 1977.
Chlorine–Its Manufacture, Property and Uses, Sconce (ed.), Reinhold/Chapman & Hall, 1962.
Developments and Trends in the Chlor-Alkali Industry, C. Jackson and S. F. Kelham, *Chemistry and Industry*, 1984, 397–402.
Modern Chlor-Alkali Technology, Vol. 3, K. Wall (ed.), Ellis Horwood, 1986.
Chlorine Bicentennial Symposium, T. C. Jeffrey *et al.*, Electrochemical Soc. Inc., 1974.

CHAPTER TEN

CATALYSTS AND CATALYSIS

J. PENNINGTON

10.1 Introduction

In the physical processes which we encounter every day, the roles of the laws of thermodynamics are frequently self-evident. Slow processes are usually associated with visible spatial or physical barriers, and we have well-founded expectations as to the direction in which changes will occur.

When we turn to chemical reactions, the laws of thermodynamics again provide the directional signposts. But here, and particularly in organic chemistry, the signposts are often obscure, and the chemist may well lack any 'feel' as to whether a proposed new reaction is thermodynamically feasible or not. This is largely attributable to the inherent stability of covalent chemical bonding, which introduces a barrier to change at the molecular level. Of course, were kinetic barriers absent, no natural or synthetic organic matter could survive in an oxygen-containing atmosphere. This lack of reactivity also implies that very high temperatures would be required for many reactions, providing little hope for selective conversions; reactions which become thermodynamically unfavourable with increasing temperature (methanol synthesis, olefin (alkene) hydration) would be impracticable.

However, the existence of life forms shows that such constraints need not limit the scope of organic chemistry. Photosynthesis, the controlled stepwise oxidation of carbohydrates and many more highly specific chemical transformations all proceed at near ambient temperatures.

Early in the 19th century, Berzelius began to recognize a common feature within a number of isolated observations, such as the behaviour of hydrolytic enzyme concentrates *in vitro* and the effects of acids, platinum metal and other materials on simple chemical reactions. In 1835, he introduced the terms 'catalytic force' and 'catalysis' to describe the property by which some materials, soluble or insoluble, could effect or accelerate a reaction, to which the catalytic material itself remained 'indifferent'. When the catalytic material is in the same phase as the main reactants, for example in solution in a liquid reaction mixture, the term 'homogeneous catalysis' is applied. The term 'heterogeneous catalysis' is used when the catalytic material exists as a distinct, usually solid, phase with liquid or gaseous reactants.

The complexity of enzyme systems limited the interaction between biochemical and chemical interpretations of catalytic mechanisms until the middle of the present century. In the meantime, industrial catalysis progressed through empirical developments, with homogeneous catalysis by sulphuric acid, alkalis and transition metal compounds. But the historical landmarks, the Contact Process for sulphuric acid (1890s) and the Haber–Bosch Process for ammonia (1913), established a very strong tradition for predominantly heterogeneous catalysis by metals and metal oxides. Organometallic chemistry has introduced some important new applications of homogeneous catalysis, but has by no means changed the face of the petrochemical industry as yet.

Not all petrochemical processes are catalytic—the steam cracking of hydrocarbons to lower olefins is a thermal process at 700 to 800°C or more. However, excluding free-radical polymerization processes, this is a rare example, though severe conditions may still be required in some catalysed processes on thermodynamic grounds or to achieve acceptable rates (several mol h^{-1} per litre of reaction volume). As we shall see in this and the following chapter, the major impact of catalysis is to provide a remarkably wide range of products from a small number of building blocks.

10.2 Definitions and constraints

10.2.1 *Essential features*

A catalyst brings about or accelerates a specific reaction or reaction type. Catalysis of a specific reaction provides a route which proceeds in *parallel* with any existing thermal, or even other catalytic (possibly unrecognized), modes of reaction within the particular mixture. Hence, useful catalysis implies that the rate of the desired reaction considerably exceeds the rates of all other possible reactions.

In the original definition, a catalyst was referred to as being 'indifferent' to the reactions being catalysed. This is often interpreted to mean that the catalytic material must be recovered unchanged, but we have to be a little more circumspect in the light of our increased knowledge of catalytic mechanisms. Mineral acid catalysts used in esterification, etherification, etc., may be partially converted to mixtures of the original acid and esterified forms (particularly with sulphuric and phosphoric acids). Soluble salts and complexes of metals may be converted into a variety of forms, often in various oxidation states. Heterogeneous catalysts will become covered in chemisorbed and adsorbed reactants, products and intermediates, often again with changes in the oxidation states of catalytically effective metal centres. In continuous processes we should see a constant compositional state, reaction rate and yield pattern develop. In batch operations, compositional changes during a reaction may cause changes in the proportions of different catalytic species, but recovery and re-use of the catalytic material should permit indefinite reproduction of rates and yields.

Even with our modified definition of 'indifferent', we still require that the catalytic material should act indefinitely once introduced. This requirement is also fulfilled by a number of essential materials added to some catalytic processes, and often referred to as co-catalysts or promoters. For example, the copper (I)-copper (II) chloride redox system used in Wacker's palladium-catalysed oxidation of ethylene to acetaldehyde[1] (section 10.7.7.3) behaves in a true catalytic manner in the single-reactor variant of the process (ethylene and O_2 introduced into the same reaction vessel).

$$C_2H_4 + \tfrac{1}{2}O_2 \xrightarrow{PdCl_2/CuCl_2} CH_3CHO$$

However, when the copper (II) chloride is used as a stoichiometric oxidant in one reaction vessel, and is then re-oxidized in a second (another commercialized variant of the process), it should possibly be regarded as a regenerable or catalytic reagent.

$$C_2H_4 + 2CuCl_2 + H_2O \xrightarrow{PdCl_2} CH_3CHO + 2CuCl + 2HCl$$
$$2CuCl + 2HCl + \tfrac{1}{2}O_2 \rightarrow 2CuCl_2 + H_2O$$

Nonetheless, it may still be treated as an integral part of the 'catalyst package' in the overall process. In the Monsanto acetic acid process[2], methyl iodide (iodomethane) is so volatile that some escapes continuously in the gas purge stream, but is efficiently recovered and returned, and is again part of the 'catalyst package' (section 10.7.8.3).

$$CH_3OH + CO \xrightarrow[CH_3I]{Rh\ complex} CH_3CO_2H$$

10.2.2 Initiators

The above comments also provide a basis for distinguishing between catalysis and 'initiation' of a reaction. Initiation effectively introduces a package of energy into a small number of reactant molecules. Thereafter, if the overall process is thermodynamically favourable, the energetic molecules, most frequently free radicals, propagate the reaction by a chain mechanism, and the chemical nature of the initiating species is no longer relevant; moreover, the number of energetic species can change with time. (An initiator may also catalyse reactions in subsequent stages.) The distinction may be of more than academic interest if an initiation/chain mechanism implies that the rate-increasing additive will be consumed at an uneconomic rate.

10.2.3 Co-reactants

In the oxidation of p-xylene (in acetic acid solution) to terephthalic acid, catalysed by cobalt and manganese cations, bromide ions are generally introduced as co-catalysts to accelerate the oxidation of the second methyl group (section 10.7.7.1). Now halide ions in carboxylic acid solution are extremely corrosive even to high-grade stainless steels. However bromides can be omitted if materials such as acetaldehyde or 2-butanone, which are oxidized rapidly to acetic acid via peroxide intermediates, are added continuously to the reaction mixture in relatively large quantities (0·25–1 mole per mole of p-xylene). The peroxides generate high concentrations of Co(III) and Mn(III) ions to initiate attack on the methylaromatic substrate. Such added substances are referred to as co-oxidants, or more generally co-reactants. They are always consumed in the process.

10.2.4 Inhibition

No simple, single-step chemical reaction can be inhibited, in the sense that the rate constant cannot be reduced. The occurrence of inhibition implies that

(a) a reactant or important intermediate has been removed by a competing reaction with, or catalysed by, the inhibitor (occasionally by complexation in a way which does not affect the analytically determined quantity of that reactant), or

(b) a catalytic species required for the reaction, whether previously recognized or not, has been destroyed or removed from the sphere of activity.

10.3 Thermodynamic relationships

10.3.1 Application

A catalyst in no way alters the *relationships* between the thermodynamic properties of reactants and products. Hence, a catalyst can only effect, or accelerate, a reaction which is thermodynamically feasible under practically attainable conditions. Furthermore, a catalyst accelerates both forward and back reactions in a reversible process, and thermodynamic calculations are essential for the selection of preferred reactions conditions.

If we consider any reversible reaction of the form

$$aA + bB + \ldots \rightleftharpoons 1\,L + mM + \ldots \tag{10.1}$$

we can relate the thermodynamic properties of the reactants and products to the equilibrium constant expressed in terms of *activities* of the elements or compounds, A, B, etc., in the reaction system.

Activities of materials may be defined with respect to one of three 'standard states', the pure element or compound in its natural form, the ideal aqueous solution or the ideal gas state all at 1 atmosphere (1 atm = 1.01325 bar; 1 bar = 10^5 Pa). They are often quoted at a temperature of 25°C (298.15 K). Aqueous solutions are of little relevance to petrochemical operations.

The major problem in the use of the pure element or compound as a reference state lies in quantifying the activity of a component in a non-ideal liquid mixture. As activities are usually determined by measuring the partial pressures of each component in the vapour space, it is quite logical to use a gas state as reference for all organic compounds. The data are also directly applicable to the many industrial gas-phase chemical processes. Several texts present tables of ideal gas-phase data for a range of organic compounds, and provide guidance in their use. Furthermore for new compounds, accurate entropies and specific heat capacities can now be calculated from structural and spectroscopic data; additionally, group contribution methods for predicting entropies and enthalpies of formation in the gas phase exist.

In the gas phase, the term 'fugacity' (f_A, etc.) usually replaces the term activity; but at modest total pressures (up to several atmospheres) this is approximately equal to the partial pressure (p_A etc.) in *atmospheres*. Hence

$$K_p^\ominus = \frac{(p_L/p^\ominus)^l(p_M/p^\ominus)^m}{(p_A/p^\ominus)^a(p_B/p^\ominus)^b} \tag{10.2A}$$

where $p^\ominus$ is a standard pressure.

Taking $p^\ominus = 1$ atmosphere, this reduces to

$$K_p^\ominus = \frac{p_L^l p_M^m}{p_A^a p_B^b} \tag{10.2}$$

If $\Delta G^\ominus$ is the Gibbs free energy change from reactants to products (individually at 1 atm. and at a temperature T in kelvin) we have

$$\ln K_p^\ominus = -\Delta G^\ominus/RT. \tag{10.3}$$

(However, even at low partial pressures, carboxylic acids may be extensively dimerized in the vapour phase, and corrections are necessary; at higher pressures the fugacities of components must be obtained from published data or by applying a 'fugacity coefficient', $\phi = f/p$, derived from a generalized or compound-specific chart.)

10.3.2 *Effect of total pressure*

In equation 10.2, each partial pressure can be represented by the product of the mole fraction (y_A, etc.) and the total pressure (P in atm.)

hence
$$K_p = \frac{(y_L^l y_M^m \ldots)P^{(l+m\ldots)}}{(y_A^a y_B^b \ldots)P^{(a+b\ldots)}}$$

If $r = a + b + \cdots$ the number of moles of reactants
and $q = 1 + m + \cdots$ the number of moles of products

$$K_p P^{r-q} = \frac{y_L^l y_M^m \cdots}{y_A^a y_B^b \cdots}$$

Hence, the mole fractions of products will increase with increasing pressure if $r > q$ (or decrease if $r < q$) in accordance with Le Chatelier's principle; this relationship proves very important in seeking high conversions for processes such as ammonia and methanol syntheses, ethylene hydration, etc.

10.3.3 *Rough calculations*

The free energy of reaction, ΔG°, is derived from the change in enthalpy, ΔH°, and the change in entropy, ΔS°, by the equation:

$$\Delta G^\circ = \Delta H^\circ - T \Delta S^\circ$$

and $\ln K_p = \Delta S^\circ / R - \Delta H^\circ / RT$ (all dimensionless terms)

For *rough* purposes, we can take ΔS° and ΔH° as approximately constant for gas phase reactions, and we can see immediately how the equilibrium constant will vary with temperature according to the sign and magnitude of ΔH°. Furthermore, for many organic reactions of interest, values of $\Delta S^\circ / R$ are of the order 15 $(q - r)$, less 10 for C—C bond formation with cyclization. We now have a very crude method for assessing feasibility based solely on (gas-phase) heats of formation.

Thus, the value of $\Delta H^\circ / R$ for the hydrogenation of an olefin (alkene),

$$RCH = CH_2 + H_2 \rightleftharpoons RCH_2CH_3$$

is typically of the order $-15\,000\,K$. (ΔH° is approximately -120 to $-130\,kJ/mol$.)

Therefore $\qquad \ln K_p \sim -15 + 15\,000/T$

At low temperatures, $\ln K_p$ will be positive, but will fall to zero at about 1 000 K, above which the reverse reaction becomes possible. In contrast, for the hydration of an olefin $(\Delta H^\ominus / R \sim -5\,500\,K)$ $\ln K_p$ becomes negative $(K_p < 1)$ at about 370 K, and the reverse reaction will be significant under all practical conditions.

10.3.4 *Thermodynamic traps*

Although a catalyst cannot affect thermodynamic relationships, it can affect the thermodynamic activities of reactants or products, particularly when present in large amounts.

A particularly misleading situation can occur when a reaction product forms a complex with the catalytic material. Thus the Gatterman–Koch reaction of benzene and carbon monoxide to produce benzaldehyde is thermodynamically unfavourable, but reaction proceeds under pressure in the presence of at least 1 mole $AlCl_3$/mole benzene. However, the benzaldehyde–$AlCl_3$ complex formed must be cleaved *chemically*, with water for example, to liberate the product. Surprisingly, the similar conversion of toluene to *p*-tolualdehyde (*p*-methylbenzaldehyde) is favourable, and separation from the catalyst system can be achieved by simple distillation; workers with Mitsubishi in Japan have used BF_3/HF[3].

The above examples are from an area of traditional chemistry. However, in seeking to translate organometallic chemistry into catalytic processes, we may often find products or intermediates in the desired process scheme as ligands on the metal. In such circumstances, it is important to recognize that their formation confirms that a mechanistic pathway exists, but does not imply that the formation of the free product or intermediate is necessarily feasible under the conditions used.

10.4 Homogeneous catalysis

10.4.1 *General features*

In homogeneous catalysis, the catalyst is usually dissolved in a liquid reaction mixture, though some or all of the reactants may be introduced as gases, or even as solids. A small number of examples exist in which the reactants and catalyst are all in the vapour or gaseous state; one example is the 'cracking' of acetic (ethanoic) acid to ketene (ethenone) and water at about 700°C, with diethyl phosphate vapour as an acid catalyst.

In solution chemistry, the nature of the catalytic material introduced is usually well defined, and we expect reproducible kinetic behaviour. Hence, considerable scope exists for accurate kinetic investigation to throw light on mechanistic aspects. Further information can be derived by examining the kinetic behaviour of isotopically labelled or otherwise substituted compounds which are closely related to the original reactants and catalyst, or may be intermediates in the reaction[4].

As was indicated earlier the main catalytic species in solution may not be identical with that introduced; it is quite common for two or more species derived from the catalyst to be present simultaneously. Changes in the catalytic species can often be followed by conventional analytical or spectroscopic methods, NMR, ESR, etc. Interest in the nature of catalytic species under actual working conditions has prompted the development of high-temperature or high-pressure cells for such techniques. It may also be possible to isolate and characterize the catalytic species, or intermediates in a catalytic cycle. Care is required in interpretation, as the true catalyst may be a minor component, or various reactions such as ligand exchange may occur during the isolation procedure. Nevertheless, such characterized species, and their stoichiometric reactions, can sometimes bring the worker very close to an understanding of the catalytic mechanism.

10.4.2 Catalyst life and poisons

In seeking to apply any new catalytic material the useful life of the catalyst is important. Loss of activity may be inherent i.e. an inevitable consequence of the chemical nature of the catalytic material, reactants and products, or a result of 'poisoning'. Some examples are shown in Fig. 10.1.

Within the first category, loss as vapour, thermal decomposition or internal rearrangement of the catalytic material may occur as reaction temperatures are increased to obtain commercially viable rates of reaction. For example, arenesulphonic acids decompose and hydrolyse slowly. Metal-centred complexes may lose ligands required to maintain activity, selectivity or solubility, or metal-ligand reactions may occur (e.g. orthometallation of the phenyl group in a triphenyl phosphine ligand). Additionally, side reactions can occur between the catalytic material and reactants or products. Sulphuric acid and, to a lesser extent, sulphonic acids slowly oxidize many organic substances, such as alcohols, with a consequent loss of catalytic activity (and the introduction of organic impurities into the products). Many such reactions are identifiable by extending studies to extreme reaction conditions. Rather more difficult to identify are interactions between catalytic materials and those reaction by-products which accumulate only after extended operation with re-cycle streams.

By catalyst 'poisons' we normally mean materials which do not form part of the defined process chemistry, but which gain entry into the reaction mixture and lead to permanent or temporary catalyst deactivation. In general, poisons are impurities in the chemical feedstocks or corrosion products from materials of construction. Many chemical feedstocks derived from petroleum contain traces of sulphur compounds, whilst chloride ion is almost ubiquitous (in the air at seaboard factory sites). These impurities, together with traces of organic impurities such as dienes and alkynes in alkenes, can again displace preferred ligands, destroy ligands (e.g. oxidation by traces of air) or compete with the reactant. Metal ions, in feedstocks or introduced by corrosion, may lead to progressive neutralization of an acid catalyst, or otherwise interfere with the catalyst performance; metal ions may also catalyse undesirable side reactions.

10.4.3 Limitations

In practice, there are a number of serious limitations to the application of homogeneous catalysis.

Firstly, no suitable catalyst may have been found in the past. For example, organometallic chemistry has introduced soluble and stable zero- or low-valency metal complexes, and derived hydrido-complexes, which catalyse hydrogenations in the liquid-phase. But with the enormous amount of accumulated general and company-specific information and experience in the

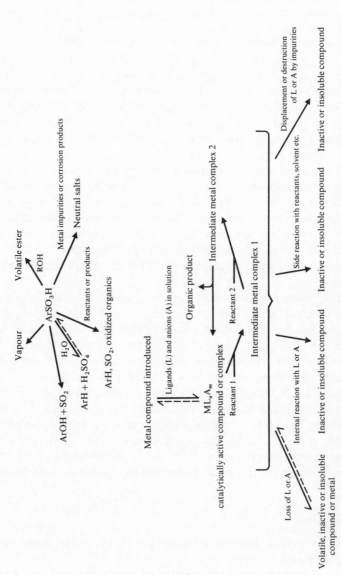

Figure 10.1 Some modes of deactivation of homogeneous catalysts during reaction or product separation.

heterogeneous catalysis of hydrogenation processes, there is little incentive for industrial organizations to evaluate novel (and expensive) homogeneous catalysts, except in a very small number of specific applications. We still have no homogeneous catalysts for many of the industrially important lower olefin (alkene) oxidation processes (e.g. acrylonitrile (propenonitrile) from propylene (propene), butadiene from butenes).

In other cases, homogeneous catalysis or mechanisms have been demonstrated, but rate, or more often thermodynamic, requirements can only be met at temperatures too high for liquid-phase operation. Gas-phase reaction over a heterogeneous catalyst permits a far wider range of reaction conditions.

However, the major problem encountered in homogeneous catalysis, particularly when the catalytic material is expensive, lies in the separation of reaction products from the catalyst such that the recovery process is efficient and does not impair catalytic activity. Thus we can afford to lose small, catalytic quantities of mineral acid, alkalis or base metal compounds. But only minute losses can be tolerated if a noble metal catalyst costs over £12 000 (say $20 000) per kg, and the product sells for about 40–70 pence (say 70–120 cents) per kg.

10.5 Heterogenization of homogeneous catalytic systems

A number of methods have been used to convert catalytic species into a heterogeneous form, and thereby simplify separation from the reaction mixture, but without changing drastically the nature of the catalytic species, or its immediate environment. To ensure that the original mechanistic pathway is unaffected, this category must exclude the substitution of a supported metal for a soluble metal compound or complex. The original modes of decay and interference by poisons are then similarly unaffected.

Two distinct approaches have been adopted: (a) the chemical attachment or tethering of the original catalytic species to a support material (applied mainly to reactions in the liquid phase), and (b) the physical absorption of the original soluble catalyst/solvent system into an inert porous material (applicable solely to gas-phase reactants and products).

Heterogenization of a catalytic system introduces mass transfer (diffusional) constraints, which may affect both activity and selectivity in comparison with the original homogeneous catalytic systems. Some comments on such effects will be made in this section, but further discussion is deferred to the section on heterogeneous catalysis (section 10.6).

Possibly the simplest example of the first approach would be the substitution of a sulphonated polystyrene-divinylbenzene polymer ion-exchange resin (in free acid form) for p-toluene sulphonic acid as the catalyst in an esterification or other acid-catalysed chemical reaction (see section 10.7.2). The ion-exchange resin may be used in bead form and packed into a tubular reactor, or may be introduced as a powder and subsequently filtered off. Furthermore, the solid resin may prove less corrosive to the metal walls of an industrial reaction vessel.

The physical nature of such resins, whereby the majority of sulphonic acid groups lie within the resin beads or particles, and are accessible only via narrow pores, introduces considerable resistance to the flow of reactants to, and products from, these catalytic 'sites'. Hence, in general, reaction rates are reduced compared with those attainable with an equivalent quantity of a soluble sulphonic acid, though 'macroreticular' resins with improved access for relatively large organic molecules are now available. As an ion-exchange material, the sulphonic acid resin efficiently scavenges the reaction mixture of traces of corrosion metal ions, thereby leading to slow neutralization of the sulphonic acid groups. Further, at high temperatures we encounter the typical modes of thermal and hydrolytic decomposition of aromatic sulphonic acids, together with the possibility of degradation of the polymeric structure itself. Nonetheless, several processes utilizing sulphonic acid ion-exchange resins as catalysts have now been commercialized in the petrochemical field.

A further example of the first approach to the heterogenization of organometallic complexes entails the chemical attachment of a ligand to an organic or inorganic support[5]. A considerable number of variations on this theme have been described. One example, developed by BP workers, utilizes the reaction of a silyl alkyl phosphine (amine, thiol) with the hydroxylic surface of a porous silica (alumina, etc.). Rhodium, for example, can then be attached by liganding to the phosphine groups. The resulting materials are capable of effecting hydrogenation and hydroformylation of alkenes. Again, diffusional limitations can affect rates and selectivities. At higher temperatures, slow loss of the catalytic metal by dissociation or displacement of ligands, or cleavage of the ligand–support linkage, may occur, and these materials have not found commercial use as yet.

To illustrate the second approach to heterogenization, we will consider the hydration of ethylene. Ethylene can be converted to ethanol with sulphuric acid in a two-stage process.

$$C_2H_4 + H_2SO_4 \rightarrow C_2H_5HSO_4$$

$$C_2H_5HSO_4 + H_2O \rightarrow C_2H_5OH + H_2SO_4$$

For a catalytic version, dilute sulphuric acid (5–10% w/w) must be used to limit ether production, oxidation and sulphate esters in the products, but corrosion of reactor materials is extremely high at the temperatures required (about 200°C).

In contrast phosphoric acid is less corrosive and non-oxidizing, while only very small amounts of ethyl esters are formed. But the lower acidity of phosphoric acid necessitates temperatures in excess of 250°C, and pressures of 60–70 atm., to achieve sensible conversions and rates. If a controlled quantity of phosphoric acid is absorbed into a siliceous material of appropriate pore-structure, the acid spreads over the internal surface of the support material and leaves an appreciable part of the pore volume unoccupied, permitting access to most of the acid solution by gaseous reactants. The material can be packed into a simple tubular reaction vessel. Obviously, the use of such a system is limited to operations in which reactants are introduced, and products removed, in a gaseous form; if liquid were to condense within the reaction vessel, a part of the phosphoric acid solution would be washed off the support material.

This type of catalytic material and mode of use provided the basis for the Shell process[6], the first direct hydration route to ethanol. The approach remains the basis for the manufacture of all synthetic ethanol and most of the isopropanol (2-propanol) produced today.

A similar approach has been adopted in a number of other reaction systems. A common example is the use of supported melts of copper chloride (with possibly other metal chlorides incorporated to reduce volatility or provide a lower melting-point eutectic mixture) for 'oxychlorination' processes, in which hydrogen chloride introduced into the reaction mixture, or formed in it, is reoxidized to chlorine *in situ* (section 10.7.7.6).

The maintenance of a supported liquid layer in gas-phase reactions is also important in other heterogeneous catalytic applications, such as the Bayer/Hoechst process for vinyl acetate manufacture[7]. However, in these systems, the catalytic metal is reduced to the metallic state, leading to significant mechanistic differences from the formally related homogeneous Wacker-type alkene oxidation/acetoxylation processes (section 10.7.7.3).

10.6 Heterogeneous catalysis

10.6.1 *Introduction*

A heterogeneous-catalyst is a solid composition which can effect or accelerate reaction by contact between its surface and either a liquid-phase reaction mixture (in which the catalytic material must be essentially insoluble) or gaseous reactants. In liquid-phase systems, one or more of the reactants may be introduced as a gas, but access of such reactants to the (fully wetted) surface of the catalyst is almost invariably by dissolution in the reaction medium and subsequent diffusion.

The existence of heterogeneous catalysis has been recognized by chemists for over 150 years, and practical applications increased dramatically from the beginning of the present century. While mechanistic aspects were poorly understood, certain basic features were established and industrialists built up a mass of empirical facts, from which a number of 'rules of thumb' could be deduced.

In the last 20 years or so many new techniques have been applied to characterize the materials themselves and to throw more light on the mechanistic aspects of heterogeneous catalysis. Nevertheless, no single theory has swept the board and found universal acceptance.

We will start, therefore, with some of the more pragmatic aspects of heterogeneous catalysis, for which little theoretical background knowledge is required.

10.6.2 *Major (primary) and minor (secondary) components*

Many heterogeneous catalysts in commercial use contain several components, often referred to as the major (or primary) component, minor (or secondary) components and the support, with which we shall deal later.

While all components contribute to the overall performance of the catalytic material, the major component is essential for any activity in the type of reaction required. If several components show individual activity, the major component is often taken to be the most catalytically active material. (Two major components may be essential in bifunctional catalysts, section 10.7.4.)

With metal catalysts, classification is usually reasonably straightforward. However, with oxide catalysts none of the components may be individually active. In most cases of this type, a true mixed oxide, containing two metals (or metalloids) within the crystal structure, is active when used alone, and is regarded as the major component. Such a component may be named by joining the names of the two constituent oxides, e.g. silica-alumina or tin-antimony oxide, but often names of the type bismuth molybdate, cobalt tungstate, copper chromite are used. Such names should not be taken literally, as the two metals may not be present in a simple stoichiometric ratio or the composition may contain microcrystals of different structures each containing differing ratios of the two metals.

Minor (or secondary) components are introduced to modify the crystal structure or electronic properties of the major component, to improve activity, selectivity or thermal stability. The situation is highly complex, but some examples of secondary components, and their effects, will be included in subsequent sections.

10.6.3 *Operational modes*

In liquid-phase reactions the heterogeneous catalyst may be introduced in powder form into the reaction mixture, and maintained in suspension by stirring or other means. The powdered catalyst must be recovered subsequently by settling, filtration, centrifuging etc. The efficiency of this recovery process usually dictates the minimum particle size usable.

Alternatively, the catalytic material may be used in a coarser form (0·5 to 8 mm or more in diameter), and packed as a *fixed bed* into a tower or tubular reaction vessel. Practical considerations would be concerned with the means of retaining the catalytic material, while avoiding high pressure drops which might cause physical damage (crushing) to the catalyst beads, granules, pellets etc. It is sometimes possible to pass the liquid upwards through the catalytic bed so that the particles are slightly lifted and separated, but are not carried away in the supernatant liquid; such *expanded-bed* operation requires careful choice of particle size and liquid velocity for successful operation. When one of the reactants is gaseous, a *trickle-bed* configuration may be adopted, in which the liquid reaction medium trickles down over the surface of the catalytic material but the gas occupies most of the space between the granules (up or down flow).

When all the reactants and products are gaseous under reaction conditions, the fixed bed configuration is frequently adopted. However, if the chemical reaction is either strongly exothermic or strongly endothermic, the diameter of a tubular fixed bed may have to be limited to 25–50 mm (1–2 inches), to permit heat transfer through the wall and avoid excessive changes in temperature along the bed. Thus, a 'reactor' for the gas phase oxidation of o-xylene (1, 2-dimethylbenzene) to phthalic (benzene- 1, 2-dicarboxylic) anhydride will actually comprise 10 000 to 15 000 parallel tubes of about 30 mm diameter, each containing coarse catalytic material, all surrounded by a circulating molten salt mixture to remove the heat (enthalpy) of reaction[8]. To simplify engineering design, a number of other modes of operation have been adopted—see Fig. 10.2.

The most common form of *fluidized bed*[9] is similar to the expanded bed referred to above. Gaseous reactants pass upwards through a bed of small catalyst particles at a velocity which induces turbulence in the bed, to given good heat transfer to the walls and immersed cooling tubes, but does not carry away excessive quantities of the catalyst. (Filters may be fitted.) A further variant, used for the catalytic cracking of petroleum fractions with catalyst regeneration by oxidation, employs a transported bed. The catalyst is fluidized, but transported through the main reactor and separated from the emerging product stream. The catalytic material is again transported (in air) through a second reactor (regenerator) if required, before its return to the main reactor. (Compare this operation with the pneumatic transport of grain, powdered coal etc.)

When we move to very high-pressure exothermic processes (such as methanol and ammonia syntheses), different engineering approaches have resulted in a variety of other reactor types[10].

10.6.4 *Chemisorption and active sites*

Heterogeneous catalysis occurs on the surfaces of solid materials. These surfaces are almost invariably covered by various species—water, carbon dioxide, and other materials (including ions) from the preparative mixture or atmosphere. The majority of these species are very weakly adsorbed, or physisorbed, via van de Waals' forces. However, for a chemical reaction

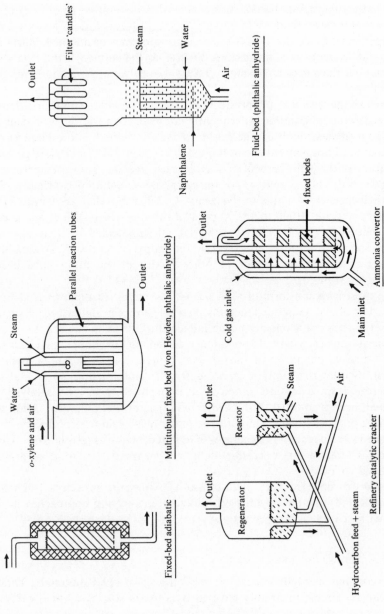

Figure 10.2 Various reactor types (schematic) for heterogeneous catalysts and gaseous reactants.

involving an otherwise stable molecule to be initiated, a significant electronic disturbance in that molecule, or even bond scission, must occur at the catalytic surface.

The adsorption process that brings about such a chemical modification, which may occur in several steps, is usually referred to as *chemisorption*. Furthermore, when two or more molecules are involved in a reaction on a catalytic surface, as in hydrogenation or oxidation processes, we usually find evidence that the major reactions occur between chemisorbed or surface species derived from each of the reactant molecules.

It was recognized early in the history of hetrogeneous catalysis that, in many instances, only a relatively small proportion of the surface was catalytically active. The pre-exponential rate factor was seen to be very small in relation to likely collision frequencies of molecules adsorbed on the surface, taking into account steric requirements, while poisoning (inhibition of the catalysed reaction) could result from surprisingly low levels of specific impurities (see below). Hence the term *active sites* was coined to describe those localities on the surface which would induce the desired chemical reaction.

10.6.5 *Physical forms and their preparation*

Solid catalytic materials can be divided into two main groups:

(a) Bulk catalytic materials, in which the gross composition does not change significantly throughout the material, such as a silver wire mesh or a compressed pellet of 'bismuth molybdate' powder. As we shall see later, the composition in the top few atomic layers may be different from that in the bulk.

(b) Supported catalysts, in which the active catalytic material is dispersed over the surface of a porous solid, such as carbon or silica.

Bulk metals can be used in traditional engineering forms, more particularly as fine wire woven into gauzes. Such forms are generally used only in high-temperature processes, such as the partially oxidative dehydrogenation of methanol (over Ag) or ammonia oxidation (over Pt–Rh) at about 500–600°C and 850–900°C respectively. Mechanical stability is of greater importance than high surface area.

However, for high activity at modest temperatures, forms which present a much higher surface area to the reactants are highly desirable. Finely divided metal powders often show very high catalytic activity, but may present separation problems, whilst agglomeration often leads to a progressive loss in activity. Hence, we find the development of methods to produce coarser particles of metals in porous form, such as platinum sponge or, far more commonly, Raney nickel. (The Raney technique has been applied to other metals such as cobalt and copper.) However, the scope for producing physical forms with a high proportion of accessible metal atoms is seriously limited.

In the case of oxides of metals and metalloids, even small particles (0·1 mm diameter or less), whether naturally occurring or produced by precipitation, gelling, crystallization or powder reactions, are usually polycrystalline or agglomerates, with appreciable porosity (after driving off water and other weakly adsorbed materials). Quite often, by allowing precipitates or microcrystals to stand for appreciable periods, further agglomeration will provide relatively large, porous granules, several millimetres across. Alternatively, larger *formed* catalysts in the shape of cylinders, spheres, rings, etc. can be produced from powder, optionally incorporating a small amount of a 'cement' or 'binder', by compression into moulds or extrusion of a slurry and drying. In all cases, preparative conditions affect surface properties. Most oxide catalytic materials provide (specific) surface areas from about 0·5 up to 700 m² per gram, more usually 10 to 300 m^2/g. The zeolites (molecular sieves), of course, also have very small channels of 0·4–1 nm (4–10 Å) diameter extending throughout the crystal structure[11].

There are two major reasons for the use of supports, to provide a stable extended surface over which an active component can be dispersed, and to confer mechanical strength. In either case, the desired property may well have to be maintained to a high temperture.

Synthetic silicas, aluminas and carbon blacks—there are many variations in form, surface area, impurity levels and crystal structure (for aluminas)—are probably the most common supports for metals. Their use with noble metals can provide a high *dispersion* (the proportion of metal atoms on to which a reactant can be adsorbed), and thereby provide a very considerable economic benefit. However, these materials, together with calcined clays and diatomaceous earths, are also commonly used as supports for less expensive metals, such as nickel and copper, for use at higher temperatures where sintering of base metals can occur.

Preparation most frequently involves absorption of a solution of a suitable metal compound into the support, followed by drying, thermal decomposition and reduction. This treatment, and the reductant employed (such as hydrazine in solution, or hydrogen gas), often has a significant effect on the physical form and dispersion of the resulting metal. There are few rules, and this area is far too complex for discussion here.

When we turn to metal-oxide catalysts, a few of low porosity, such as zinc oxide, benefit from supporting the finely divided material. However, a great many metal-oxide systems of acceptable porosity prove too soft for commercial handling and use; often only small additions (15–30%) of a hard 'support' will confer adequate strength. Preparative techniques include absorption of salt solutions into the support (followed by calcination to oxides), precipitation of hydroxides in the presence of support material, mixed powder compression, slurry extrusion and coating or spraying on to low-porosity support materials (quartz, glazed ceramic). The preparative method and conditions again significantly affect performances.

10.6.6 Support interactions

While the practical reasons referred to above represented the main driving force for the use of supports, this is by no means the whole story. As all the surface is rarely active, the support material can act as a heat sink or source for active sites or zones in exothermic and endothermic reactions, while increasing the total area for heat transfer between solid material and the gas (or liquid) phase.

The support material may have catalytic properties in its own right, either useful or otherwise; hence exposed areas of the support can influence the selectivity of the overall process, and special pretreatments may be required to enhance or inhibit such behaviour, as appropriate. More importantly, however, the support may induce crystallographic modification of a thin layer of overlying metal or oxide, or electron transfers may occur between the support and the supported catalytic material, leading to significant changes in the structure and electronic (acidic/basic or electrophilic/nucleophilic) properties of the latter material, and hence its catalytic behaviour. When the primary component is a metal, the term strong metal-support interaction, often abbreviated to SMSI, has recently appeared in the literature.

Such chemical interactions have prompted extensive investigations of the influence of support materials on overall catalytic behaviour in many systems, and have extended considerably the range of mainly oxide materials used as supports in commercial catalysts. In many systems the so-called support material is thus an important part of the complete catalyst formulation, fulfilling a dual role, though usually retaining crystallographic identity except at boundaries with other components.

10.6.7 Catalyst structure

The surface area, pore structure and chemical composition of the surface are important parameters of any support material or solid catalyst. Even when mechanistic interpretation is not a primary aim, many techniques are now applied in industrial laboratories to establish correlations between these parameters and the performance of specific catalysts. Such techniques may also be used on a routine basis to monitor the reproducibility of purchased materials and catalyst preparation methods.

The pore volume of a solid material can be estimated from the actual density and the true density of the constituents. More frequently, the absorption of a suitable liquid into a (pre-evacuated) sample is used as a guide.

Surface areas are usually obtained by the BET Method[12]. The adsorption isotherm (at $-196°C$) for nitrogen onto the thoroughly de-gassed surface is measured; a 'knee' at monolayer coverage provides a value for the specific surface area, while the complete curve gives information on the pore size distribution, particularly for micropores of less than 50 nm diameter. Low-temperature gas-solid chromatographic (GSC) techniques have also been applied[13]. Information on the size distribution of larger pores can be obtained by forcing mercury into the pores, a technique known as *mercury porosimetry*. The *tortuosity* of larger pores through formed or shaped catalysts (particularly cylinders) can be estimated from the pressure differential required to establish a gas flow through a suitably mounted sample.

With a supported catalyst system, the information derived by the above methods relates to the composition as a whole. A number of methods have been applied to measure the sizes of metal crystallites on the support surface, and hence estimate the dispersion or surface area of the metal itself. These include electron microscopy, X-ray diffraction and Mössbauer spectroscopy. More direct measurements of metal surface area have been obtained from adsorption isotherms for hydrogen, carbon monoxide and other adsorbates, by assuming that all the exposed metal surface adsorbs such gases. GSC methods can again be applied.

A host of spectroscopic techniques have now been applied to obtain compositional, structural and valency information for the topmost layers of catalyst compositions. These include traditional methods, such as UV/visible reflectance, IR and Raman spectroscopy. X-ray fluorescence spectroscopy (XRF) has been applied to catalytic samples or surface rubbings. A potentially more informative group of modern techniques includes low-energy electron diffraction (LEED), X-ray photoelectron spectroscopy (also known as electron spectroscopy for chemical analysis, ESCA) and Auger electron spectroscopy (AES)[14]. LEED provides lattice information, and led to the discovery that the surfaces of some metals and most semiconductor oxides have structures which differ from those projected by simple extension of bulk geometry. ESCA and AES provide moderately quantitative compositional information, and indications of valencies or bonding states. The former has shown that some components may be present in excess, or be deficient, at the catalyst surface. However, *no* single technique can be regarded as definitive.

10.6.8 *General kinetic behaviour*

Rate data have often been expressed in terms of simple powers of reactant concentrations (liquid phase) or partial pressures (gas phase), such as

$$-\frac{dB}{dt} = k(B)^{\beta}(C)^{\gamma}\dots.$$

Such equations are frequently acceptable for design purposes and, if β, γ etc. are integers or simple fractions and remain constant, for interpretative work also. However, orders of reaction rarely remain constant with variations in temperature. More widely applicable rate equations were derived by Hinshelwood and others to take into account competition between reactants, and possibly intermediates and products, for active sites[4].

The form of the best equation can, of course, provide some mechanistic guidance; when the process is multistep, with desorption/readsorption of intermediates, more mechanistic information may be derivable, but the kinetic description can become extremely complex.

Normally, the dependence of the rate constant on temperature takes the traditional Arrhenius form, $k = A \exp(-E/RT)$. However, some systems, particularly oxide catalysts in gas-phase processes, can show a change in activation energy with temperature; this may reflect a multistage process or changes in the composition (or oxidation state) of the catalytic surface, which may also be time-dependent. If the chemical process is strongly exothermic, such as the oxidation of o-xylene (1, 2-dimethylbenzene) over vanadia catalysts, heat-transfer limitations lead to appreciable variations in the catalyst and gas temperatures along the catalytic bed. Considerable effort has been devoted to the mathematical modelling of interactions between kinetic and heat-transfer parameters for such systems, aimed at improved control and reactor designs.

A common feature of heterogeneous catalysis is an increase in rate with increasing subdivision of the catalytic material. This arises from increasing accessibility of the surface, and reduction in diffusional constraints between reactants and catalytic sites. In general, continued subdivision will eventually lead to a levelling off in the reaction rate at a value dictated solely by adsorption and chemical processes. For simple, selective reactions (olefin hydrogenation, oxidation of sulphur dioxide), the ratio of the reaction rate with a practical catalyst form to the maximum rate attainable by mechanical subdivision is often referred to as the *effectiveness factor*.

When parallel reactions or further reaction of the desired product occur, as in most oxidation processes, larger catalyst particles can also affect selectivity. Differences in diffusion rates of individual reactants lead to changes in their ratios within deep pores, which also favour further reaction of a product in the pore network. Therefore, some catalysts have been developed with a 'shell' structure, in which only the outermost zone of a porous support is loaded with active material or an impervious central core is coated with the catalytic components. All such effects are, of course, very important in scale-up for commercial processes, for which relatively large granules may be desirable.

Not all pore diffusional effects are necessarily undesirable. The molecular sieving action of zeolites, as catalysts or supports, can lead to preferential reaction of small or linear molecules (such as n-paraffins) in complex mixtures, or modify the product distribution in other ways (cage effects). In the latter context, the zeolite ZSM-5 shows exceptionally high activity for the cyclization and aromatization of hydrocarbons[29].

10.6.9 Catalyst deactivation and life

In general the activity of any catalyst falls off with time. Ultimately the conversion of reactants or production rate reaches an unacceptable level, or a compensatory temperature increase degrades selectivity. Long life is especially important for expensive catalytic materials; deactivated precious metals may be recoverable, but the credit will only partially offset the cost of new material. With fixed beds, even of cheap catalysts, considerable labour costs are incurred in discharging and recharging the reaction vessel, while loss of production during these operations will mean a loss of income (and possibly customers).

Loss of catalytic activity can occur in several ways, but firstly we will consider simple physical loss. Particulate catalysts used in agitated liquid phases or fluidized beds are liable to both wear and fracture through collisions of the particles with each other, the vessel walls and fittings. The process is usually referred to as *attrition*. The finest particles formed tend to escape the main separation or filtration equipment, and continual make-up of the catalyst charge is required.

Losses can also arise from dissolution or volatilization of catalytic components. Phosphoric acid is slightly volatile in the gas phase hydration of ethylene, leading to its depletion in the catalytic material at the reactor inlet and deposition in downstream equipment. Similar problems occur with alkali metal acetates and, in the presence of steam, alkali metal hydroxides. Some compensation may be possible by introducing a small quantity of the appropriate material with the reactants. Nickel and palladium catalysts are subject to slow dissolution in unsaturated fatty acids if the hydrogen concentration falls too low locally, for example within deep pores. Surprisingly, even the precious metals in platinum-rhodium gauzes used for ammonia oxidation slowly volatilize, owing to the formation of uncharacterized gaseous oxides at high temperatures.

A far more common mechanism for catalyst deactivation in high-temperature gas phase reactions is *sintering* of the support or active material, leading to a reduction in the effective surface area. Many metals and oxides begin to sinter at a temperature equal to about half the melting point in kelvin (the *Tamman* temperature), though surface diffusion of atoms can occur at still lower temperatures. Thus silica should be stable at temperatures up to about 500°C, but structural changes still occur at lower temperatures under high partial pressures of water vapour. The presence of certain components in the gas phase (often oxygen or water) can again accelerate sintering or growth of metal crystallites. Fortunately, alloys of the principal metal with a second, often catalytically inactive, metal may provide more thermally stable dispersions; examples include supported Pt–Ir, Pt–Re, Pd–Au and Ni–Cu. Small changes in catalytic behaviour may also result.

As in the case of homogeneous catalysis, poisons can also lead to deactivation of heterogeneous catalysts. Soluble or volatile metal or nitrogen compounds can destroy acid sites, while carbon monoxide and sulphur compounds almost invariably poison nickel and noble metal hydrogenation catalysts by bonding strongly with surface metal atoms. These considerations often lead to the selection of less active, but more poison-resistant, catalysts for industrial use.

Slow catalyst deactivation may also result if certain impurities, present in commercial grades of chemicals used for catalyst preparation, migrate to the catalyst surface in use.

Finally, the formation of deposits on the surface and pore blockage may introduce a physical barrier between reactants and catalytic sites. In acid-catalysed reactions of olefins at modest temperatures, insoluble or involatile polymers, derived from the reactant olefin or impurities, may accumulate slowly. However, *coke laydown* is a much faster process in the high temperature, gas-phase reactions of hydrocarbons over acid- and metal-acid catalysts. The coke can be burned off, but only at higher temperatures; hence the adoption of continuous regeneration in catalytic cracking, the incorporation of additives to accelerate burn-off, and the advantages of using the more thermally stable zeolites as catalysts or supports.

10.6.10 *Studies on surface chemistry*

More specific information on the nature of active sites can often be deduced from adsorption isotherms and heat of adsorption distribution measurements (by calorimetry or GSC)[13] with individual reactant gases or vapours. These can be backed up with many of the spectroscopic techniques referred to above, to characterize the nature of the adsorbed layer and chemisorbed species, and their effects on the structure and oxidation state of the surface itself[14].

Acidity is frequently an important parameter in the chemisorption process or general catalytic behaviour. The extent (and heat) of adsorption of various bases, such as ammonia, pyridine and hydrocarbons (e.g. triphenylmethane and perylene), again backed up with spectroscopic techniques, and a titrimetric (H_0 indicator) method for n-butylamine adsorption, have all proved informative. Similarly, basic sites can be quantified by adsorption of acidic materials. (Total surface hydroxyl can often be assessed by NMR and deuterium displacement for comparison). Kinetic studies on partially neutralized samples can point to the acid/base strength required to effect a particular reaction. These are specific examples of a general approach, in which the effects of selected poisons can throw light on the useful catalytic area and mechanism.

Isotopic labelling of reactants, or a switch to labelled reactant during reaction, can give information on kinetically limiting steps, symmetry patterns in intermediates, the involvement of lattice oxygen, etc. Microreactors and pulse techniques are usually most suitable for such work.

The co-adsorption of reactants at below normal reaction temperature, with temperature-programmed desorption of reactants, intermediates and products, has also received increasing attention in recent years. GSC techniques again find use in this context.

Simulation of catalytic reactions on metal films allows closer spectroscopic examination of the reacting system. A more recent step from convention is the use of molecular beams.

10.6.11 Theoretical approaches

Early ideas on heterogeneous catalysis assumed the need for appropriate spacing of surface atoms to effect bond breaking and making, the *geometric concept*. However, chemical bonds in liquids and gases are by no means rigid, while the use of crystal data for the solid can be misleading as the surface structure may differ or show some ionic mobility. Nevertheless, the undoubted involvement of multiple sites (ensembles) in some processes, or side-reactions, can sometimes allow useful catalyst modification based on simple geometric considerations.

The *electronic, semiconductor*, or *band theory*, as developed during the 1950s and early 60s, treats the catalytic surface of an oxide as a general entity. Reactions on the surface are effected by transfer of electrons into and out of the acceptor bands of a *p*-type semiconductor (or vice-versa for *n*-type donor bands). The theory showed modest success in rationalizing the hydrogenation and oxidation performance of some oxides, but is now considered inadequate for general use. However, as electron transfers *are* involved in such processes, the idea of acceptor bands warrants incorporation into many mechanistic interpretations, and semiconductor theory can provide useful guidelines for the selection of minor components to improve performance.

A more sophisticated approach starts from molecular orbital theory, which in itself proves too tedious for all but simple molecules and complexes. However, a combination of *electrostatic theory* and *crystal field theory* has proved valuable in relating changes in the stabilization energy of *d*-orbitals with changes in coordination patterns and the number of *d*-electrons. In particular, no changes in energy are predicted for metal ions with zero, five or ten *d*-electrons, pointing to a twin-peaked pattern of activity. Such a pattern has indeed been observed in many catalytic reactions over single oxides, but little real progress can be made to define the chemical reactions on the surface. Furthermore, the majority of practical oxide catalysts contain several metal species.

10.7 Applications and mechanisms

10.7.1 Introduction

The following sections are by no means comprehensive, but are intended to illustrate the scope of catalysis. Table 10.1 presents a partial summary of petrochemical applications.

These highlight some chemical advantages of heterogeneous catalysis.

(a) A very wide range of reaction temperatures is usable in gas phase reactions, allowing many kinetically difficult reactions to be effected and thermodynamic boundaries to be crossed.
(b) Heterogeneous catalysts can often be tailored empirically to give a wider range of products from an individual feedstock.

However, extensive use of heterogeneous catalysts has shown up a number of problem areas.

(a) Where a reaction can be effected with both homogeneous and heterogeneous catalysts, the latter are often less active, and the compensatory use of higher temperatures may degrade selectivity.
(b) Heterogeneous catalyst systems are less well defined, and the structural and electronic properties are subject to disturbance by trace impurities. Hence reproducibility may prove problematical.

The main emphasis here is on mechanisms, and few process details are included (see Chapter 11 for the latter). Whenever possible, related homogeneous and heterogeneous catalytic processes have been brought together for comparison. However, many of the mechanistic equations, diagrammatic representations and catalytic cycles have been greatly simplified, and details often continue to attract considerable debate.

The absence of a section devoted to polymerization processes may be noted. However, the majority of these, including the production of low density polyethylene and the polymerization of vinyl chloride, vinyl acetate, acrylonitrile, butadiene and styrene for example, involve initiated free-radical chain reactions which are considered to lie outside the scope of the present chapter. Despite their growing importance, only brief reference is made to olefin polymerizations by metal complexes, as kinetic and other characteristics may depart significantly from those associated with traditional catalytic processes.

10.7.2 Acid catalysis

Acid catalysis finds very wide industrial application in refinery operations and chemicals production. Both Brönsted acids (proton donors) and Lewis acids (electron acceptors) are used; the latter are often precursors for Brönsted acids or related species (traditional Friedel–Crafts' systems);

$$AlCl_3 + H_2O \rightleftharpoons H^+[HOAlCl_3]^-$$
$$AlCl_3 + EtCl \rightleftharpoons Et^+[AlCl_4]^- (\rightleftharpoons C_2H_4 + H^+[AlCl_4]^-)$$
$$BF_3 + HF \rightleftharpoons H^+BF_4^-$$

The author has adopted the negative of the H_0 scale as a convenient means of comparing the acidity of widely different acid systems, though rates of acid-

Table 10.1 Some industrial catalytic systems

Reaction	Homogeneous	Heterogeneous (Major components only)
Acid-catalysed		
Esterification	$ArSO_3H$;$Ti(OR)_4$, etc.	Acid ion-exchange resin
Cumene (isopropybenzene) hydroperoxide rearrangement	H_2SO_4	
Methanol to gasoline		Zeolite (ZSM-5)
Hydration of olefins	H_2SO_4	Supported H_3PO_4
Alkylation of benzene	$AlCl_3$-HCl	Supported H_3PO_4
Alkylation of isobutane	H_2SO_4 or HF	Silica-alumina; Zeolites
Catalytic cracking		
Hydrogenation		
General olefin, aromatic		Supported Ni or noble metal
Desulphurization		Co-Mo oxides
Dehydrogenation		
Butane → butenes		Chromia-alumina
Ethylbenzene → styrene (phenylethene)		Fe_2O_3-Cr_2O_3-K_2CO_3
Isopropanol → acetone (propanone)		ZnO
Dual-function (acid-catalysed reactions with hydrogen transfer)		
Isomerization of alkanes		Pt-acidified Al_2O_3
Catalytic reforming		Pt-Re-acidified Al_2O_3
Other metal-catalysed		
Oligomerization of C_2H_4	$AlEt_3$; $Ni(O)$ complexes	Supported chromium (II) oxide; supported Ziegler
Polymerization of C_2H_4		

Oxidations

Reaction	Catalyst
p-Xylene (1,4-dimethylbenzene) → terephthalic (benzene-1,4-dicarboxylic) acid	Co-Mn bromides
C_2H_4 → acetaldehyde (ethanal)	$PdCl_2$-$CuCl_2$
C_2H_4 + HOAc → vinyl acetate (ethenyl ethanoate)	Supported Pd(-Au)-KOAc
C_2H_4 → ethylene oxide (oxiran)	Supported Ag(-Cl)
CH_3OH → HCHO	Ag (partial ox.); Fe-Mo oxides
C_3H_6 + NH_3 → acrylonitrile (propenonitrile)	Bi-Mo; U-Sb; Sn-Sb oxides
C_4H_8 → butadiene	Ferrite spinels
o-Xylene (1,2-dimethyl benzene) → phthalic (benzene-1,2-dicarboxylic) anhydride	Supported V_2O_5-TiO_2
C_2H_4 + 2HCl → $ClCH_2CH_2Cl$	Supported $CuCl_2$ melt

Carbon monoxide chemistry

Reaction	Catalyst
$CO + 3H_2 \rightleftharpoons CH_4 + 3H_2O$	Supported Ni
$CO + H_2$ → paraffins (alkanes), etc.	Iron oxide (promoted)
$CO + 2H_2O \rightarrow CH_3OH$	Cu-ZnO (-Cr_2O_3, etc.)
$CO + H_2O \rightleftharpoons CO_2 + H_2$	Fe_3O_4; Cu-ZnO
Olefin + CO/H_2 → aldehyde	Co; Rh-phosphine
$CH_3OH + CO$ → acetic (ethanoic) acid	Rh-CH_3I
$CH_3OAc + CO$ → acetic (ethanoic) anhydride	Rh(Ni)-I-base
($CH_3OAc + CO/H_2 \rightarrow CH_3CH(OAc)_2$	Rh(Ni)-I-base)

catalysed reactions are rarely simply related to acidity on this $- H_0$ scale. Some values for strong (anhydrous) acid systems are listed below.

System	$- H_0$
HF	11·0
H_2SO_4	12·6
CF_3SO_3H	14·6
FSO_3H	15·1
$AlCl_3/HCl$ or $GaCl_3/HCl$	15 to 16
SbF_5/HF	up to 25

The value of $- H_0$ associated with the hydrated proton is about 1·7, and water acts as a base in systems of higher acidity. Alcohols and ethers, ketones and esters also act as bases at progressively higher values of $- H_0$.

Earlier reference has been made to sulphonic-acid resins and supported mineral acids as heterogeneous catalysts. The chlorination of alumina also provides a strongly acidic surface similar to supported aluminium trichloride. However a number of metal oxides also show acidic properties without treatment.

Differences in coordination number become significant in mixed oxide systems, where we find many examples of enhanced Lewis or Brönsted acidity, even for binaries such as SiO_2-MgO[15]. However, the best known and possibly most remarkable system of this type is silica–alumina. Typical materials have acid titres of the order 1 milliequivalent/g, but only a small proportion of the acid sites show $- H_0$ values of 3 or more.

A reduction in alumina content enhances acid strength up to a point, but a reduction in physical strength and thermal stability limits the utility of such materials. A breakthrough came with the synthesis of the low-alumina-content zeolites, by the growth of highly regular silica–alumina crystals around bulky quaternary ammonium ions. These materials provide nearly 1 milliequivalent/g of sites of $- H_0$ from 3 to 10 or more, while showing markedly improved thermal stability[11].

10.7.2.1 *Esterification and ester hydrolysis.* A considerable number of commercial esterifications are still carried out by conventional acid catalysis. For practical esterifications, a slightly modified version of the Goldschmidt (proton sharing) equation remains useful, if not exact.

$$\text{Rate} = \frac{k[H^+ \text{total}][RCO_2H][ROH]}{K[ROH] + [H_2O]}$$

However, acid catalysts promote side reactions, and 'uncatalysed' esterifications become very slow at high conversions. Hence, for the production of high-quality esters, a large family of metal compounds are now used as esterification catalysts. These include the soluble alkoxides and carboxylates, and insoluble oxides and oxalates of titanium, aluminium, zinc, cadmium, antimony, tin and many other metals. The phenomenon is often referred to as amphoteric catalysis, and is related to the coordinating capability, i.e. Lewis acidity, of the metal towards the acyl C=O group; electron withdrawal permits nucleophilic attack by alcohol (or alkoxide) in a similar, but less effective, manner to protonation. Temperatures required are of the order 200°C or more; essentially all such processes are carried out in the liquid phase, under pressure if necessary.

10.7.2.2 *Rearrangement of oxonium ions.* In the acid-catalysed cleavage of cumene hydroperoxide (to phenol and acetone), an important step is aryl transfer from carbon to oxygen in the intermediate oxonium ion:

$$\underset{\underset{\text{Me}}{|}}{\overset{\overset{\text{Me}}{|}}{Ph-C-O\overset{+}{O}H_2}} \xrightarrow{-H_2O} \underset{\underset{\text{Me}}{|}}{\overset{\overset{\text{Me}}{|}}{Ph-C-\overset{+}{O}}} \longrightarrow \underset{\text{Me}}{\overset{\text{Me}}{>}}\overset{+}{C}-OPh$$

Alkyl transfers from O to C (Stevens rearrangement), carbenes and methyl carbonium ions have all been postulated to explain the formation of lower olefins from methanol and dimethyl ether over heterogeneous acid catalysts[16]; the reaction is autocatalytic, e.g.

$$Me\overset{+}{O}H_2 + MeOH \rightleftharpoons Me_2\overset{+}{O}H + H_2O \rightleftharpoons Me_2O + H_3\overset{+}{O}$$

$$\left. \begin{array}{l} Me\overset{+}{O}H_2 + Me_2O \rightarrow Me_3\overset{+}{O} + H_2O \\ Me_3\overset{+}{O} \rightarrow EtMe\overset{+}{O}H \rightleftharpoons C_2H_4 + Me\overset{+}{O}H_2 \end{array} \right\} \text{ slow}$$

$$Me\overset{+}{O}H_2 + EtOMe \rightarrow Et\overset{+}{O}Me_2 \rightarrow PrMe\overset{+}{O}H \text{ etc. (faster)}$$

$$(\rightarrow \text{propylene})$$

These reactions represent the first steps in the conversion of methanol to hydrocarbon fuels over Mobil's ZSM-5 catalyst; further reactions of the olefins are described in section 10.7.2.5.

10.7.2.3 *Formation of carbonium ions from olefins (alkenes).* Many industrial reactions of olefins involve protonation to give a carbonium ion, which is subject to nucleophilic attack, followed by proton transfer from the product to olefin. The ease of protonation follows the stability of the carbonium ion formed in the sequence tertiary > secondary > primary. Additional proton exchanges can occur at any stage in the overall process, leading to double-bond shifts in the olefinic feedstock and mixed products in some cases. (At high temperatures, products with terminal substituents may also be detectable).

10.7.2.4 *Hydration and etherification.* The direct hydration of ethylene has been discussed (section 10.5). Propylene can also be hydrated in the gas phase over supported phosphoric acid (*c.* 180°C), or with an ion-exchange resin catalyst at about 140°C, with liquid water and gaseous propylene. The use of an ion-exchange resin as a catalyst has also been commercialized for the hydration of n-butenes, though the sulphuric acid two-stage process still predominates. The use of very weak acid systems at much higher temperatures ($> 250°C$) has also been studied.

In all the above processes, the corresponding symmetrical ethers are co-produced by the reaction of product alcohol with the carbonium ion. Ion-exchange resins protonate iso-butene in methanol (at below $100°C$) to produce methyl $tert$-butyl ether (MTBE).

Despite the reversibility of these reactions, many acidic oxides catalyse the dehydration of alcohols but show no significant activity for olefin hydration. The high partial pressure of water thermodynamically necessary leads to excessive surface coverage by water, with a marked fall in effective acidity.

10.7.2.5 *Carbon–carbon bond formation and cleavage.* In olefin hydration processes, minor amounts of dimers and polymers are produced by the mechanism

$$-\overset{|}{\underset{|}{C}}-\overset{|}{\underset{|}{C}}{}^{+} \ \rightleftharpoons \ \longrightarrow \ -\overset{|}{\underset{|}{C}}-\overset{|}{\underset{|}{C}}-\overset{|}{\underset{|}{C}}-\overset{|}{\underset{|}{C}}{}^{+}$$

Under anhydrous conditions, traditional mineral acid and Friedel–Crafts' systems (liquid phase), as well as supported phosphoric acid (gas phase), can be used to produce dimers and trimers through to relatively high molecular weight viscous liquid polymers from C_3 and C_4 olefins. These same catalyst systems are also used in the alkylation of aromatic hydrocarbons.

Branched alkanes are also alkylated by lower olefins in the presence of concentrated sulphuric acid or anhydrous HF at near-ambient temperatures; an additional reaction, *hydride transfer*, is involved. If we consider the reaction of propylene and iso-butane (in excess), the chain reaction sequence is as follows:

$$\text{MeCH}{=}\text{CH}_2 + \text{H}^+ \rightarrow \text{Me}_2\text{CH}^+ \qquad \text{(protonation)}$$

$$\text{Me}_2\text{CH}^+ + \text{Me}_3\text{CH} \rightarrow \text{Me}_2\text{CH}_2 + \text{Me}_3\text{C}^+ \qquad \text{(hydride transfer)}$$

$$\text{Me}_3\text{C}^+ + \text{CH}_2{=}\text{CHMe} \rightarrow \text{Me}_3\text{C} \cdot \text{CH}_2\overset{+}{\text{C}}\text{HMe}$$

$$\text{Me}_3\text{CCH}_2\overset{+}{\text{C}}\text{HMe} + \text{Me}_3\text{CH} \rightarrow \text{Me}_3\text{CCH}_2\text{CH}_2\text{Me} + \text{Me}_3\text{C}^+$$

(alkylation)

All the above reactions are reversible. Hence, at higher temperatures with zeolites, cleavages of olefins and isomerization and trans-alkylation of alkylaromatics can occur; in the presence of alkenes and alkylaromatics as hydride acceptors, alkanes can also take part.

If we generate a hydrocarbon of appropriate structure with both unsaturation and a protonated centre, then intermolecular addition, cyclization, can occur. C_5 and C_6 ring systems are most favoured thermodynamically. Further hydride transfers to olefins can lead to aromatization of C_6 ring systems. Thus, under appropriate conditions, lower olefins can be converted to mixtures of alkanes and aromatic hydrocarbons by acid catalysis alone. Such reactions form the basis for aromatics production from intermediate olefins in the homologation of methanol over Mobil's ZSM-5 zeolite (section 10.7.2.2). Further hydride transfers from polyaromatic hydrocarbons lead to 'coke'.

10.7.2.6 *Koch reaction.* Carbonium ions react with carbon monoxide (under pressure) to form acyl cations; the overall reaction from isobutene gives 2, 2-dimethylpropionic (2, 2-dimethyl propanoic) acid, for example.

$$R^+ + CO \rightleftharpoons RCO^+ \xrightarrow{H_2O} RCO_2H + H^+$$

Traditionally, concentrated sulphuric acid has been used as the reaction medium, necessitating dilution with water to give the carboxylic acid, and reconcentration of the mineral acid. The more direct reaction of olefin, water and carbon monoxide in the presence of BF_3 requires much higher temperatures and pressures, but has possibly become the preferred system.

10.7.2.7 *Carbonium ion rearrangements.* At low temperatures, strong acids ($-H_0$ about 10) induce methyl shifts in branched alkanes (a hydride acceptor must be present to form the carbonium ion).

Similarly, methyl shifts on aromatic rings are relatively facile. Under the severe conditions used in refinery processes, more dramatic rearrangements occur towards thermodynamically favoured highly branched products.

10.7.3 Hydrogenation

The two critical steps in alkene hydrogenation by metal complexes are now moderately well defined; recent reports of alkane activation essentially confirm that the final step is also reversible[17]. (Di-hydrogen addition can occur at various stages to complete the cycle).

There have been few reports of carbonyl hydrogenation with metal complexes. However, alkoxy-metal intermediates have been proposed in hydrogen transfers from alcohols.

Similar mechanisms are postulated for commercial alkene/arene, carbonyl and nitrile hydrogenations on metal surfaces; in particular, individual metal atoms are involved. In contrast hydrogenolysis, the cleavage of C—C or C—O (N, S, etc.) bonds, appears to need two or more adjacent sites and can sometimes be reduced by alloying the main component (addition of copper to nickel, for example). The stability of supported metal (especially platinum) catalysts permits their use at high temperatures, to promote hydrogen transfers between alkanes, alkenes and arenes or dehydrogenation processes.

However, a number of metal oxides also show high hydrogenation/dehydrogenation activity at higher temperatures, and find extensive commercial use in dehydrogenation processes. The oxides of chromium, iron and zinc, among others, are common catalyst components for the dehydrogenation of alkanes, alkenes (e.g. butane or butene to butadiene), ethylbenzene and secondary alcohols (for ketones). Alkali 'promoters' are often added to eliminate side reactions caused by Brönsted acid sites, but the slightly reduced surfaces show Lewis acid behaviour and semiconductor properties. A possible scheme for the dehydrogenation of isopropanol over zinc oxide involves proton transfers, thus:

$$
\begin{array}{ccc}
Me_2CH - O - H & Me_2C{=}O & H_2 \\
\downarrow & \xrightarrow{\hspace{2cm}} & \xrightarrow{\hspace{2cm}} \\
O^{2-} - M - O^{2-} & \bar{O}H - M - O\bar{H} & O^{2-} - M - O^{2-} \\
& \downarrow & \uparrow \\
& 2e\ \text{to acceptor band} & 2e
\end{array}
$$

(effective reaction is $2OH^- + 2e \rightarrow 2O^{2-} + H_2$)

Metal-hydride bonding may also be involved. More readily reducible mixed oxides, such as copper chromite, possibly present a more metallic surface, but with an electron acceptor band of the semiconductor type.

Metal sulphides are also used in hydrogenation processes. Thus nickel sulphide ($Ni_3 S_2$) permits reduction of alkynes and dienes to alkenes. Various tungstates and molybdates (particularly of cobalt) are used in sulphided form for the hydrogenolysis of sulphur compounds, and the saturation of sulphur-containing refinery streams.

10.7.4 Dual-function catalysis

For hydrocarbon reactions, metals (particularly platinum and its alloys) are frequently applied to acidic supports to catalyse hydrogen transfers. Thus platinum on a chlorinated alumina support accelerates the acid catalysed isomerization of n-alkanes (at about 150°C). In *hydrocracking*, the metal catalyses hydrogenation of heavy aromatic and polyaromatic components; the resulting cycloparaffins (cycloalkanes) undergo zeolitic cracking, with olefin hydrogenation, to give paraffinic naphthas (mainly C_5 to C_8). In the *catalytic reforming* of naphthas the presence of a Pt–Ir or Pt–Re alloy on an acidic alumina (with halogen in the feed), leads to fast dehydrogenation, cyclization and aromatization of the paraffinic hydrocarbons without the cage effect which promotes such reactions on the ZSM-5 zeolite.

Metallic components have also been added to a variety of heterogeneous oxide catalysts, to introduce additional hydrogenation, dehydrogenation and hydrogen transfer processes during aldolization, ketonization or Tishchenko reactions. Examples include acetone (propanone) to 4-methyl-pentan-2-one, ethanol to acetone and methanol to methyl formate (methyl methanoate), e.g.

$$CH_3OH \xrightarrow{Cu(ZrO_2)} HCHO + H_2 \qquad 2HCHO \xrightarrow{(Cu)ZrO_2} HCO_2CH_3$$

10.7.5 Olefin (alkene) polymerization and dismutation on metals

Hydroboration has become a very useful synthetic tool in academic chemistry. The addition of an olefin to a boron hydride is essentially irreversible.

Although there is a fund of interesting chemistry for alkylboranes, no simple catalytic possibilities exist.

Aluminium hydrides show somewhat greater versatility. Although the addition of olefins is not readily reversible, exchange occurs between the aluminium alkyl groups and olefins. Olefin insertions into the metal–alkyl extend the alkyl chain; displacement by olefin gives relatively low molecular weight polymers (up to $c.$ C_{20} linear alpha-olefins from ethylene).

$$\rangle AlEt + nC_2H_4 \longrightarrow \rangle Al(C_2H_4)_nEt \xrightarrow{C_2H_4} \rangle AlEt + CH_2 = CH(C_2H_4)_{n-1}Et$$

Phillips supported chromium (II) catalyst for high density polyethylene (HDPE) possibly behaves in a similar manner, but the olefin insertion reaction is faster by several orders of magnitude. In Ziegler catalyst systems, the aluminium alkyl reductively alkylates the primary component, most frequently a titanium compound, to give the true catalytic species, which similarly undergoes very fast olefin insertion.

$$AlR_3 + \equiv Ti\text{-}X \rightarrow XAlR_2 + \equiv TiR \xrightarrow{nC_2H_4} \equiv Ti(C_2H_4)_nR (X = Cl \text{ or } OR')$$

Aluminium and other metal alkyls also activate tungsten and molybdenum compounds (particularly oxychlorides), to generate homogeneous or supported olefin dismutation catalysts. It is now believed that an initial M-alkyl (M = W, Mo) group is converted to a metal alkylidene group by α-hydrogen abstraction. Coordinated olefin now gives a metallocyclobutane (isolable in some cases)[18]:

$$
\begin{array}{ccc}
M{=}CR_2 & M{-}CR_2 & M \quad CR_2 \\
+ & \rightleftharpoons \quad | \quad | \quad \rightleftharpoons & \| + \| \\
R_2''C{=}CR_2' & R_2''C{-}CR_2' & R_2''C \quad CR_2
\end{array}
$$

However, some oxides, such as supported Re_2O_7 (at 50–120°) or molybdenum oxide (at higher temperatures), show dismutation activity without alkyl treatment. Presumably, partial reduction of the surface by olefin exposes suitably liganded (M—O—M=CR_2) centres. Further reduction (slow anion migration to the surface) leads to deactivation, but activity is restored by re-oxidation.

Finally, nickel (0) complexes and supported nickel show somewhat similar properties to aluminium alkyls in ethylene oligomerization to alpha-olefins. However, sparsely liganded (phosphine) nickel (0) complexes form dually bonded σ, π (allyl) C_4 ligands from butadiene, which undergo a variety of insertion reactions with olefins or butadiene and ene reactions between ligands. The nickel acts as a template around which simple and puckered ring systems can build up, with phosphine (or other ligand) size and basicity as selectivity control parameters.

10.7.6 *Base catalysis*

The traditional alkaline catalysts are still used for aldol condensations; cross-Cannizzaro reactions occur when formaldehyde is one of the reactants, e.g. for 'pentaerythritol':

$$CH_3CHO + 4HCHO \rightarrow (HOCH_2)_3CCHO + HCHO$$
$$\rightarrow C(CH_2OH)_4 + HCO_2H$$
$$\text{pentaerythritol}$$

Alkali metal alkoxides catalyse the alcoholysis of esters, by a mechanism analogous to basic hydrolysis. Additionally, alkoxides catalyse the reaction of alcohols with carbon monoxide to give formate esters:

$$RO^- + CO \rightleftharpoons RO_2\bar{C} \underset{RO^-}{\overset{ROH}{\rightleftharpoons}} HCO_2R$$

There is now increasing commercial interest in the dimerization of olefins over supported alkali metals, via a carbanion mechanism. Propylene selectively produces 4-methylpent-1-ene and alkylaromatics are alkylated on the side-chain (α-carbon) with these materials.

$$C_3H_6 + Na \xrightarrow[-\frac{1}{2}H_2]{} Na^+C\bar{H}_2—CH=CH_2 \xrightarrow{C_3H_6} C\bar{H}_2\cdot CHMeCH_2CH=CH_2$$
$$\text{(accepts proton from } C_3H_6)$$

10.7.7 *Oxidations*

10.7.7.1 *Catalysis in liquid-phase free-radical oxidations.* The conventional liquid-phase oxidation of hydrocarbons and their derivatives with air involves a free-radical chain mechanism:

$RH + O_2 \rightarrow R^\cdot + HO_2^\cdot$	Slow initiation
$R^\cdot + O_2 \rightarrow RO_2^\cdot$	Propagation
$RO_2^\cdot + RH \rightarrow RO_2H + R^\cdot$	
$RO_2H \rightarrow RO^\cdot + HO^\cdot$	Branching
2 radicals $\rightarrow$ non-radical products	Termination
(R = alkyl, aralkyl, acyl, etc.)	

Organic compounds which decompose to give free radicals (e.g. peroxides) may be added to accelerate initiation, and some metals in high oxidation states may also fulfil this role,

$$RH + M^{(n+1)+} \rightarrow R^{\cdot} + M^{n+} + H^+$$

With methyl aromatic substrates, the presence of bromide ions also aids hydrogen abstraction without significant formation of halogenated by-products.

$$M^{(n+1)+} + Br^- \rightarrow M^{n+} + Br^{\cdot}$$
$$ArCH_3 + Br^{\cdot} \rightarrow ArCH_2^{\cdot} + Br^- + H^+$$

However, the major role of most metal compounds, in solution or not, is to accelerate decomposition of the hydroperoxide intermediate in a branching mode:

$$RO_2H + M^{n+} \rightarrow RO^{\cdot} + M^{(n+1)+} + OH^-$$
$$RO_2H + M^{(n+1)+} \rightarrow RO_2^{\cdot} + M^{n+} + H^+$$

In general, metal concentrations of only 10–100 parts per million are required in the oxidation of alkanes and their oxygenated derivatives, somewhat higher levels for alkyl aromatics.

When the desired product is an alkyl or aralkyl hydroperoxide or peracid, careful control of metal concentration is required to minimize hydroperoxide cleavage. When the reactant is an alcohol, the hydroperoxide is the hydrogen peroxide adduct of an aldehyde or ketone, i.e.

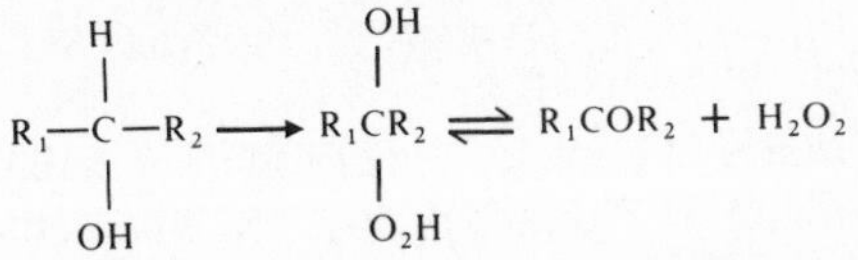

$$R_1\!-\!\underset{\underset{OH}{|}}{\overset{\overset{H}{|}}{C}}\!-\!R_2 \longrightarrow R_1\underset{\underset{O_2H}{|}}{\overset{\overset{OH}{|}}{C}}R_2 \rightleftharpoons R_1COR_2 + H_2O_2$$

Hence hydrogen peroxide may be a recoverable product from such oxidations under mild conditions.

Many of the above reactions of metal ions are reversible, and the kinetically favoured direction varies with the metal and type of radical. Hence, individual metals may cause acceleration or inhibition with different substrates, often accompanied by changes in product distribution.

Cobalt (III) appears to be unique, in that higher concentrations (0·5 to 2% w/w) permit high rates in aerial oxidation of hydrocarbons while maintaining specificity in the point of attack; free radicals, in the conventional sense, appear to be absent. Further evidence of a strong association between cobalt ions and reactive species is the exceptionally high reactivity of 'Co^{3+}' regenerated in situ compared with typical cobaltic compounds prepared by other means.

Finally, in the Bashkirov oxidation of normal paraffins to secondary alcohols, the boric oxide introduced is often referred to as a catalyst. This material does indeed catalyse the decomposition of hydroperoxides, but by a non-radical route to alcohol and oxygen, and therefore slows down the overall rate of reaction.

10.7.7.2 *Liquid-phase epoxidation.* Epoxides are produced in liquid-phase reactions of olefins with hydroperoxides in the presence of (soluble or insoluble) compounds of molybdenum, tungsten or vanadium.

$$\text{\large$\sum$}C = C\text{\large$\big\langle$} + \; RO_2H \longrightarrow \text{\large$\sum$}\overset{O}{\overset{/\backslash}{C - C}}\text{\large$\big\langle$} + ROH$$

The hydroperoxides (and hydrogen peroxide) form well-characterized peroxy-complexes with oxides of the above metals, but the mechanism whereby the olefin coordinates and reacts with these complexes is still unclear. (No catalyst is required if a peracid is used as the epoxidizing reagent.)

10.7.7.3 *Wacker-type oxidations.* Ethylene is oxidized to acetaldehyde in the presence of an aqueous solution of palladium (II) chloride and copper (II) chloride[1,21]. The initial reaction is believed to follow a sequence of the type

$$\| + \left[Pd^{II} Cl_n H_2O \right] \xrightarrow{-Cl^-} \left[\|\rightarrow Pd^{II} Cl_{n-1} H_2O \right] \xrightarrow{-H^+} \left[HOCH_2CH_2Pd^{II} Cl_{n-1} \right]$$

Although both chloride and hydrogen ions are essential for catalyst stability, the rate of reaction shows a negative order on both. β-Hydrogen abstraction then follows, nominally to give palladium metal by complete decomposition,

$$\left[HO - CH_2CH_2 - Pd^{II} Cl_{n-1} \right] \longrightarrow CH_3CHO + Pd^0 + H^+ + (n-1)Cl^-$$

However, in the presence of copper (II) chloride, the palladium is reoxidized without significant metal precipitation, possibily via a chloride bridged Cu–Pd species. The copper (I) chloride is reoxidized to copper (II) chloride in situ, or in a separate stage.

Higher olefins react more slowly than ethylene with the Pd(II)–Cu(II) chloride system, and some chloro-substituted by-products are formed in addition to the expected ketone. Attempts have been made to overcome these drawbacks by replacing the Cu(II) chloride by low chloride or chloride-free redox systems for the palladium, based for example on Fe(III) sulphate or vanado-phosphomolybdate complexes. Further systems, involving the formation of, and hydrogen abstraction from, β-hydroxyalkyl ligands on Rh(III)[22] or Tl(III) in lieu of Pd(II), have also been studied. However, none of these alternative systems has been commercialized to date.

Substitution of acetic acid for water as solvent in the Wacker process leads to the formation of vinyl acetate (ethenyl ethanoate) from ethylene by an essentially identical mechanism (called 'acetoxylation'). This liquid phase system (chlorides in acetic acid) is exceedingly corrosive. However, the use of supported palladium catalysts in the liquid phase provides modest rates without chloride. Similarly, solid-catalyst systems (e.g. Pd/Te on carbon) are effective for the conversion of butadiene to 1,4-diacetoxybut-2-ene, which forms the basis of one commercialized route to 1,4-butanediol.

The commercial route to vinyl acetate is a gas-phase process, utilizing palladium or a palladium-gold alloy supported on alumina or a spinel[7]. The catalyst also carries potassium acetate, and sometimes other metal acetates, to provide a supported liquid phase (KOAc–HOAc) in which the reaction takes place. The role of palladium may be described by the reaction scheme

$$\| + Pd^0 + HOAc \longrightarrow \| \longrightarrow Pd \overset{(H)}{\underset{OAc}{\diagup}} \longrightarrow Pd \overset{(H)}{\underset{CH_2CH_2OAc}{\diagup}}$$

$$\longrightarrow Pd \overset{(H)}{\underset{(H)}{\diagup}} + CH_2{=}CHOAc \quad (\beta\text{-hydrogen abstraction})$$

Hydrides on the palladium are continually removed as water by oxygen present; no redox component is necessary. However, other oxidizing components were included in heterogeneous catalysts for the uncommercialized oxidation (Pd-phosphomolybdate) and 'oxycyanation' (olefin/HCN/O$_2$ over Pd-vanadia) of ethylene to acetaldehyde (ethanal)/acetic (ethanoic) acid and acrylonitrile (propenonitrile) respectively.

10.7.7.4 *General gas-phase oxidation over metals.* Only silver, gold and the noble Group VIII metals do not form bulk oxides in the presence of air at high temperatures, and of these only silver, as gauze or supported metal, finds application in a number of selective gas-phase oxidations of organic substances. (Pt and other noble metals catalyse total combustion for exhaust gas clean-up).

Silver has a relatively strong affinity for oxygen, which is activated by anion formation. For a univalent metal, each electron transfer involves another adjacent metal atom; on silver, only O_2^- and O^{2-} ions have been reported.

$$Ag + O_2 \rightarrow Ag^+ + O_2^-$$
$$4\,Ag\,(adjacent) + O_2 \rightarrow 4\,Ag^+ + 2O^{2-}$$

(The positive charges are partially delocalized).

In most oxidation processes, dioxygen ions are considered undesirable, showing peroxidic behaviour with high activity and low selectivity. However, in the oxidation of ethylene to ethylene oxide, it is now believed that the O_2^- ion is essential:

$$C_2H_4 + O_2^- + Ag \rightarrow C_2H_4O + O^{2-} + Ag^+$$

Furthermore, the O^{2-} ions formed in this step, together with any produced by chemisorption of oxygen onto groups of adjacent silver atoms, lead to combustion products. This possibly explains why partial ($c.$ 25%) blocking of the surface with chloride ions specifically inhibits, but cannot totally eliminate, the combustion process.

10.7.7.5 *Gas-phase oxidation over metal oxides.* There is one particularly important finding common to essentially all metal oxide catalysts which provide rapid, moderately selective oxidations of organic substrates. If the supply of oxygen is interrupted, oxidation of the organic substrate continues, initially with little change in rate and selectivity. The oxide system becomes reduced, and tens, or even hundreds, of atomic layers below the surface become oxygen deficient. What are the implications?

The Redox Model of Mars and Van Krevelen describes the overall process as

$$MO_x + substrate \xrightarrow{r_1} MO_{x-n} + products$$

$$MO_{x-n} + \frac{n}{2}O_2 \xrightarrow{r_2} MO_x$$

Under steady-state conditions, r_1 equals r_2, and the oxide must be in a reduced state. Furthermore, the degree of reduction and catalytic performance will change with operating conditions.

Oxide anions (O^{2-}) must be mobile within the oxide structure, requiring both an appropriate spacing between the metal atoms in the crystal structure and stability of the metal atom lattice. Such structural stability occurs in vanadia provided that no more than two-thirds of the V(V) atoms are reduced to V(IV); further reduction causes changes in both structure and catalytic behaviour. The partially reduced oxides (and sometimes the fully oxidized states) therefore have vacancy (coordinatively deficient) structures.

Oxide ions also provide nearly all the oxygen present in the products and by-products, with little involvement of other chemisorbed oxygen species such as O_2^-. Most selective oxide catalysts show little or no oxygen isotopic exchange, again pointing to minimal involvement of such species. Hence substrate oxidation and oxygen reduction do not necessarily occur at the same site. In bismuth molybdate, the molybdenum and bismuth may respectively perform these separate functions. This feature requires electron transfers between metal atoms in the lattice; there must be partial covalency in the M—O—M bonds and near-degeneracy between some of the electronic states in the two metal atoms. The $4d$ states of molybdenum and the $6p$ band of bismuth are nearly degenerate, for example. It is also evident that other components can significantly affect both structural and electronic characteristics, and most commercial catalysts are indeed very complex. (One example is $K_{0.1}Ni_{2.5}Co_{4.5}Fe_3BiP_{0.5}Mo_{12}O_{55} + 17.5\%$ w/w SiO_2). Finally to achieve both reduction of atmospheric oxygen, yet selective oxidation of the organic substrate, requires a restricted range of metal–oxygen bond strengths.

Up to this point, we have only looked at general requirements of the oxide material, without considering the nature of the chemisorption and oxidation processes themselves. The situation proves complicated as a particular organic substrate can give different products over related families of oxides.

In the oxidation or ammoxidation of propylene on bismuth molybdate (and U–Sb, Sn–Sb oxides), all the evidence indicates that the first step is hydrogen abstraction to give a symmetrical intermediate. The first stages can possibly be written:

$$H—CH_2—CH=CH_2 \qquad\qquad \left[CH_2{=}\!{=}CH{=}\!{=}CH_2 \right]^-$$

$$\searrow O^{2-} - Mo - O^{2-}\ O^{2-} - \quad\longrightarrow\quad HO^- - Mo - O^{2-}\ O^{2-} -$$

Electrons must now drain away to permit formation of a C—O(N) bond before further hydrogen removal can occur. (The manner of nitrogen incorporation to form acrylonitrile is still under review[23]. The formation of acrolein (propenal), though until recently of lesser commercial importance, will be pursued).

$$\left[CH_2{=}\!{=}CH{=}\!{=}CH_2 \right]^- \qquad\qquad CH_2{=}CH—CH_2 \qquad\qquad CH_2{=}CHCHO$$

$$HO^- - Mo - O^{2-}\ O^{2-} - \longrightarrow HO^- - Mo - O^{2-} - \longrightarrow HO^- - Mo - HO^-$$

Elimination of water and diffusion of two O^{2-} anions from the bulk complete the cycle. When the feedstock in an n-butene, the electron shift prompts the transfer of a second proton to surface oxygen, rather than C—O bond formation, leading to butadiene.

Little or no acrylic (propenoic) acid results from overoxidation of acrolein on bismuth molybdate. The addition of cobalt, nickel or vanadium oxides is necessary to induce a further transfer of oxygen (or OH) to carbon. However, if we oxidize propylene over tin or cobalt molybdates alone, the major product is acetone. The surfaces of these materials have high hydroxyl concentrations, with Brönsted acidity, leading to isopropoxy groups from which the α-hydrogen is readily abstracted[24]. n-Butenes give 2-butanone and acetic acid. In contrast, the incorporation of phosphorus into bismuth molybdate does not modify the initial stage, but butadiene formed from n-butenes is chemisorbed more strongly, and further stepwise oxidation to maleic anhydride occurs. (Modified vanadia catalysts are preferred for maleic anhydride production from C_4's).

Surprisingly perhaps, oxidation of C_3 and C_4 olefins over bismuth oxide alone leads mainly to oxidative dimerization to C_6 or C_8 dienes, and small amounts of cyclic hydrocarbons. The surface is possibly highly reduced, providing many Lewis acid centres for olefin coordination but few oxide ions for hydrogen abstraction and transfer. (The supported gallium oxide catalyst, developed by BP, leads to further hydrogen abstractions and significant yields of aromatics from lower olefins).

Finally vanadia catalysts are used extensively for oxidations of aromatic hydrocarbons. With benzene, the mechanism for ring breakage is not well defined, and the desorption of maleic anhydride itself appears to be rate-controlling. For the oxidation of o-xylene, the use of supported vanadia-titania catalysts limits ring cleavage. A well-defined major product sequence, o-xylene → o-tolualdehyde → o-toluic acid → phthalide → phthalic anhydride, defines the main series of hydrogen abstraction/oxygen transfer processes which follow the initial coordination and hydrogen abstraction step.

10.7.7.6 Halogen-mediated oxidations and oxychlorination. A number of liquid-phase oxidation processes appear to be mediated by *in situ* oxidation of halide, halogen addition to an olefin, and solvolysis of the resulting intermediate. This comment applies particularly to Halcon's tellurium bromide system (in aqueous acetic acid) for glycol acetates from olefins. Halogenated organic products are co-produced, but are separated for recycle. (The process was commercialized as part of a manufacturing route to ethylene oxide, but apparently failed.) The mechanism is essentially:

$$TeBr_4 \rightleftarrows TeBr_2 + Br_2$$

$$TeBr_2 + 2H^+ + 2Br^- + \tfrac{1}{2}O_2 \rightarrow TeBr_4 + H_2O$$

Similarly, in the high-temperature gas-phase oxychlorination of alkanes over supported copper chloride melts, generation of the free chlorine appears to be necessary.

$$2CuCl_2 \rightarrow 2CuCl + Cl_2$$

$$Cl_2 + CH_4 \rightarrow CH_3Cl + HCl$$

$$\underline{2CuCl + 2HCl + \tfrac{1}{2}O_2 \rightarrow 2CuCl_2 + H_2O}$$

$$\text{overall } CH_4 + HCl + \tfrac{1}{2}O_2 \rightarrow CH_3Cl + H_2O$$

In contrast, the oxychlorination of ethylene occurs at much lower temperatures than those required to generate chlorine. Hence, direct transfer of halogen to coordinated olefin, with simultaneous reduction of Cu^{II} to Cu^{I}, has been proposed.

A direct halogen transfer and solvolysis mechanism has been put forward for the liquid phase 'oxycyanation' of butadiene to 1, 4-dicyano-but-2-ene by the system copper-iodine—HCN—air.

10.7.8 Carbon monoxide chemistry

10.7.8.1 *Heterogeneous catalysis.* In 1923, Fischer and Tropsch showed that carbon monoxide and hydrogen gave complex mixtures of hydrocarbons and oxygenated derivatives over iron catalysts at about 400°C and high pressures. Later cobalt and promoted iron catalysts gave predominantly linear hydrocarbons at much lower temperatures (200–300°C) and pressures (1–20 atm.). Under similar conditions, supported nickel gave mainly methane, but high-molecular-weight paraffins were formed on ruthenium.

Hydrocarbon formation is now believed to proceed via atomic carbon (carbide) to bridging methylene groups, and a chain polymerization process[25]:

$$\begin{array}{ccccc} H & CH_2 & CH_2 & & CH_3 & CH_2 & & CH_3CH_2 \\ \backslash & /\backslash & /\backslash & & \backslash & /\backslash & & \backslash \\ M-M-M & \longrightarrow & M-M-M & \longrightarrow & M-M-M \end{array}$$

β-hydrogen abstraction and α-hydrogen addition give alkene and alkane respectively, while incorporation of unreduced carbon monoxide leads to oxygenates. The reverse, depolymerization, process is implicated in hydrogenolytic cleavage of hydrocarbons on these metals, with appreciable co-production of methane. The overall mechanism is retraced in the steam reforming of hydrocarbons over nickel to produce synthesis gas (CO/H_2 mixtures).

$$C_nH_{2n+2} + nH_2O \rightarrow nCO + (2n+1)H_2$$

Developments in the 1930s led to zinc chromite for methanol synthesis, with pressures of about 300 atm-required for thermodynamic reasons at about 400°C. Reduction on a single metal centre seems a likely mechanism. The higher *initial* activity of copper was known for many years, but the commercial success of ICI's lower-temperature, lower-pressure catalyst stems mainly from their success in stabilizing the metal dispersion.

10.7.8.2 *Hydroformylation.* Experiments with supported cobalt catalysts led to the hydroformylation (or OXO) process for the conversion of olefins and synthesis gas to aldehydes. Homogeneous catalysis followed, and is now used exclusively. The generally accepted mechanism involves the following reactions:

$$Co_2(CO)_8 + H_2 \rightleftharpoons 2HCo(CO)_4 \rightleftharpoons 2HCo(CO)_3 + 2CO$$

$$(CO)_3CoH + CH_2 = CHR \rightleftharpoons (CO)_3CoCH_2CH_2R \text{ (and } (CO)_3CoCHRCH_3)$$

$$CO + (CO)_2Co{\overset{\diagup CO}{\underset{}{—}}}\dot{C}H_2CH_2R \rightleftharpoons (CO)_3CoCOCH_2CH_2R$$

$$\overset{H_2}{\rightarrow} (CO)_3CoH + HCOCH_2CH_2R$$

The order on carbon monoxide is *negative*, but a high partial pressure is required to maintain catalyst solubility and limit olefin hydrogenation.

CO-insertion (or alkyl transfer to CO) occurs widely in organometallic chemistry; with cobalt, the addition of extra ligands drives the equilibrium to the right. (Substitution of water or alcohol for hydrogen in the process leads to carboxylic acid or ester, but the mechanism may then differ).

The selectivity to linear product is improved with a trialkyl phosphine ligand, but the major product is the corresponding *alcohol*. The use of triphenylphosphine rhodium hydrocarbonyl permits milder reaction conditions and gives a still higher selectivity to linear aldehyde, by essentially the same mechanism.

10.7.8.3 *Methanol carbonylation*. BASF employ a cobalt catalyst, with iodide promoter, for the carbonylation of methanol to acetic acid under severe conditions (250°C, over 300 atm.). The mechanism is ill defined. Monsanto showed that rhodium again allows much milder conditions. In the following scheme the oxidative addition of methyl iodide to the rhodium(I) dicarbonyl diiodide anion appears to be rate-controlling[26]:

$$MeOH + HI \rightleftharpoons MeI + H_2O$$

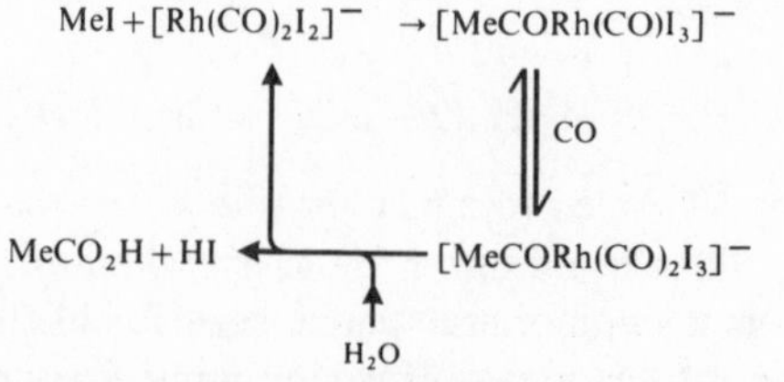

$$MeI + [Rh(CO)_2I_2]^- \rightarrow [MeCORh(CO)I_3]^-$$

$$MeCO_2H + HI \longleftarrow \qquad [MeCORh(CO)_2I_3]^-$$

Similar intermediates are possibly involved in the Halcon/Eastman route to acetic anhydride from methyl acetate[27], and BP Chemicals' acetic acid/acetic anhydride co-production process, both now commercialized. However, all these cyclic mechanisms may be incomplete, as individual steps may themselves be complex or catalysed by other species present, and the author has omitted many (minor) side-reactions.

10.7.8.4 *New developments*. If hydrogen is introduced during the carbonylation of methanol with a soluble cobalt/iodide catalyst, acetaldehyde is produced. Similarly, ethylidene diacetate, formally an adduct of acetaldehyde and acetic anhydride, is formed by the reaction of methyl acetate with carbon monoxide and hydrogen in the presence of rhodium catalysts. In both these processes, hydrogenolysis of the metal-acyl is probably involved, linking together parts of the methanol carbonylation and hydroformylation cycles. At very high pressures (> 100 atm.), ethylene glycol is formed from synthesis gas with soluble rhodium and ruthenium catalysts, presumably via a hydroxymethyl ligand and CO-insertion.

The reaction of methanol, carbon monoxide and oxygen ('oxycarbonylation') with soluble palladium and copper salts gives dimethyl carbonate and/or dimethyl oxalate (dimethyl ethanedioate), for hydrogenation to ethylene glycol. Postulated mechanisms include the sequence

$$(CO)_2Pd(OMe)_2 \rightarrow Pd(CO_2Me)_2 \rightarrow Pd^0 + (CO_2Me)_2$$

Heterogeneous Fischer–Tropsch chemistry is coming under closer scrutiny, and there is increasing interest in the use of traditional catalysts and supported carbonyl clusters of Group VIII metals for the possible production of lower olefins or alcohols from synthesis gas. The heterogenization of homogeneous catalysts for hydroformylation, carbonylation, etc. is also attracting much attention. Conversely, several homogeneous systems have now been found to effect the water-gas shift reaction, though heterogeneous catalysts have yet to be challenged for industrial use.

10.8 The future

The ultimate objective of the industrial chemist is the ability to design catalyst systems, homogeneous or heterogeneous, which will

(a) effect known reactions with high selectivities;
(b) extend the range of possible chemical reactions; and
(c) provide commercially viable rates under the mildest conditions consistent with thermodynamic requirements.

Enzymes provide selectivity, and effect certain reactions which have proved difficult to duplicate in more conventional chemical systems, but with very specific reactants. While the active centre is often organometallic, the protein structure is important for steric and conformational control (co-operative distortion of ligands to control redox potential, etc.), but is subject to 'denaturation' at temperatures too low for acceptable rates in other than speciality uses (some large tonnage in the food industry). Furthermore, the systems remain too complex for any thought of catalyst design along these lines, though model systems are being studied.

Homogeneous catalysis provides the maximum opportunity for mechanistic interpretation. The blossoming field of organometallic chemistry and catalysis also seems to provide the greatest scope for achieving novel reaction chemistry[28]. However, this very feature reflects the fact that organometallic chemistry is still young, and we are probably still on an early part of the learning curve. A considerable part of the effort directed at catalytic processes therefore retains the purely speculative or empirical approach.

Heterogeneous catalysis entails reactions between organic molecules and the surface of inorganic materials. Although a great deal of work has been carried out, and continues, to characterize the surfaces of materials, we remain far from being able to predict surface properties for all but the simplest. Even then, the interaction of an organic molecule with the surface may be characterizable, but is rarely predictable for any reaction, other than coordination. Nonetheless, the amount of information reported is exceedingly large, and by appropriate classification we can pick out short lists of possible candidates as starting points for a particular reaction. Theoretical considerations may point to those additives most likely to affect performance, but in the final analysis we follow a largely empirical approach yet again.

Only a few years ago, it was suggested that chemical technology was approaching maturity, but it now appears that a considerable number of possible new processes, many based on novel catalyst systems, are vying for consideration. However, the objective of catalyst design still seems remote.

References

1. J. Smidt, C&I, 1962, 54–61.
2. J. F. Roth et al., Chemtech, 1971, 1, 600–5.
3. S. Fujiyama and T. Kasahara, Hydrocarbon Processing, 1978, (November), 147–9.
4. E. S. Lewis (ed.), Investigation of Rates and Mechanisms of Reactions (Techniques of Chemistry, Vol. VI), 3rd edn., Part I, Wiley-Interscience, 1974.
5. D. C. Bailey and S. H. Langer, Chem. Revs., 1981 81, (2), 109–48.
6. C. R. Nelson and M. L. Couter, Chem. Eng. Progress, 1954, 50 (10), 526–31.
7. C&I, 1968, 1559–63.
8. O. Wiedemann and W. Gierer, Chem. Eng., 1979, (Jan. 29), 62–3.
9. (a) J. J. Graham, Chem. Eng. Progress, 1970, (Sept.), 54–8.
    (b) Kirt-Othmer Encyclopedia of Chemical Technology, Petroleum (Refinery Processes Survey), 3rd edn., Vol. 17, Wiley, 1982.
    (c) M. L. Riekena et al., Chem. Eng. Progress, 1982, (April), 86–90.
10. Kirk-Othmer Encyclopedia of Chemical Technology, Ammonia, 3rd edn., Vol. 2, Wiley, 1978.
11. (a) J. M. Thomas et al., New Scientist, 1982, (18 Nov.), 435–8.
    (b) C. Naccache and Y. B. Taarit, Pure & Appl. Chem., 1980, 52, 2175–89.
    (c) Ref. 29.
12. S. Brunauer, P. H. Emmett and T. Teller, J. Amer. Chem. Soc., 1938, 60, 309 (but see modern texts for current techniques and interpretation).
13. N. C. Saha and D. S. Mathur, J. Chromatography, 1973, 81, 207–32.
14. J. T. Yates Jr., C&EN, 1974, (Aug. 26), 19–29.
15. K. Tanabe et al., Bull. Chem. Soc. Japan, 1974, 47 (5), 1064–6.
16. (a) Y. One and T. Mori, J. Chem. Soc. Faraday Trans. I, 1981, 77, 2209–21.
    (b) C. D. Chang and C. T-W. Chu, J. Catalysis, 1982, 74, 203–6.
    (c) C. D. Chang, C. T.-W. Chu and R. F. Socha, J. Catalysis, 1984, 86, 289.
17. R. H. Crabtree et al., J. Amer. Chem. Soc., 1982, 104, 107–13.

18. A. K. Rappe and W. A. Goddard, *J. Amer. Chem. Soc.*, 1982, **104**, 448–56.
19. (a) A. W. Shaw *et al.*, *J. Org. Chem.*, 1965, **30**, 3286–9.
    (b) British Patent 932, 342 to British Petroleum Company, 1963.
20. *Oxidation of Organic Compounds*, Vols. I and II, Advances in Chemistry Series, Nos. 75 and 76, American Chemical Society, 1968.
21. (a) P. M. Maitlis, *The Organic Chemistry of Palladium* (2 vols.), Academic Press, 1971.
    (b) J. E. Bäckvall *et al.*, *J. Amer. Chem. Soc.*, 1979, **101**, 2411–6.
22. H. Mimoun *et al.*, *J. Amer. Chem. Soc.*, 1978 **100**, 5437.
23. J. D. Burrington *et al.*, *J. Catalysis*, 1983, **81**, 489–98.
24. Y. Takita *et al.*, *J. Catalysis*, 1972, **27**, 185–92.
25. W. A. Herrmann, *Angew. Chemie, Int. Ed.*, 1982, **21**, 117–30.
26. D. Forster, *Advances in Organometallic Chemistry*, 1979, **17**, 255–67.
27. (a) J. L. Ehrler and B. Juran, *Hydrocarbon Processing*, 1982, (Feb.), 109–13.
    (b) M. Schrod and G. Luft, *Ind. Eng. Chem. Product Research Dev.*, 1981, **20**, 649–53.
28. (a) D. L. King *et al.*, *Hydrocarbon Processing*, 1982 , (Nov.), 131–6.
    (b) A. Aquilo *et al.*, *Hydrocarbon Processing*, 1983, (March), 57–65.
    (c) E. L. Muetterties and M. J. Krause, *Angew. Chemie, Int. Ed.*, 1983, **22**, 135–48.
    (d) D. F. Shriver, *Chem. Brit.*, 1983, (June), 482–7.
29. S. L. Meisel, *Chemtech*, 1988, (Jan.), 32–37.

Bibliography

Thermochemical Kinetics, 2nd edn., S. W. Benson, John Wiley and Sons, 1976.

Thermochemistry of Organic and Organometallic Compounds, J. D. Cox and G. Pilcher, Academic Press, 1970.

The Chemical Thermodynamics of Organic Compounds, D. R. Stull, E. F. Westrum and G. C. Stinke, John Wiley and Sons, 1969.

Catalysis and Inhibition of Chemical Reactions, P. G. Ashmore, Butterworths, 1963.

Principles and Applications of Homogeneous Catalysis, A. Nakamura and M. Tsutsui, John Wiley and Sons 1980.

Catalysis, Vols I–VII (ed. P. H. Emmett), Reinhold, 1954–60.

Physical and Chemical Aspects of Adsorbants and Catalysts (ed. B. G. Linsen), Academic Press, 1970.

Introduction to the Principles of Heterogeneous Catalysis, J. M. and W. J. Thomas, Academic Press, 1967.

Design of Industrial Catalysts, D. L. Trimm, Elsevier, 1980.

Heterogeneous Catalysis in Practice, C. N. Satterfield, McGraw-Hill, 1980.

CHAPTER ELEVEN

PETROCHEMICALS

J. PENNINGTON

11.1 Introduction

11.1.1 *Layout*

With over 90% of synthetic organic materials presently produced from petroleum or natural gas, the main technical questions are how we derive feedstocks or building blocks suitable for chemical processes from such raw materials, and what sort of chemistry follows?

After the introduction and a brief section on refinery processes, this chapter describes the major petrochemical cracking and reforming operations, which provide the more reactive feedstocks for chemical conversions. In these, energy requirements play an increasingly important role, and some space has, therefore, been devoted to this topic. Some immediate downstream products have also been included in these sections.

Thereafter, the operations of the petrochemical industry, or more accurately bulk organic chemical industry, centre on chemistry. High yield is a major target, but there is great emphasis on catalysis, to permit the use of the cheapest feedstocks available and eliminate expensive reagents. Presently, the industry is at a turning point, and future flexibility with respect to primary raw materials is of great concern. Production routes to acetic (ethanoic) acid and anhydride may now provide a pointer to future possibilities, and have been considered in a separate section. The remaining chemical products are dealt with in carbon number sequence, to bring together non-synthetic and synthetic routes.

Petrochemicals are predominantly intermediates for polymers, and reference to these materials is essential to put the present topic into perspective, although plastics and polymers are not considered in detail here.

11.1.2 *The beginnings*

A major factor leading to initial developments in the petrochemical field was the enormous increase in popularity of the motor-car in the United States during the 1920s and 30s. Henry Ford paved the way to mass ownership, and the first large finds of petroleum provided cheap gasoline. However, the separation of a gasoline cut from petroleum left refiners with increasing amounts of high-boiling fractions; thermal cracking gave additional gasoline, together with hydrocarbon gases. These were initially burned as refinery fuel, but an excess soon prompted the development of chemical processes for their use, mainly within the U.S. oil industry. Subsequently, the development of wet natural gas supplies, i.e. gas containing ethane, propane etc., which were available in the U.S.A., also attracted chemical interest, particularly in the use of ethane as a precursor for ethylene. As a result, the U.S. production of organic chemicals from these sources grew from about one hundred tonnes in 1925 to a million tonnes in 1940.

11.1.3 Into the 70s

The wartime and postwar increases in demand for fuel, and the introduction of new synthetic materials, led to a very rapid growth in the U.S. petrochemical industry into the early 70s, still largely (c. 80%) based on cheap refinery and natural gases.

In Europe, fuel oil requirements have dominated refinery operations. Gasoline (petrol) always represented less than 20% of petroleum consumption, and refinery cracking operations never figured prominently until the 70s. However, naphtha (paraffinic, or low octane, motor spirit) became surplus to requirements, and the advent of progressively larger naphtha crackers into Europe after World War II (and later in Japan) allowed these countries to enter petrochemicals. From the 1950s, growth in Europe was higher than in the U.S.A., and the West European consumption of lower olefins caught up in 1973 at about 10 million tonnes (Mt) of ethylene and 5 Mt of propylene.

Until 1950 the coal-tar industries were the major source of aromatics in both Europe and the U.S. Thereafter, they contracted slowly, while demand for aromatics increased. The chemical industry again turned to petroleum, which eventually supplied some 90% of benzene (4–5 Mt) and other aromatic hydrocarbons worldwide.

The third major doorway into petrochemicals is via synthesis gas. Methanol synthesis is the major organic downstream operation, but is often integrated with still larger ammonia production facilities. U.S. methanol production in 1973 was about $2\frac{1}{2}$ Mt (cf. 14 Mt of ammonia), with a similar figure for Western Europe.

11.1.4 The present

After a short slump, following oil price increases in the early 70s, the recovery in 1976–7 was interpreted as heralding continued expansion. Construction of additional capacity commenced, inevitably in very large blocks; a typical cracker will produce 5×10^5 tonnes per year (500 kilotonnes per annum) of ethylene, and the total output from the complex may exceed 1 Mt per annum.

In the event, demand increased up to 1979–80, but generally fell, or at best levelled off, following a further sharp upward movement in oil prices. Despite cancellation of several planned expansions and new projects, a number of new plants were still coming on-stream throughout the world, including the less well-developed countries of the Far East, Middle East, Africa and Central/South America. The fall in export potential contributed to something like a 60% excess of nominal capacity over actual requirements in both Europe and the U.S.A. during 1982, leading to a number of plant shutdowns, both temporary and permanent. The 1982 U.S. production figures for major chemicals were only slightly higher than for 1973.

However, just as the industry was beginning to stabilize, the start of 1986 brought a dramatic fall in crude oil prices. Following a period of readjustment, demand has again begun to surge ahead, except in one or two areas (ammonia, for instance) and the resurrection of old plants and plans for expansions and new plants are the order of the day. A few industry observers are getting jittery, lest we are initiating yet another go-stop cycle.

The situation remains rather too fluid to be able to quote production capacities with any certainty. However, some indication of reasonably recent production figures have been included, where available (mainly from U.S. sources). For comparison with the figures for 1973, 1987 saw U.S. production reach 15·9 Mt ethylene, 8·4 Mt propylene, 5·3 Mt benzene, 14·7 Mt ammonia and 3·3 Mt methanol.

11.1.5 Individual feedstocks and routes

Despite the high tonnages of petrochemicals, the chemical industry as a whole consumes rather less than 10% of available petroleum and natural gas hydrocarbons as feedstocks, with possibly a further 4–5% as fuel. For comparison, the current consumption of gasoline alone in Western Europe exceeds 120 Mt per annum, while the U.S. figure is nearly 300 Mt per annum. Hence, prices of individual hydrocarbon feedstocks are largely determined by other forces; the most economic feedstock/route combination has frequently changed with time, and may differ in different parts of the world. Furthermore, while a specific route may be preferred for new plants, older plants for which the capital is largely written off may well remain economically viable. Finally, special situations may prompt individual solutions. For example, Rhone–Poulenc in France derive the carbon monoxide for a very modern acetic acid plant, based on Monsanto's methanol carbonylation process, from the partial oxidation of methane to acetylene and synthesis gas. They will, therefore, continue to manufacture vinyl acetate by the otherwise obsolescent route from acetylene rather than ethylene (see section 11.3.3.1). Therefore, we find a surprising variety of co-existing technologies for many downstream products.

11.2 Crude oil, gas and refinery operations

11.2.1 *Crude oil and natural gas*

Petroleum, or crude oil, is an extremely complex mixture derived like coal from prehistoric vegetation. The components range from gaseous to semi-solid or solid hydrocarbons, with compounds of sulphur, nitrogen, oxygen and various metals as impurities. Distillation gives roughly the fractions shown in Table 11.1. The hydrocarbons are almost entirely *saturated* paraffins, cyclo-paraffins (naphthenes) and aromatics/polyaromatics; the proportions vary enormously from one source to another.

Methane dissolved in or trapped above the oil is referred to as associated gas; ethane, propane etc. are usually also present. Some of this gas separates at the well-head, and must be disengaged. While separate distribution of the gas presents no problem for land-based oil wells in developed countries, the facilities required to eliminate flaring in the North Sea and Middle East entailed considerable capital expenditure.

Gas may also be found unassociated with oil. The gas may be 'wet', containing appreciable amounts of ethane and other hydrocarbons, but gas from the (southern) North Sea is 'dry' i.e. almost entirely methane.

11.2.2 *Refinery operations*[1]

The terms 'cracking' and 'reforming' are often used without qualification to describe a variety of refinery and primary petrochemical operations. It is hoped

Table 11.1 Petroleum fractions

Methane ⎱ Ethane ⎰	Associated gas	(Natural gas)
Propane ⎱ Butanes ⎰	Liquefied petroleum gases (LPG)	(Natural gas liquids—NGL)
C5(> 0°C)–70°C	Light gasoline or light naphtha	Feedstocks for motor spirit, also known as straight-run gasoline (US).
70°C–170°C	Naphtha (mid-range)	
170°C–250°C	Kerosine (UK)*	Vaporizing oil, jet fuel
250°C–340°C	Gas oil	Diesel fuel, light fuel oil
340°C–500°C	Heavy distillates (heavy gas oil in U.S.A.)	Feedstocks for lubricants and waxes, or heavy fuel oils.
Fluid residues (from light crudes)		
Semi-solid residues	Bitumen or asphalt	

* Depending upon projected use, fractions within this or a similar boiling range may be included within (full-range) naphtha or (light) gas oil classifications.

that Table 11.2 will provide a general guide to what is meant in a particular context.

11.2.2.1 *Catalytic cracking.* The old thermal cracking processes gave way to catalytic cracking—the 'cat-cracker'—during the 1940s in the U.S.A. The original clay catalysts have now been replaced by zeolites, operating at about 500°C and 2 atm. Carbon is laid down on the catalyst surface, and must be burned off at about 700°C; in the modern fluid (fluidized-bed) catalytic cracker (FCC unit), the particulate catalyst is transported continuously through the hydrocarbon reaction and regeneration zones. In the process, higher alkanes are cleaved, aromatics partially dealkylated and some naphthenes converted to aromatics by hydrogen transfer to olefins. The product mixture has a broad carbon number spread with some 60% by weight in the C_5 to 220°C cut and 15% as gases, mainly C_3 and C_4 alkenes and alkanes. The latter may be used in other refinery or chemical processes.

11.2.2.2 *Hydrocracking.* Hydrocracking is a more limited (and expensive) operation, aimed at converting still higher boiling fractions to naphtha. The process operates at about 450°C and 150–200 atm of hydrogen, with a metal (often palladium) on zeolite catalyst. Heavy aromatic and polyaromatic components are saturated and cleaved, to give a mainly paraffinic product (i.e. naphtha) suitable for reforming or petrochemical use.

Table 11.2 Refinery and petrochemical cracking and reforming operations

Refinery	
Thermal cracking	Obsolete, thermal decomposition of middle/higher fractions to increase gasoline range hydrocarbons.
Thermal cracking/coking (delayed coking etc.)	Thermal decomposition of very heavy fractions to give mainly gases and a high coke yield.
Catalytic cracking	Accelerated decomposition, with some aromatization, of middle/higher fractions over solid acidic catalysts.
Hydrocracking	Accelerated hydrogenolysis/decomposition of heavy fractions to paraffinic hydrocarbons over metal/acid catalysts.
Thermal reforming	Obsolete, thermal rearrangement and aromatization of naphthas under high pressure.
Catalytic reforming (platforming etc.)	Metal/acid-catalysed rearrangement and aromatization of naphthas.
Petrochemical	
Steam cracking	Thermal cracking of C_2 + hydrocarbons to olefins in the presence of steam.
Thermal/autothermal cracking	More general term, including methane to acetylene—autothermal, with partial combustion.
Steam reforming	Nickel-catalysed formation of synthesis gases from hydrocarbons.

11.2.2.3 *Catalytic reforming*. Older thermal reforming processes similarly gave way to catalytic reforming. A typical catalyst comprises platinum or a platinum-rhenium alloy (0·5–1%) on an acidic alumina, and is used at about 500°C and 7–30 atm pressure with a 80°–230°C naphtha feedstock. The most important reactions occurring are cyclization and dehydrogenation (known as 'dehyᵣdrocyclization') of alkanes, and dehydrogenation of naphthenes to aromatics. Various isomerization, alkylation, dealkylation, cleavage and hydrogen transfer reactions occur simultaneously. A typical feed and product composition might be:

	Feed	Product
		(c. 75% yield by volume)
Paraffins	60	32
Naphthenes	25	2
Aromatics	15	66
Individual aromatics (%)		
Benzene		4
Toluene		18
Xylenes and ethyl benzene		23
Higher		21

The rather low benzene content is fairly common and, with benzene in greatest demand, other sources of this material are required. Hydrogen and saturated C_{1-4} hydrocarbons (10–14% w/w) are also co-produced in reforming operations.

11.2.2.4 *Other refinery processes*. Some of the more intractable residues of refinery distillations are subjected to thermal coking, to provide small amounts of gas/liquid fuels while producing a more readily handled 'coke' fuel.

Olefins from cat-cracker operations (or even steam cracking of natural gas liquids in the U.S.A.), may be used to alkylate branched alkanes, particularly isobutane (c. H_2SO_4 or HF catalyst at near ambient temperature). Alternatively, propylene and butenes may be oligomerized, and the resulting C_6 + olefins hydrogenated to give 'polygasoline'.

The term 'hydrotreating' tends to be used to cover all refinery hydrogenation processes. These may range from hydrogenation of olefinic mixtures (liquid cracker products for example) under mild conditions, to the hydrodesulphurization of heavy fractions and the hydrogenation of (poly-) aromatic components.

11.2.3 *Energy consumption*

According to the number of operations carried out on a refinery site, some 7 to 10% of the fuel value of crude oil introduced is consumed in the separation and interconversion of product streams. If we momentarily take naphtha as a reference point, energy is required for the catalytic reforming process to provide higher octane gasoline. The subsequent separations of individual aromatic hydrocarbons consume still more energy; hence the 'chemical values' of these materials exceed their fuel values.

In later sections, the author has assigned a fuel value to naphtha. However, even this material requires the production, transportation and distillation of crude oil at the very least. Thus, any strict assessment of energy utilization for a particular product should ultimately include inputs to bring the raw material out of the ground and for all subsequent operations.

11.3 Lower olefins (alkenes) and acetylene (ethyne)

The paraffinic hydrocarbons in natural resources show very low reactivity under moderate conditions. Early petrochemical processes drew upon the greater reactivity of olefins, and when refinery supplies failed to meet demand means of converting alkanes to alkenes were developed. During the later period, low prices of naphtha (in Europe and Japan) and natural gas (in the U.S.A., and more recently Europe) also promoted petrochemical routes to acetylene.

The resulting processes all require high temperatures and are, therefore, both energy- and capital-intensive, so that very large scales of operation are desirable for efficiency and economic viability.

11.3.1 Cracking processes

11.3.1.1 Steam (thermal) cracking for lower olefins. A typical naphtha for cracking has a carbon number range from 4 to 12 or more. The uncatalysed cracking reaction is carried out within tubes in a furnace enclosure at near atmospheric pressure (less than 3 atm). Steam makes up 30–45% w/w of the total feed to improve heat transfer, reduce the partial pressure of hydrocarbons (thermodynamically desirable) and remove carbon by the reaction

$$C + H_2O \rightarrow CO + H_2$$

Cracking temperatures were traditionally 750–850°C, but temperatures up to 900°C (high severity cracking) are becoming more common, in conjunction with shorter residence times (about 0·1 second).

Hydrocarbon dehydrogenation and cleavage reactions of the type

$$A \rightleftharpoons B + C + D \dots$$

are always strongly endothermic, with ΔH values in the range 120–180 kJ for each additional molecule produced. Figure 11.1 shows the temperature dependence of $\log 10\, K_p$ (atm) values per g-atom of carbon for the formation of a number of relevant hydrocarbons from the elements. Either methane or benzene is the most stable hydrocarbon over the temperature range of interest, but *no* hydrocarbon is stable with respect to the elements between 800 and 4000 K. Hence, the successful formation of olefins in significant yields implies that kinetic and mechanistic factors are very important.

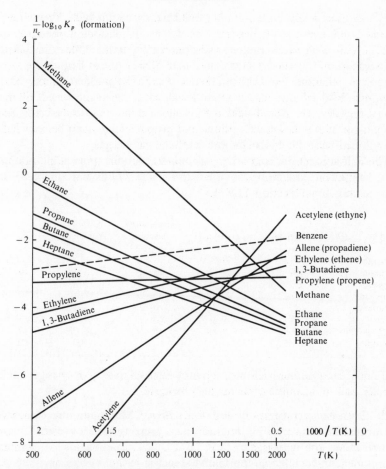

Figure 11.1 Temperature dependence of log K_p per gram atom of carbon for the formation of a series of hydrocarbons from their elements.

The reaction mechanism, and its implications with respect to the product distribution, has been described in some detail by Cattanach (Tedder *et al.*, 1975)[2]. There is now evidence that coke formation on the tube walls may to some degree be metal-catalysed, and a current development within BP is the application of an inert coating to inhibit this process. With increasing interest in the steam cracking of still higher fractions (such as gas oils) such a development could become even more important.

The table on p. 356 shows typical overall yield patterns (%w/w), with ethane/propane recycle, for a number of cracker feedstocks; *n*-butane gives high ethylene yields, isobutane more propylene and methane. In all cases, the cracker product stream is cooled rapidly to below 400°C to minimize further reactions. After further cooling and separation of condensed hydrocarbons and water, the gases (H_2, C_1—C_4) are compressed, scrubbed with aqueous alkali to remove

CO_2 and other acidic contaminants, and dried over solid beds. Thereafter, the C_2 and C_3 alkanes and alkenes are separated by distillations at pressures up to 35 atm., with refrigerated condensers for the early columns in the train. Selective hydrogenation to remove acetylenes and dienes (most frequently over a supported palladium catalyst) and further distillative purification may also be required. With ethane and propane feedstocks, the unconverted alkane is always recycled; the overall yield of ethylene on ethane may exceed 80%. In the cracking of higher feedstocks, ethane and propane may again be recycled, or recombined with the hydrogen and methane as fuel gas.

The C_4 fraction from the cracking of naphtha contains appreciable quantities of 1, 3-butadiene, and represents the major source of this material in Western Europe and Japan (section 11.9.1).

Product	Ethane	Propane	n/iso-Butane mixtures	Naphthas	Gas oil
Hydrogen	5–6	1–2	1	1	0.5
Methane	10–12	20–25	15–25	13–18	10–12
Ethylene	75–80	40–45	20–30	25–37	22–26
Propylene	2–3	15–20	15–25	12–16	14–16
Butadiene ⎱		1–2	0–2	3–5	3–5
Other C_4's ⎰	1	3–4	15–22	3–5	3–6
20(C_5's)–220°C	1	5–10	5–14	18–28	17–22
Fuel Oil	—	—	—	4–8	18–22

Finally, the gasoline fraction, frequently referred to as 'pyrolysis gasoline', is usually rich in aromatics, particularly benzene.

11.3.1.2 *Cracking processes for acetylene (ethyne)*. Minor amounts of acetylene are formed in steam cracking, but rarely justify separation. However, if cracking temperatures are increased to over 1100°C acetylene becomes a significant component, and a different approach to reactor design is necessary. The Wulff process[3] furnaces contain stacks of tiles, and operate on 1-minute heating-cracking cycles. Fuel gas or oil is burned in air to heat the tiles to about 1200°C, and the feed is then switched to naphtha and steam, the temperature falling to about 1000°C to complete the cycle. The operation of 36 furnaces, with periodic flow reversals and coke burn-offs, is controlled by an IBM computer at Marathon Oil's plant in West Germany.

At still higher temperatures (over 1500°C), methane can be partially converted into acetylene. An electric arc process has been operated, but the most successful and widespread approach is the BASF or Sachsse partial oxidation process. Typically methane (in excess) and oxygen are fed to a special burner, in which the temperature falls from a peak of about 2500°C to 1300°C in a very short time, at which point the gases enter a quench. A typical product stream (water-free basis) would contain some 8% (molar) acetylene, 25% carbon monoxide, 55% hydrogen and small amounts of methane, carbon dioxide and other gases. After acetylene extraction, the remaining gases may be integrated into a synthesis gas system. Société Belge d'Azote and Montecatini have commercialized their own, similar processes.

Finally, BASF have also developed a submerged flame autothermal cracking process for crude oil[5]. Partial combustion raises the temperature locally, to the point at which cracking to both ethylene and acetylene occur. A typical off-gas composition comprises approximately 1 mol acetylene, 1·15 mol ethylene, 7·5 mol carbon monoxide and 4·5 mol hydrogen, with minor amounts of C_3 and higher products.

11.3.2 Energy balances and economics

A standard economic analysis, as described in section 5.11, is obviously necessary when a company considers alternative primary raw materials and cracking process technologies. However, raw materials show significant variations in local availability and relative pricing, and primary petrochemical cracking operations give a variety of co-products which vary in value according to local demand. Hence, the relative economics for such alternatives vary quite markedly from one location to another. A rather more fundamental basis for comparison is the total net primary energy, including the energy value of the hydrocarbon (or other) feedstock(s), required per unit quantity of the major product. This figure determines the minimum non-renewable energy consumed in making all subsequent derivatives, and hence the status of materials and routes in attempts to conserve energy sources (see section 11.3.1).

The following simplified examples are illustrative, rather than definitive. A full analysis requires detailed data for each individual operation. The author has used lower, or net, heats of combustion at 25°C for the energy (fuel) values of raw materials i.e. where all combustion products are in the gaseous state. Steam, whether generated or consumed, has been assigned an energy value of 42 kJ/mol, unless at high superheat. Electricity has been included at the primary fuel requirement based on a generation efficiency of 33%, which could be improved upon by future combined heat and power generation schemes. (Energy use in cooling water systems has been ignored).

Some improvement in efficiency is almost invariably achievable with additional capital equipment. However the metals used and fabrication also entail energy consumption, which must be compared with the energy saved over the life of that equipment. The employment of labour also entails additional energy expenditure, but the relevant amounts involved are minor and the topic has not been pursued.

11.3.2.1 *Ethylene (ethene) and other olefins (alkenes)*. We will start with the simpler cracking of ethane according to the following equation and theoretical energy balance (in MJ/mol ethylene), based on energy (fuel) values as defined above.

$$C_2H_6(+ \text{ heat}) \rightarrow C_2H_4 + H_2 \qquad (\Delta H_r^\ominus = 0\cdot137)$$
$$1\cdot428 + 0\cdot137 \qquad 1\cdot323 + 0\cdot242 \qquad (\text{total } 1\cdot565)$$

Although absent from the equation, generation of the co-fed steam requires energy, and extra heat is necessary to raise the temperature of the reactants to over 800°C. Some of this extra energy is recovered (by steam generation) from the furnace flue gases and the reactor outlet gases, but, with additional steam required for product separation and purification, a net steam consumption results. The overall yield of ethylene on ethane in a modern U.S. plant is about 85% molar, and an approximate energy balance might be as follows:

Inputs:	Ethane (1·18 mol)	1·68
	Furnace fuel	0·53
	Steam consumed (net)	0·23
	Electricity	0·01
		2·45 MJ/mol

By-products:	Gases (H_2, CH_4)	0·33
(credited at	C_3, C_4	0·07
fuel values)	Pyrolysis gasoline	0·05
		0·45 MJ/mol

| Net input for ethylene | | 2·00 MJ/mol (71GJ/tonne) |

Thus the net energy* utilization is about 1·5 times the fuel value of ethylene; this is indicative of significant inefficiency in the overall process. For comparison, a traditional costing ($, 1980) based on the same figures would take the following form. (The U.S. dollar is increasingly used as a basis for comparative costings in Europe and elsewhere).

Capacity	500 000 tonnes/year	
(Actual production	350 000 tonnes/year)	
Fixed capital	$220 million (63 c/kg produced)	
Ethane	(17·25 c/kg)	21·8 c/kg
Other primary energy	(0·2 c/MJ)	5·5
Other chemicals		1·3
By-product gases	(0·2 c/MJ)	(2·4)
C_3 and higher	(say 20c/kg)	(1·9)
Net direct (or variable) costs		24·3 c/kg
Depreciation, maintenance, and labour (fixed costs)		16·0
Production cost		40·3 c/kg
20% return on fixed capital		12·6
Target sale price		52·9 c/kg C_2H_4

---

*Note: 'Energy' is used in a more limited sense in this chapter and is essentially only the chemical energy referred to in chapter 7.

Such a shortfall in production applied during 1981/2 when the U.S. price of ethylene fell to 44 cents/kg. At the present time, a lower feedstock price and full production should have offset any increases in capital and labour rates, while demand has pushed the price of ethylene to over 70 cents/kg.

(As an alternative to cracking, Union Carbide have patented improved catalysts for the selective oxidation of ethane to ethylene, which may ultimately prove more energetically and economically favourable.)

The analysis of propane or naphtha cracking is rather more complex, owing to the variety of products formed, and changes in the distribution with operating conditions. However, some reasonably typical figures with ethane recycle (MJ/mol ethylene) are as follows:

Propane (64·2g)	2·98	—
Naphtha (84g)	—	3·62
Other fuel	0·58	0·80
Steam consumed	0·26	0·32
Electricity	0·01	0·01
Inputs	3·94	4·75
Gases ($CH_4, H_2, tr C_2$)	1·00	0·80
Propane	(recycled)	0·03
Propylene	0·37	0·66
C_4's	0·08	0·38
Pyrolysis gasoline	0·35	0·55
Fuel oil	—	0·08
By-and co-products (credited at fuel values)	1·80	2·50
Net input for ethylene	2.14 MJ/mol (76GJ/tonne)	2.25 MJ/mol (80GJ/tonne)

The energy requirements appear to be higher than for ethane cracking. But all energy losses have been assigned to ethylene by crediting the by- and co-products at fuel values. The deliberate production of propylene and other olefins as chemical feedstocks, and gasoline fractions of higher octane value than naphtha, would have incurred additional energy utilization and losses. Thus the useful co-products should be credited at a higher energy value, referred to as the chemical value in later analyses.

From an economic viewpoint, the capital costs of naphtha-crackers are higher, but the recent fall in the price of world-traded naphtha has significantly reduced the historical advantage enjoyed by U.S. production based on ethane-cracking.

11.3.2.2 *Acetylene (ethyne)*. Before evaluating the petrochemical routes to acetylene, we should look first at the long-established process for producing acetylene via calcium carbide and water:

$$CaO + 3C \rightarrow CaC_2 + CO$$

$$CaC_2 + H_2O \rightarrow C_2H_2 + CaO \text{ (or } Ca(OH)_2 \text{ with excess water)}$$

Essentially all the energy is consumed in calcium carbide manufacture. In the electrothermal process, the reaction is carried out at temperatures in excess of 1850°C. If we ignore the lime (recovered), and convert other consumption/by-product figures to energy values, we obtain (per mol acetylene equivalent):

Coke (3·5g-atoms C/mol)		1·4	
Electricity—actual input	1·0		
Primary fuel value	—		3·0
		2·4 MJ/mol	4·4 MJ/mol
CO produced (1·32 mol, fuel value)		0·4	
Net input		2·0 MJ/mol (77GJ/tonne)	4·0 MJ/mol (154 GJ/tonne)

Hence, with electricity generated by conventional means, the net primary energy consumption is very high at over 3 times the fuel value of acetylene. (Any energy losses in the conversion of coal, as mined, into coke have been ignored.)

BASF sought to develop a process in which high temperatures were generated by the partial combustion of additional coke within the carbide-producing furnace. If the co-produced CO were required for chemical purposes, this approach would seem to be more energetically attractive. As a further alternative, AVCO in the U.S.A. have demonstrated that acetylene can be produced directly from coke and hydrogen in an electric arc, but primary energy inputs are presently excessive.

Returning to petrochemical raw materials, some figures for a Wulff naphtha cracking operation are as follows (in MJ/mol acetylene):

Inputs:	Naphtha feed (113 g/mol)	5·07
	Furnace fuel	1·69
	Steam (net, for separations)	1·03
	Electricity (primary fuel)	0·11
		7·90 MJ/mol
By-products:	Gases (5 mol, H_2, CH_4, CO, C_2H_6)	1·97
	C_3 and higher	0·65
	Carbon	0·11
		2·73 MJ/mol
	Net	5·17 MJ/mol
Co-products:	C_2H_4(0·94 mol, @2 MJ/mol) chemical value	1·88
	Input for acetylene, net	3·29 MJ/mol (127GJ/tonne)

The 'chemical value' assigned to ethylene is based on the net energy requirement for ethylene production by steam cracking.

For the partial oxidation of methane, a generalized 'design case' gives the following set of approximate figures (again in MJ/mol acetylene).

Inputs: Natural gas   5·45
                 Additional fuels/steam   0·83
                 Oxygen (4 mol/mol)   0·40
                 Electricity (primary fuel value)   0·35
                                                      ─────
                                                      7·03 MJ/mol

By-products: Methane (in synthesis gas)   0·49
                 C_2 and higher   0·27
                                                      ─────
                                                      0·76 MJ/mol

                                    Net   6·27 MJ/mol

Co-products: Synthesis gas (10 mol, $c.$ $2H_2$:1CO);
                 fuel value   2·55
                 chemical value (say 1·2x)   3·06
                                                      ─────   ─────
                 Input for acetylene, net   3·72 MJ/mol 3·21 MJ/mol
                                                     (143 GJ/tonne) (123 GJ/tonne)

Somewhat similar net figures are obtained for the autothermal cracking of crude oil if the ethylene coproduct is credited at 2 MJ/mol. Even with chemical values attached to co-products, the petrochemical processes consume over 3 MJ/mol for the production of acetylene. Furthermore, the capital costs for these processes are appreciably higher than for ethylene crackers. Thus, while the production costs of acetylene show considerable variations from one location to another, it is invariably dearer, by 30% to over 100%, than ethylene.

11.3.3 Lower olefins (alkenes) versus acetylene (ethyne)

The major outlets for ethylene, propylene, butadiene (section 11.9.1) and acetylene are shown in Fig. 11.2.

11.3.3.1. General considerations. A number of organic products are potentially producible from either acetylene or a lower olefin, for example:

$$C_2H_2 + H_2O \quad\quad \rightarrow CH_3CHO \quad\quad \leftarrow C_2H_4 + \tfrac{1}{2}O_2$$
$$C_2H_2 + CH_3CO_2H \rightarrow CH_2 = CHO_2CCH_3 \leftarrow C_2H_4 + \tfrac{1}{2}O_2$$
$$+ CH_3CO_2H$$
$$C_2H_2 + HCN \quad\quad \rightarrow CH_2 = CH{\cdot}CN \quad\quad \leftarrow C_3H_6 + NH_3 + 1\tfrac{1}{2}O_2$$

The reactions of acetylene are simpler, and were important early commercial routes to the products indicated, acetaldehyde, vinyl acetate and acrylonitrile.

We have just seen that the net energy requirements for acetylene are higher than for ethylene, and the difference is reflected in the energy requirements for all downstream products. To give an indication of how energy balances can continue to be applied the following table has been compiled from fairly modern sources. Energy inputs of 2·0 and 3·0 MJ/mol have been used for ethylene and acetylene respectively.

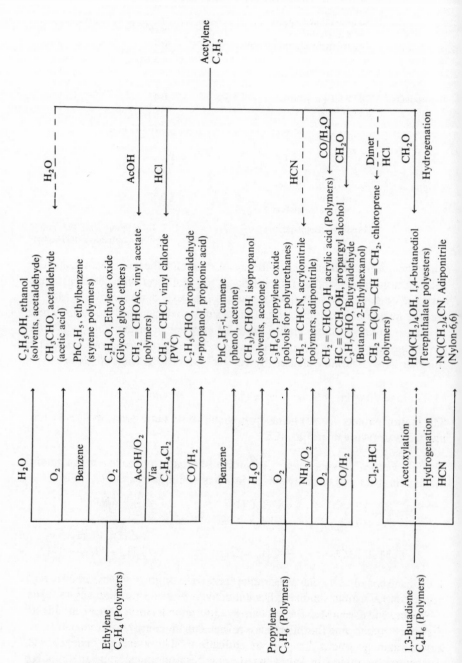

Figure 11.2 Major outlets for ethylene, propylene butadiene and acetylene. Brackets indicate further uses of the intermediate product.

	Acetaldehyde		Vinyl acetate	
	C_2H_2	C_2H_4	C_2H_2	C_2H_4
C_2H_2 (1·05 mol)	3·15	—	3·15	—
C_2H_4	— (1·05 mol)	2·10	— (1·08 mol)	2·16
Oxygen	—	0·06	—	0·07
Steam	0·21	0·12	0·46	0·48
Electricity	0·08	0·02	0·05	0·08
Acetic acid			say 2·55	2·55
Total input	3·44 MJ/mol	2·30 MJ/mol	6·21 MJ/mol	5·34 MJ/mol
	(78 GJ/tonne)	(52GJ/tonne)	(72GJ/tonne)	(62·1GJ/tonne)

However, the main driving force to discover and develop olefin based routes was the dramatic increase in scale of steam cracking operations in the post-war years. Early increases in demand for synthetic ethanol and ethylene oxide/glycol were followed by a rapid escalation in polyethylene production, now the major outlet for ethylene. In contrast, the group of products for which acetylene was the preferred feedstock grew more modestly, and a disparity in required production scales and prices developed. The steam cracking of naphtha in Europe and Japan also gave propylene in large quantities, for which new outlets had to be found. Thus acetylene was slowly squeezed out, including later displacement by butadiene for chloroprene manufacture, and the ratio of ethylene production to acetylene production is now 100-fold in the U.S.A., if somewhat less elsewhere.

11.3.3.2 *Vinyl chloride (chloroethene)*. The production of vinyl chloride naturally attracted the attention of those companies already involved in chlorine production and chlorination processes. Many chlorination processes are substitutive, i.e.

$$RH + Cl_2 \rightarrow RCl + HCl$$

The quantities of by-product hydrogen chloride produced were often an embarrassment and the relatively facile production of vinyl chloride from acetylene and anhydrous HCl, over a supported mercuric chloride catalyst, provided a valuable additional outlet.

As the market for vinyl chloride grew, major petrochemical companies without existing chlorination capabilities looked to other possible routes. With falling ethylene prices, the route pursued was the addition of chlorine to ethylene, to give ethylene dichloride (EDC, 1, 2-dichloroethane), and the thermal cracking of EDC at about 550°C to vinyl chloride and HCl.

$$CH_2 = CH_2 + Cl_2 \rightarrow ClCH_2CH_2Cl \rightarrow CH_2 = CHCl + HCl$$

A snag was that these companies now had to dispose of hydrogen chloride. The most elegant solution was to react this with acetylene in a separate stage, to give what became known as the 'Balanced Process'. The overall stoichiometry is then

$$C_2H_4 + C_2H_2 + Cl_2 \rightarrow 2CH_2 = CHCl$$

This scheme became a driving force for naphtha cracking to mixtures of ethylene and acetylene, as exemplified by the Wulff process.

Nevertheless these special crackers proved expensive, and many gave serious operational problems. The 'final solution' came with the development of the ethylene oxychlorination process. The reaction

$$CH_2 = CH_2 + 2HCl + \tfrac{1}{2}O_2 \rightarrow ClCH_2CH_2Cl + H_2O$$

is carried out at 250–300°C over supported copper chloride (melt) catalysts. This process may be combined with chlorine addition and EDC cracking to provide a balanced operation, if required; the overall stoichiometry then becomes

$$2CH_2 = CH_2 + Cl_2 + \tfrac{1}{2}O_2 \rightarrow 2CH_2 = CHCl + H_2O$$

As a guide, a rough energy balance for a modern vinyl chloride pro-duction/polymerization operation is as follows (MJ/mol or GJ/tonne as indicated)

Ethylene (1·05 mol)	2·1	MJ/mol
Chlorine	0·8	
Processing inputs	0·9	
Vinyl chloride	3·8	MJ/mol

which gives

Vinyl chloride monomer	62	GJ/tonne
Processing inputs	10	
Polyvinyl chloride	72	GJ/tonne

The polymer referred to here is rigid PVC. The flexible material contains appreciable amounts of 'plasticizers' (mainly phthalate esters), which obviously affects the overall energy utilization.

11.3.4 *Polyethylene (polyethene) and polypropylene (polypropene)*

Low-density polyethylene (LDPE) is produced by a free radical process, initiated with traces of oxygen or peroxides, at temperatures of 200°C to over 300°C and pressures up to 3000 atm. Over half of this low-melting point, flexible product is used for packaging film manufacture.

High density polyethylene (HDPE) is now produced mainly by polymer growth on microscopic particles of catalytic materials (chromium or Ziegler systems) suspended in a non-solvent or carried along in the gas-phase, at only

70–125°C and 15–40 atm. The product has a higher melting point and is more rigid, and finds major uses in containers, moulded items and pipes. Polypropylene, with both molding and fibre uses, ethylene-propylene copolymers and the newest product line, linear low-density polyethylene (LLDPE), are produced by similar methods. A C_4/C_6 terminal olefin co-monomer is used in the latter to provide a tough, flexible film forming product.

Yields in all these processes are in the range 97–99% of theory overall, giving feedstock energy inputs of 72–73 GJ/tonne with ethylene at 2 MJ/mol. Processing requirements add 10–12 GJ/tonne for the high pressure (LDPE) process and 6–7 GJ/tonne for the low pressure (HDPE, LLDPE) processes.

11.3.5 Production and use statistics

The U.S. production of ethylene passed the 16 Mt per annum mark in 1988. Although well over half the capacity can use naphtha and heavier feedstocks, ethane and other natural gas liquids are reportedly still used for over 60% of ethylene produced.

In Western Europe, ethylene production is running at over 13 Mt per annum. Naphtha and a small quantity of heavier fractions have been the traditional cracker feedstocks. However, the commissioning of the 500 kt per annum Esso-Mossmoran ethane cracker and the 100 + kt per annum switch to ethane by BP Chemicals a few years ago, and other scheduled or proposed expansions based on ethane or LPGs, herald significant changes in the future feedstock pattern.

Japanese ethylene production was about 4·6 Mt in 1987, almost entirely based on naphtha.

The following table gives an indication of the major end uses of ethylene in the U.S.A.

LDPE (high pressure)	15%	
HDPE	23%	
LLDPE	13%	
Ethylene oxide	13%	
Dichloroethane	10–11%	(for vinyl chloride)
Ethylbenzene	7%	(for styrene)
Vinyl acetate	3%	
Acetaldehyde	2%	
Ethanol	1%	

Outside the U.S.A. the pattern varies, with LLDPE capacity continuing to lag behind, despite rapid expansions. In Western Europe, vinyl chloride production (c. 4·25 Mt per annum) is higher than in the U.S.A. But 50% or more of ethylene still goes into polyethylene in all zones, and over 70% of all ethylene ends up in polymeric materials worldwide.

The ratio of propylene to ethylene production is just over 0·5 in the U.S.A., slightly higher in Europe, but over 0·7 in Japan, reflecting differences in cracker feedstocks. The proportion converted to polypropylene is over 40% in Japan, compared with about 35% in Europe and the U.S.A. Some other figures for propylene usages in the U.S.A. are:

Acrylonitrile	17%
Propylene oxide	10%
Cumene	8%
Oxo feedstock	7%
Isopropanol	5%

In Western Europe, percentages into acrylonitrile, the OXO process (butanols and 2-ethylhexanol) and propylene oxide are all higher, with isopropanol rather less important.

The U.S. production of acetylene fell from a peak of over 450 kt per annum in the late 1960s, and has levelled off at about 130 kt per annum. The largest producers use methane oxidation, with carbide in second place. Relatively minor amounts of vinyl chloride and vinyl acetate are still produced from acetylene, but the most important use is for 1,4-butanediol (section 11.9.8).

The displacement of acetylene was somewhat slower in Europe. With recent ups and downs, the author has lost track of the status of the various production facilities, but there are continuing outlets into vinyl acetate and BASF's growing C_4 diols production.

Acetylene is relatively more important in Japan, with (carbide-based?) production reportedly of the order of 150 kt per annum.

11.4 Synthesis gas, ammonia and methanol

(See also section 2.2.2.2 part (b))

The production of mixtures of carbon monoxide and hydrogen from coal or coke was the basis for 'town gas' manufacture from well before the turn of the century, and the processes were adapted to provide appropriate feeds for ammonia and methanol synthesis in Germany and elsewhere.

$$C + H_2O \rightarrow CO + H_2 \quad \text{(endothermic)}$$
$$C + \tfrac{1}{2}O_2 \rightarrow CO \quad \text{(exothermic)}$$
$$CO + H_2O \rightleftharpoons CO_2 + H_2 \quad \text{(water gas shift)}$$

Low prices for natural gas in the US and naphtha in Europe prompted a change in feedstock. With no natural gas then available in the U.K. 'steam reforming' of naphtha was also adopted for 'town gas' generation and provided about 90% of supplies by 1968. However the European production of synthesis gas for chemical processing is presently based almost solely on steam reforming of natural gas.

11.4.1 *Process descriptions*

11.4.1.1 *Synthesis gas processes.* The methanation of carbon monoxide

$$CO + 3H_2 \rightarrow CH_4 + H_2O$$

over supported nickel catalysts has long been known to be very thermodynamically favourable at temperatures up to $400°C$. In fact, the reaction is reversible, the free energy change passing through zero at about $620°C$. The endothermic reverse reaction, the 'steam reforming' of methane, is carried out with a H_2O/CH_4 molar ratio of about $3:1$ at $800-850°C$, within catalyst filled tubes in a fuel-fired furnace[7]. The feedstock must first be freed from sulphur compounds. Although thermodynamically undesirable, the process may operate at 10–40 atm. to take advantage of gas supply pressures and minimize subsequent compression. An important secondary reaction is the water-gas shift reaction, which forms CO_2 and increases the H_2/CO ratio to c. $5:1$ molar (the equilibrium constant is roughly 1 at reformer temperature). The CO_2 may be scrubbed out, but if higher concentrations of CO are required, this CO_2 and possibly additional CO_2 recovered from the furnace flue gas may be recycled to the reformer inlet. Alternatively, if very high H_2/CO ratios are required, the exit gas is subjected to further water-gas shift reaction over iron or copper catalysts, and the CO_2 is scrubbed out and rejected.

The steam reforming of naphtha is, in essence, the reverse of the Fischer–Tropsch reaction, and is carried out under very similar conditions to methane reforming.

$$(CH_2)_n + nH_2O \rightarrow nCO + 2nH_2$$

The feedstock must again be desulphurized, most frequently by 'hydrodesulphurization' over a cobalt-molybdenum oxide catalyst. The nickel reforming catalyst is usually doped with alkali or supported on magnesia to reduce carbon deposition. The stoichiometry leads to somewhat lower H_2/CO ratios, but CO_2 recycle or addition and further shift reactions can again be applied if required.

If the gas mixture is destined for ammonia synthesis, the initial steam reforming of natural gas or naphtha may be effected at lower temperatures, with incomplete conversion of the hydrocarbon, and the process stream then subjected to 'secondary reforming'. Air is introduced and the combined stream passed through a compact catalyst-filled reaction vessel. Here, partial combustion raises the temperature to complete the reforming process, while consuming oxygen to leave the requisite quantity of nitrogen. Multiple shift reaction and CO_2 removal stages, and a final methanation step to eliminate traces of carbon oxides, would generally follow.

Finally, any hydrocarbon can be converted to CO/H_2 mixtures by partial oxidation, e.g.

$$CH_4 + \tfrac{1}{2}O_2 \rightarrow CO + 2H_2$$

$$(CH_2)_n + \frac{n}{2}O_2 \rightarrow nCO + nH_2$$

When naphtha prices were high, there was considerable activity on developing and piloting improved partial combustion processes for heavier oil fractions and coal (especially in W. Germany and the U.S.A.). But with lower efficiencies, the need for expensive gas purification and the handling problems of coal and residual ash, the work has lost impetus for the time being.

If required, relatively pure carbon monoxide can be recovered by cryogenic separation (i.e. condensation at very low temperatures), use of selective scrubbing solutions (for example $CuAlCl_4$ in toluene, the COSORB process) or solid adsorbents, and by other methods. Some CO recovery techniques are also applicable to the CO-rich (c. 70% molar) off-gases from basic oxygen furnaces and the leaner off-gases from air blast furnaces produced in steel manufacture, though operations of this type are still limited.

11.4.1.2 *Ammonia synthesis*[8]. Although inorganic, ammonia is currently dependent on hydrocarbon feedstocks, with production figures of a similar order to ethylene from plants making 1000 tonnes/day or more. The reaction

$$3H_2 + N_2 \rightleftharpoons 2NH_3$$

is exothermic to the extent of over $45\,kJ/mol\ NH_3$. Ammonia formation becomes quite unfavourable at the temperatures of about 450°C required to achieve suitable reaction rates over iron catalysts, and pressures of 250–300 atm. are necessary to achieve conversions of about 20–25% per pass. Thus, a hydrogen-rich, CO-free synthesis gas will be combined with extra nitrogen, if necessary, pressurized by means of a very large centrifugal compressor, and passed through a series of adiabatic catalyst beds, with inter-cooling by heat exchange or cold gas injection. The ammonia may be condensed by cooling under pressure, and the gases recycled, with a purge to limit the build-up of 'inerts'.

11.4.1.3 *Methanol synthesis.* The thermodynamics of methanol synthesis

$$CO + 2H_2 \rightleftharpoons CH_3OH \qquad \Delta H° = -100\,kJ/mol.$$

show some similarities to ammonia synthesis. The following table shows the effect of temperature on the equilibrium constant and the calculated pressure (uncorrected for departures from ideality) for an equilibrium concentration of 10 mol % methanol in a 1 $CO:2H_2$ mixture.

Temperature (°C)	$K_p(atm^{-2})$	Pressure for 0·1 mole fraction (atm.)
250	$2·09 \times 10^{-3}$	21·6
300	$2·85 \times 10^{-4}$	58·3
350	$5·29 \times 10^{-5}$	136
400	$1·24 \times 10^{-5}$	281

Any carbon dioxide present can react directly with hydrogen to give methanol

$$CO_2 + 3H_2 \rightarrow CH_3OH + H_2O$$

or undergo the water-gas shift reaction. However, larger amounts detract from the overall rate and equilibrium conversion, as do any inert components. Other minor side-reactions give higher alcohol impurities.

The earlier zinc-chromium oxide catalysts require reaction temperatures of about 400°C, and hence pressures of 300 atm and above are commonly used. For very large-scale operation (500 tonnes/day or more), the compression/reaction system is very similar to that described for ammonia synthesis. However, large methanol plants are not required in less well-developed areas. The characteristics of compressors are such that smaller high-pressure plants must use reciprocating machinery, which is costlier to instal and maintain.

The introduction of copper into methanol catalysts by ICI in the late 1960s, with other modifications to provide acceptable catalyst life, revolutionized this situation by permitting reaction temperatures of about 250°C. Thus pressures down to 50 atm. became usable, allowing the retention of centrifugal compressors, with smaller economic penalties, for modest scales of operation. Methanol yield also improved by 2–3%. Other process licensing companies have now developed copper-containing catalysts, and these form the basis for essentially all new methanol plants, but with somewhat higher pressures (up to 100 atm.) in the larger plants. This approach has allowed Lurgi to design multitubular reactors, cooled by steam generation in the shell, to provide improved temperature control and energy recovery.

11.4.2 Energy balances and economics

For synthesis gas mixtures, or syngas for short, energy inputs have been expressed per mole of reducing gas, that is the total quantity of carbon monoxide and hydrogen. This quantity is unaffected by CO_2 addition, recycle or removal and any water-gas shift operations, though the fuel value of the mixture changes with composition.

For hydrogen production, a high steam/hydrocarbon ratio is used in the reformer, followed by shift reactions and CO_2 rejection. Some mid-70s data for such operations with natural gas (NG) and naphtha feedstocks are given below.

| Feedstock | H₂-rich syngas | | H₂ for ammonia | Linde H₂ process | |
	Natural Gas	Naphtha	Natural Gas	Natural Gas	Naphtha
Reformer feed	217	220⎱	330–5 net⎱	345 net	350 net
Reformer fuel	146	150⎰	⎰		
CH₄ in syngas	(16)	(15)	(In purge gases recycled as fuel)		
	347	355			
Electricity	2	2	2	3	4
Steam generated	(36)	(27)	(Used in NH₃ production)	(42)	(25)
Net energy input	313	330	(c. 300)	306	329

For comparison, recent data for ammonia production (Haldor Topsøe and Uhde)[10] has been used to back-calculate the inputs per mole of hydrogen required ($\frac{2}{3}$ NH$_3$), and data for Linde hydrogen production units, which use solids adsorption separation methods after the reformer, are included. All figures are kJ/mol.

The most efficient processes evidently require little more than 300 kJ/mol reducing gas. The total inputs for the manufacture of ammonia are then about 0·5 MJ/mol or 30 GJ/tonne. For comparison, the inputs for a modern coal-based ammonia plant would be 50–56 GJ/tonne, with higher capital costs as well.

For methanol synthesis, older catalysts were often fed with gas mixtures containing H$_2$, CO and CO$_2$ in molar proportions of approximately 3:1:0·3. These were slightly hydrogen-rich compared with the stoichiometric require-ments for the parallel routes from CO + 2H$_2$ and CO$_2$ + 3H$_2$ to methanol. Some figures for the production of a synthesis gas of the above composition (at 8–10 atm.), from the late 1960s are as follows (in kJ/mol).

Syngas for methanol production

Feedstock	Natural Gas (at 10 atm.)	Naphtha
Reformer feed	221	217
Reformer fuel	125	152
CH$_4$ in syngas	(20)	(16)
	326	353
Electricity	3	3
Steam generated	(34)	(36)
Net energy input	295 kJ/mol	320 kJ/mol
CO$_2$ recovery: steam	12	
compression (electricity)	4	
	311 kJ/mol	

To show the initial impact of the ICI low-pressure catalyst, the figures for syngas generation from naphtha have been taken and some additional data from early comparisons of the high-pressure (HP) and low-pressure (LP) processes, with no other changes. The energy required for syngas compression was reduced. The higher selectivity of the ICI catalyst reduced not only the quantity of syngas required to form 1 mole of methanol (from 3·21 to 3·13 moles), but also the quantity that had to be purged from the methanol synthesis gas-recycle loop to avoid the build-up of impurities. Finally, purification of the crude methanol was easier. This notional exercise gives the figures below (in MJ/mol methanol).

	Notional synthesis (HP)	(LP)	Recent Lurgi data (LP)	(LP)
Feedstock	*Naphtha*	*Naphtha*	*Naphtha*	*Natural Gas*
Reformer feed	0·781	0·729	0·680	(Not
Reformer fuel	0·547	0·510 ⎱	0·277 net	reported)
Purge gases	(0·144)	(0·134) ⎰		
Net feed/fuel	1·184	1·105	0·957	0·938
Steam				
reformer	(0·130)	(0·121) ⎱	Steam generation in the	
syngas compr.	0·166	0·119 ⎬	reformer and methanol	
purification	0·107	0·096 ⎰	synthesis balances usages	
Net use	0·143	0·094	0	0
Electricity	0·010	0·009	v. small	0·017
Total energy input	1,337	1,208	0·957	0·955 MJ/mol
	(42)	(38)	(30)	(30) (GJ/tonne)

Thus, the ICI catalyst could reduce energy requirements by about 10%. Furthermore, the lower compressor duty reduced capital costs for the methanol synthesis plant by about 20%, and a slightly smaller reformer was required. For small scales of operation, the elimination of reciprocating machinery meant even larger notional savings.

In the meantime, each company has tackled the question of energy efficiency within the integrated synthesis gas generation/methanol production system. Lurgi's somewhat more expensive tubular synthesis reactor permits one of the most efficient operations, and figures on a comparative basis are listed in the above table[11]. Lurgi's feedstock and utilities requirements per tonne of methanol are expressed in more conventional form in the following table; it is now common to enter feedstock and fuel requirements in terms of their lower heats of combustion, rather than weight or volume.

Feedstock	*Naphtha*	*Natural Gas*	*Heavy residue (oxidation)*
Feed, GJ/tonne	21·26 ⎱	29·3	36·8
Fuel	8·67 ⎰		
Electricity, kwh	0	50	130
Feedwater, tonne	0·82	0·76	0·84
Cooling water, m³	40	45	75
Chemicals and catalysts,			
$ per tonne	1·15	1·02	0·6

The above table also contains some Lurgi data for methanol production from heavy oil residue (via partial oxidation to syngas), to illustrate the higher energy utilization for this approach. (Some capital costs for New Zealand's natural gas-based methanol-to-gasoline operation, which includes two 2200 tonnes per day methanol synthesis plants, have been published[18].)

If we require synthesis gas mixtures of still higher CO content, or pure CO, it is often convenient to produce synthesis gas mixtures with H_2/CO ratios of 2 or more, and apply partial or total separation methods. However, if a $3H_2:1CO$ mixture is generated by the steam reforming of natural gas at 310 kJ/mol, and only the CO is required, we obtain the energy balance

4 mol syngas	1240	kJ/mol CO
separation energy	50–100	
less Hydrogen fuel value	(726)	

564–614 kJ/mol CO (20–22 GJ/tonne CO)

The energy requirement for the CO produced is at least double its fuel value (283 kJ/mol or 10 GJ/tonne). We now see the importance of bringing together a number of processes which consume synthesis gas, hydrogen and carbon monoxide in proportions which match the output of a large reformer or partial oxidation unit.

11.4.3 Urea (carbamide), formaldehyde (methanal), amino resins and polyacetal

Ammonia and methanol are starting points for amino-resins, which are therefore syngas derivatives. Urea-formaldehyde resins are used in particle board (chipboard) and plywood manufacture; 'UF-foam' is also widely used for cavity wall insulation in the U.K. Melamine-formaldehyde resins are possibly more familar from the dinnerware and heat-resistant wood finishes produced.

Urea is manufactured from ammonia and carbon dioxide at about 200°C and pressures from 150 atm. upwards. Melamine (2, 4, 6-triaminotriazine) is formed by heating urea to a high temperature (over 350°C) at modest pressures (1–7 atm.) over an alumina or silica-alumina catalyst. The predominant overall reaction is:

$$6\,NH_2CONH_2 \longrightarrow \text{(melamine)} + 3\,CO_2 + 6\,NH_3$$

Urea

Melamine

Two process variants are used for the manufacture of formaldehyde from methanol, partial oxidative dehydrogenation or oxidation. In the first, methanol vapour is mixed with a stoichiometrically deficient quantity of air and passed over a silver catalyst at temperatures of 400–600°C. The two reactions

$$CH_3OH + \tfrac{1}{2}O_2 \rightarrow HCHO + H_2O \text{ (exothermic)}$$

and

$$CH_3OH \rightarrow HCHO + H_2 \text{ (endothermic)}$$

occur in parallel. In the oxidation process, methanol and excess air are passed over an iron molybdate catalyst at 350–450°C. Overall selectivities have improved from about 89% to 93% over the years.

In resin formation, the major reaction is of the type

$$-NH_2 + \overset{H}{\underset{\underset{O}{\parallel}}{\overset{\diagdown}{C}}}\overset{H}{\diagup} + H_2N - \rightarrow -NH-CH_2-NH- + H_2O$$

A ratio of formaldehyde molecules to amino groups of about 0·8 is typical. Excluding the final processing inputs, energy requirements for urea-formaldehyde resins are now quite low at about 40 GJ/tonne.

Formaldehyde (anhydrous) is also the major reactant for polyacetal resins, important for engineering components. The chemical structure is essentially $RO(CH_2O)_nR'$, with co-monomers or 'end-capping' reagents incorporated to prevent slow reversion to formaldehyde.

11.4.4 Production and use statistics

Production figures for ammonia in the U.S.A. and Western Europe are about 15 Mt and 11 Mt per annum respectively.

Urea and nitric acid production consume about 25% and 15% respectively, but, like most ammonia derivatives, are predominantly used in fertilizer outlets. Thus, in addition to the obvious uses of hydrocarbon fuels and electricity in farm machinery, we find a very considerable extra input of fossil fuel resources into modern agriculture worldwide. Plastics and synthetic fibre outlets account for less than 15%, including the urea formaldehyde (0·65 Mt in the U.S.A., higher in Europe) and melamine resins.

Of the methanol produced in the U.S.A. (3·3 Mt in 1987), about 35% was used for formaldehyde production. However, there are two, presently smaller, outlets showing significant growth, the production of over 1·5 Mt per annum MTBE (methyl t-butyl ether, requiring nearly 20%), preferred over both methanol and t-butanol as a (lead-free) gasoline octane-improver, and the carbonylation routes to acetic acid and anhydride (over 10%). Similar changes are occurring in Western Europe, but at least half of the 4 Mt per annum methanol produced is still used for formaldehyde.

11.5 Acetic (ethanoic) acid and anhydride

11.5.1 Acetic acid production

The various routes to acetic acid warrant a brief review, to illustrate the chemical industry's response to change and provide an introductory guide to future options.

The earliest route to acetic acid was the bacterial souring of wine, which eventually became the basis of vinegar manufacture. Wood distillation yielded stronger solutions of acetic acid (15 to over 90%). Surprisingly, some 10 000 tonnes per annum of acetic acid were still produced by this means in the U.S.A. in the mid-60s.

However, the major routes to acetic acid, developed during and after the First World War, were based on the oxidation of acetaldehyde derived from either acetylene or fermentation ethanol. The latter could well return to favour in countries such as Brazil (section 11.7.1). After the Second World War, fermentation ethanol gave way to synthetic ethanol, via the direct hydration of ethylene. (Synthetic ethanol made by the sulphuric acid process had already made some inroads in the U.S.A.). From 1960 onwards, the Wacker oxidation of ethylene added a further option for acetaldehyde manufacture.

When we turn to energetic aspects, acetylene hydration and ethylene oxidation have already been discussed in section 11.3.3.1. The appropriate energy input for fermentation ethanol is somewhat difficult to establish (see section 11.7.1), but for the hydration of ethylene and subsequent oxidation of synthetic ethanol (in modern plants) we have:

Ethylene (1·03 mol)	2·06
Processing energy	0·51
Input for ethanol	2·57 MJ/mol (56 GJ/tonne)
Ethanol (1·03 mol)	2·65
Processing energy	0·29
Input for acetaldehyde	2·94 MJ/mol (67 GJ/tonne)

In practice, crude ethanol can be fed to oxidation, with some overall energy savings (possibly 0·2 MJ/mol), but there is little doubt that the Wacker process is energetically advantageous.

The oxidation to acetic acid is carried out by feeding acetaldehyde and air or oxygen into acetic acid containing a soluble manganese compound at 40–60°C. (With an ester solvent, and no catalyst, peracetic (perethanoic) acid is formed. Although hazardous, this material finds use in a number of epoxidation processes). An approximate energy balance, based on the most favourable cracking and Wacker oxidation data, gives:

Acetaldehyde (1·04 mol)	2·39
Oxygen (0·55 mol)	0·06
Steam and power	0·10
Input for acetic acid	2·55 MJ/mol (42·5 GJ/tonne)

(A number of companies have sought to oxidize ethylene directly to acetic acid, with some success, but no process has been commercialized.)

Processes for the liquid-phase oxidation of hydrocarbons were conceived to by-pass the cracker. Until the early 70s, fixed costs dominated primary petrochemical operations, and represented a significant part of the price increment in each successive chemical conversion; acetic acid was well down the line. With very low feedstock prices, selectivity was of lesser concern. *n*-Butane was selected as a preferred feedstock by Celanese and Union Carbide in the U.S.A., providing acetic acid in rather more than 50% selectivity (carbon basis), with only minor amounts of formic acid (*c.* 10% w/w) and propionic acids as co-products; some 2-butanone could be withdrawn if desired. (The Union Carbide plant was shut down some years ago, while the Celanese plant is temporarily shut down for major reconstruction, following a serious fire.) BP Chemicals' 'DF Process' oxidizes light naphtha to give acetic acid as the major product, but with quite significant quantities of formic and propionic acids, acetone and other products if required. All these oxidation processes operate at temperatures of 170–200°C and with air at pressures of 50–60 atm. Over the years, optimization of operating parameters, the use of catalysts and improved hydrocarbon recoveries have provided progressive improvements in selectivities, with scope for modest changes in product ratios in the DF Process. Of course, a multiplicity of distillations are required to separate and purify the oxidation products.

Turning to energy considerations, part of the loss of selectivity is to carbon dioxide, and the recovery of energy from this oxidation/partial combustion process is important. Fortunately, the designers of the two 90 + kilotonnes per annum DF plants, commissioned in 1967 and 1972 before the oil crisis, also had the foresight to incorporate features which permit efficient energy recovery from the system as a whole. For example, air compressors are driven by gas turbines, with steam generation by exhaust gases, while off-gas turbines generate electricity.

The final route considered is the carbonylation of methanol. British Celanese reported experiments in the late 1920s, but German workers were more persistent. BASF finally commissioned a small plant in 1960, operating at about 250°C and well over 300 atm. initially, with a cobalt/iodide catalyst system. Nevertheless production was expanded several times, and the process licensed to Borden in the U.S.A. In 1970, Monsanto commissioned their own 150 000 tonnes per annum methanol carbonylation plant at Texas City, with a rhodium/iodide catalyst system at 150–200°C and only 30–50 atm. In the last 8–9 years, plants based on Monsanto technology (now owned and licensed by BP Chemicals) have come onstream in a number of countries, including BP Chemicals' own unit alongside the DF Process in the U.K.

The reaction is described by the simple equation:

$$CH_3OH + CO \rightarrow CH_3CO_2H$$

and the selectivity is very high. Therefore, with the most efficient reforming, methanol synthesis and gas separation processes, it should be possible to produce acetic acid with an energy utilization of under 2 MJ/mol (33 GJ/tonne). Furthermore, as a further coal-based option, the process is seen as providing the flexibility needed for the future, and has prompted wider attention to the feasibility of producing other C_2, and higher, products from syngas.

11.5.2 Acetic anhydride production

The mechanism of acetaldehyde oxidation is relatively complex. Considering only molecular species, the main steps appear to be

$$CH_3CHO \overset{O_2}{\to} CH_3COO_2H \overset{CH_3CHO}{\rlap{\rule[0.5ex]{2em}{0.4pt}}} CH_3COO_2\underset{CH_3}{CHOH}$$

$$2\ CH_3CO_2H \qquad \leftarrow \qquad (CH_3CO)_2O + H_2O$$

peracetic acid

If the oxidation is carried out with a metal catalyst and a hydrocarbon diluent, and the mixture is separated rapidly, an appreciable proportion of the acetic anhydride can be recovered. Thus mixtures of acetic anhydride (50–70% w/w) and acid are obtained in a total selectivity of over 90%.

An alternative commercial route is the cracking of acetic acid to ketene, in the presence of a phosphate ester at over 700°C. After separating the gaseous ketene from condensed water, it is absorbed into acetic acid.

$$CH_3CO_2H \to CH_2 = C = O + H_2O$$

$$CH_2 = C = O + CH_3CO_2H \to (CH_3CO)_2O$$

The selectivity in the cracking step is some 90%, giving an overall selectivity to anhydride based on acetic acid of about 95%.

Because acetic acid is co-produced both in the acetaldehyde oxidation process and in acetylations with acetic anhydride, many companies operate both processes simultaneously.

However, Halcon have now developed a process, catalysed by rhodium (or nickel) with iodine and other promoters, for the carbonylation of methyl acetate (or dimethyl ether) to acetic anhydride. Like the ketene route, this technology fits in well with acetylation processes.

$$CH_3O_2CCH_3 + CO \to (CH_3CO)_2O$$

$$\text{substrate} - OH + (CH_3CO)_2O \to \text{substrate} - O_2CCH_3 + CH_3CO_2H$$

$$CH_3OH + CH_3CO_2H \to CH_3O_2CCH_3 + H_2O$$

In 1983, Eastman Kodak commissioned a 250 kt per annum acetic anhydride plant based on this technology at Kingsport, Tennessee. The sting in the tail is that the whole complex is based on the gasification of coal[13].

In contrast, BP Chemicals' newest 200 kt per annum carbonylation plant in the U.K., for the simultaneous production of acetic acid and anhydride, is based on natural gas.

11.5.3 *Production and use statistics*

The reported U.S. production figures for 1987, 1·46 Mt of acetic acid and *c.* 0·80 Mt anhydride, give some idea of the importance of these products. Some 0·75 Mt of the acetic acid was used for vinyl acetate manufacture. The production of cellulose acetate involved a net consumption of over 200 kilotonnes per annum acetic acid, but accounted for most of the anhydride as an intermediate. Finally, acetate esters and the net consumption of acetic acid solvent in terephthalic acid manufacture each account for 150–200 kilotonnes per annum. European figures are roughly half those for the U.S.A.

11.6 C_1 products

For methanol and formaldehyde, see section 11.4.

11.6.1 *Formic (methanoic) acid and derivatives*

The modest U.S. requirements for formic acid (*c.* 30 kilotonnes per annum) appear to be met mainly by co-product material from butane oxidation (when available) and imports. Outlets are into a variety of speciality areas.

In Europe, demand is swollen to over 100 kilotonnes per annum by use for the preservation of damp silage (grass) and other animal feedstuffs—a means of saving the energy otherwise required for drying. Co-product material from naphtha oxidation is now exceeded by BASF's recently expanded (200 kilotonnes per annum capacity) methyl formate (methyl methanoate) production from syngas. The selective reaction

$$CH_3OH + CO \rightarrow HCO_2CH_3$$

is catalysed by sodium methoxide at 80–100°C and 30–60 atm. BASF can hydrolyse about 2/3 of the ester directly to formic acid (100,000 tonnes per annum capacity) by a recently improved process. A Finnish plant is based on by-product CO.

N, N-dimethylformamide (DMF) is an important solvent for acrylic fibres and polyurethane leather production (and for separating acetylene from ethylene). Methyl formate reacts readily with dimethylamine.

$$HCO_2CH_3 + (CH_3)_2NH \rightarrow \underset{\text{DMF}}{HCO_2N(CH_3)_2} + CH_3OH$$

Alternatively the amine can be introduced directly into methanol/carbon monoxide reaction systems. European production (possibly 60–70 kilotonnes per annum) is much higher than in the U.S.A. (about 15 kilotonnes per annum).

Mitsubishi have developed a process for methyl formate by the dehydrogenation of methanol. However, the major objective is the production of separate hydrogen and carbon monoxide streams for other uses from imported methanol.

$$2CH_3OH \rightarrow HCO_2CH_3 + H_2$$
$$HCO_2CH_3 \rightarrow CO + CH_3OH$$

11.6.2 Hydrogen cyanide

The reported demand for hydrogen cyanide in the U.S.A. is over 250 000 tonnes per annum. Possibly one-third is formed as by-product in acrylonitrile manufacture; the remainder is produced by the oxidation of methane/ammonia mixtures over platinum at about 1100°C. The major use (c. 60%) is in the manufacture of methyl methacrylate (methyl 2-methylpropenoate), with sodium cyanide second. The above figures cannot include the 150–200 kilotonnes of HCN produced and used captively by du Pont for adiponitrile production.

Acrylonitrile manufacture possibly contributes a higher proportion in Western Europe, though first-intent production is practised; Degussa dehydrogenate methane/ammonia mixtures over platinum at 1200–1300°C.

11.6.3 Chloromethanes

Passage of methanol and hydrogen chloride over a Lewis acid catalyst at about 350°C provides methyl chloride.

$$CH_3OH + HCl \rightarrow CH_3Cl + H_2O$$

This product can be chlorinated to give the more highly substituted derivatives.

However, more operations are now based on the oxychlorination of methane, over copper chloride catalysts, at temperatures of over 500°C.

$$2CH_4 + Cl_2 + \tfrac{1}{2}O_2 \rightarrow 2CH_3Cl + H_2O$$
$$2CH_3Cl + Cl_2 + \tfrac{1}{2}O_2 \rightarrow 2CH_2Cl_2 + H_2O, \text{etc.}$$

Mixtures are obtained with proportions dependent on the initial ratio of chlorine to methane. Some carbon tetrachloride is also produced by the chlorinolysis (i.e. C—C cleavage) of higher carbon number materials, and chlorination of carbon disulphide.

The total U.S. production of chloromethanes is over 0·9 Mt per annum. Chloroform (trichloromethane) and carbon tetrachloride (tetrachloromethane) are precursors for fluorocarbons; their elimination from the aerosol can market has been partially offset by increased use as refrigerants and intermediates for fluoroplastics. Methylene dichloride (dichloromethane) has picked up some of the aerosol business, and is used in paint removal and degreasing. Methyl chloride (chloromethane) is used for silicones and the waning production of tetramethyl lead.

11.7 C$_2$ products

11.7.1 Ethanol

Synthetic ethanol is produced almost entirely by the direct hydration of ethylene over supported phosphoric acid catalysts at 270–280°C and c. 70 atm.

The energy utilization for this material has been noted (c. 2·6 MJ/mol or 57 GJ/tonne, section 11.5.1), but it is worthwhile looking at the energy inputs for fermentation ethanol, for future reference in connection with alternative feedstocks and fuels. The fermentation of molasses requires processing inputs of about 1·4 MJ/mol ethanol (30 GJ/tonne). In a European operation, this energy would be supplied from conventional fossil sources, and is slightly higher than the fuel value of the ethanol produced (27 GJ/tonne). In the U.S.A., the pretreatment and fermentation of grain requires about 40 GJ/tonne for processing, while the energy inputs for grain production are equivalent to about 16 GJ/tonne ethanol, giving a total similar to that for synthetic ethanol. The main difference is that, by appropriate location of a fermentation plant, an associated power plant could generate a significant proportion of the energy inputs from agricultural wastes, though field machinery would still need a liquid fuel and net gains (from solar energy) are modest. Nevertheless, Brazil is seeking to increase fermentation ethanol production capacity to 9 million tonnes per annum—still only about 4% of fuel requirements.

The U.S. production of synthetic ethanol has fallen below 200 kilotonnes per annum, roughly corresponding with its usage as a chemical intermediate (for ethyl esters, amines etc.). Fermentation has increased rapidly, to take the total production to c. 3 Mt, of which 80% is now used in fuels, with cosmetics and other solvent outlets accounting for the remainder. The total European market for ethanol is somewhat over 0·6 Mt per annum.

11.7.2 Acetaldehyde (ethanal)

The conversion of ethanol to acetaldehyde is effected by dehydrogenation over copper at 250–300°C (preferred in the U.S.A.) or by (partially) oxidative dehydrogenation over silver at 450–500°C (mainly in Europe).

However, the Wacker process, for the direct oxidation of ethylene in aqueous solutions of Pd/Cu chlorides, has now become predominant. Two variants exist; in one, both ethylene and oxygen are introduced into a single tower reactor at about 100°C/10 atm.; in the second, ethylene is oxidized by the catalyst solution in one reactor, and the catalyst solution is reoxidized with air in a second vessel. Rhone–Poulenc (France) and BP have patented potential processes for the 'homologation' of methanol to acetaldehyde:

$$CH_3OH + CO + H_2 \rightarrow CH_3CHO + H_2O$$

U.S. production has fallen rapidly, possibly to below 300 kilotonnes per annum, as a result of incursions by carbonylation routes to acetic acid and anhydride. Synthetic pyridine derivatives, chloral (2,2,2-trichloroethanol) and pentaerythritol (2,2-di (hydroxymethyl) propane-1,3-diol), a 'polyol' for resins, are significant outlets for the remainder.

In Europe, other feedstocks also take a larger share of the acetic acid market. Some acetaldehyde producers have sought to hold their share indirectly by the production of ethyl acetate (ethyl ethanoate), via the low-temperature Tishchenko reaction catalysed by aluminium ethoxide:

$$2CH_3CHO \rightarrow CH_3CO_2\overset{.}{C}H_2CH_3$$

Obsolescent aldol condensation routes from acetaldehyde to n-butanol and 2-ethylhexanol could possibly make a resurgence in technologies based on acetylene (coal) or fermentation ethanol in the long term.

11.7.3 Ethylene oxide (oxirane) and glycol (ethane-1, 2-diol)

Ethylene oxide is manufactured by the silver-catalysed oxidation of ethylene at 200–250°C, with yields of well over 70% in modern plants. In the presence of alkalis, ethylene oxide reacts with water to give glycol and/or polyglycols, with alcohols and (alkyl) phenols to give the corresponding polyglycol ethers (nonionic detergents) and with ammonia to give ethanolamines.

$$C_2H_4 + [O] \longrightarrow \underset{CH_2-CH_2}{\overset{O}{\triangle}}$$

$$ROH \xrightarrow{C_2H_4O} ROCH_2CH_2OH \xrightarrow{C_2H_4O} RO(CH_2CH_2O)_2H \text{ etc.}$$

$$NH_3 \xrightarrow{C_2H_4O} H_2NCH_2CH_2OH \xrightarrow{C_2H_4O} HN(CH_2CH_2OH)_2 \text{ etc.}$$

The U.S. production of ethylene oxide is over 2·5 Mt, of which nearly 60% is hydrolysed to ethylene glycol (ethane -1, 2-diol). About half of the glycol is used in the production of polyethylene terephthalate, and a high proportion of the remainder is the basis for antifreeze. Di- and triethyleneglycols are used in resins and for gas drying, ethanolamines in detergents and for gas separations.

European production figures for ethylene oxide are about 1·5 Mt per annum, with less than half now hydrolysed to glycol.

Several companies have sought direct routes from ethylene to ethylene glycol, to achieve higher selectivity, while Ube (Japan) and Union Carbide (U.S.A.) are jointly developing a syngas based route:

$$2CO + 2CH_3OH + \tfrac{1}{2}O_2 \xrightarrow[CuCl_2]{PdCl_2} \underset{\text{dimethyl oxalate}}{CH_3O_2CCO_2CH_3} + H_2O$$

$$\xrightarrow{H_2} HOCH_2CH_2OH(+ 2CH_3OH)$$

The carbonylation or hydroformylation of formaldehyde may also provide a future synthetic route[15].

11.7.4 Vinyl acetate (ethenyl ethanoate)

The last U.S. plant for the production of vinyl acetate from acetylene has just shut down, though this simple reaction, over zinc acetate at about 200°C, is still in use in Europe and Japan. The production of vinyl acetate by the 'acetoxylation' of ethylene entails passage of ethylene, acetic acid vapour and oxygen over a supported palladium catalyst at about 160–200°C and 10 atm. Energetic aspects have been dealt with in section 11.3.3.1.

U.S. production of vinyl acetate is over 1·1 Mt, while Western Europe and Japan each produce over half a million tonnes. In the US and Europe, polymers are used mainly for adhesives (e.g. plywood) and paints. In Japan, however, nearly 70% of vinyl acetate polymer is hydrolysed to polyvinyl alcohol (with recovery of acetic acid), for use in fibres.

Halcon have reported the reaction of methyl acetate with carbon monoxide and hydrogen to ethylidene diacetate $CH_3CH\diagdown\begin{smallmatrix}O_2CCH_3\\O_2CCH_3\end{smallmatrix}$, an adduct of acetaldehyde and acetic anhydride. This material can be thermally decomposed to vinyl acetate, providing a route based on syngas.

$$2CH_3CO_2CH_3 + 2CO + H_2 \rightarrow CH_3CH(O_2CCH_3)_2 + CH_3CO_2H$$

11.7.5 Chloroethylenes (chloroethenes) and chloroethanes

The production of ethyl chloride from ethylene and hydrogen chloride, mainly for tetraethyl lead manufacture, is expected to fall rapidly. Ethylene dichloride (1, 2-dichloroethane) and vinyl chloride (chloroethene) have already been discussed in section 11.3.3.2. These two materials are also starting points for the production of more highly chlorinated C_2 products. The following scheme illustrates routes used for vinylidene dichloride (1, 1-dichloroethene) and 1, 1, 1-trichloroethane;

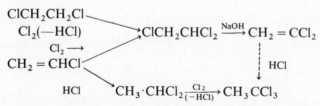

Tri- and per- chloroethylene (tri- and tetra- chloroethene) are produced together by the oxychlorination of 1,2-dichloroethane. An alternative route to the latter material is the high temperature (550–700°C) chlorinolysis of propane (or other hydrocarbons), e.g.

$$C_3H_8 + 8Cl_2 \rightarrow Cl_2C = CCl_2 + CCl_4 + 8HCl$$
$$2CCl_4 \rightarrow Cl_2C = CCl_2 + Cl_2$$

In the U.S.A., the two major products are now perchloroethylene (215 kilotonnes per annum), used for dry cleaning, and 1,1,1-trichloroethane (315 kilotonnes per annum), for cold cleaning and vapour degreasing of metals. In Europe, 1,1,1-trichloroethane production is now on a similar scale to the U.S.A.

11.8 C₃ products

11.8.1 *Isopropanol (2-propanol) and acetone (propanone)*

Isopropanol (2-propanol) is now manufactured largely by direct hydration of propylene over supported phosphoric acid catalysts at about 180°C and 50–60 atm. Deutsche Texaco has commercialized the use of an ion exchange resin as catalyst, and the old sulphuric acid process is still in use.

A considerable part (*c.* 70%) of the market for acetone (2-propanone) is supplied by the cumene-phenol process (section 11.11.2). In the U.S.A., the remainder is produced mainly by the dehydrogenation of isopropanol, with either a copper or zinc oxide catalyst. In Europe, acetone is a co-product from naphtha oxidation.

Of the 0·6 Mt of isopropanol produced in the U.S.A. in 1987, some 25% was converted to acetone. Other major uses are for solvents and chemical derivatives (e.g. isopropyl esters). European production is believed to be about 0·5 Mt per annum.

The U.S. and European production figures for acetone are approximately 1 and 0·5 Mt per annum respectively. The largest single chemical use (*c.* 35%) is in methyl methacrylate manufacture. Of the aldol derivatives, methyl isobutyl ketone [MIBK] (4-methylpentan -2-one) takes some 10–15%, but several others, such as diacetone alcohol and isophorone, have individual sales of 10 to 25 000 tonnes per annum.

11.8.2 *Propylene oxide (1-methyloxirane) and glycol (propane-1, 2-diol)*

Until about 1970, the chlorohydrin route to propylene oxide predominated worldwide.

$$Cl_2 + H_2O \rightleftharpoons HCl + HOCl$$
$$C_3H_6 + HOCl \rightarrow CH_3CHOHCH_2Cl$$

$$2CH_3CHOHCH_2Cl + Ca(OH)_2 \rightarrow 2CH_3CH\overset{O}{\overset{\diagup\diagdown}{}}CH_2 + CaCl_2 \cdot 2H_2O$$

During the 70s, Oxirane built 4 indirect oxidation plants in the U.S.A. (400 kilotonnes per annum), Holland (now 200 kilotonnes per annum) Spain (30 kilotonnes per annum) and Japan (90 kilotonnes per annum). The stoichiometry for the U.S. and Dutch operations is approximated by the equations:

$$2(CH_3)_3CH + 1\tfrac{1}{2}O_2 \rightarrow (CH_3)_3COOH + (CH_3)_3COH$$

$$(CH_3)_3COOH + C_3H_6 \rightarrow (CH_3)_3COH + C_3H_6O$$

Liquid isobutane is oxidized to a mixture of t-butylhydroperoxide and t-butanol at about 120°C and 25–35 atm., with no catalyst present. A toluene solution of the hydroperoxide is then contacted with excess propylene gas at about 160°C, in the presence of a molybdenum compound. Although some hydroperoxide decomposes, the selectivity on propylene is very high (98%). The t-butanol is used in gasoline or for isobutene production.

The operations in Spain and Japan use ethylbenzene as the hydrocarbon feedstock for oxidation. The co-produced alcohol, 1-phenylethanol, is dehydrated to styrene (phenylethene, section 11.11.1.)

The U.S. production of propylene oxide is nearly 1 Mt per annum, with a similar figure for Europe. The chemistry is similar to that of ethylene oxide; the glycol, polyethers and polyether derivatives of other polyols, such as glycerol and trimethylol propane, are used extensively in polyurethanes and polyester resins.

11.8.3 Acrylonitrile (propenonitrile)

The ammoxidation of propylene is carried out by feeding the olefin, air, ammonia and steam over a fixed or fluidized bed of catalyst at between 420 and 500°C. Several binary oxide systems, Bi-Mo, U-Sb (Sohio) and Sn-Sb (BP Chemicals), form the basis of commercial catalysts. Selectivities are now about 70% on propylene, with acetonitrile and HCN as byproducts.

$$C_3H_6 + NH_3 + 1.5O_2 \rightarrow CH_2 = CHCN + 3H_2O$$

U.S. and Western European production figures are similar at about 1.2 Mt per annum, with just under 0.6 Mt per annum produced in Japan. Use in fibres and as a co-monomer with styrene and butadiene in resins (ABS and SAN) accounts for most of the production.

11.8.4 Acrylates and acrolein (propenal)

A variety of processes have been used for the production of esters of acrylic (propenoic) acid, including the solvolysis of acrylonitrile with c. H_2SO_4 and an alcohol. The 'Reppe' process, commercialized by BASF and associated companies, is based on the reaction of acetylene with carbon monoxide, nickel carbonyl and an alcohol. However, the last U.S. operator of this process commissioned a propylene oxidation plant in October, 1982.

Several companies have oxidized propylene to acrolein (propenal) on a modest scale, mainly for the production of methionine. All but Shell (copper catalyst) used catalysts similar to those for acrylonitrile production. Two-reactor or dual-catalyst systems have now been introduced to oxidize acrolein further to acrylic acid, most frequently over cobalt and molybdenum oxides. Growth in the traditional coating and textile applications of acrylic acid and acrylates is being extended by the use of polyacrylic acid as an absorbent (in diapers!); demand now exceeds 400 kilotonnes per annum in the U.S. and 250 kilotonnes per annum in Western Europe.

11.8.5 *Allylic (propenyl) derivatives*

Allyl alcohol has been produced by the selective reduction of acrolein and the isomerization of propylene oxide. However, most allylic derivatives are produced via allyl chloride, obtained by the chlorination or oxychlorination of propylene.

A more significant use of allyl chloride is for the production of epichlorohydrin (3-chloro-1, 2-epoxypropane), an important precursor for epoxy-resins.

11.8.6 *n-Propanol, propionaldehyde (propanal) and propionic (propanoic) acid*

In the U.S.A. all these materials are produced via the hydroformylation of ethylene. Union Carbide chose the rhodium/phosphine catalyst system for their latest expansion, but older plants use cobalt catalysts.

$$C_2H_4 + CO + H_2$$
$$\downarrow$$
$$C_2H_5CHO \text{ Propionaldehyde (Propanal)}$$

$$H_2 \diagdown \text{Oxidation}$$

$$C_2H_5CH_2OH \qquad C_2H_5CO_2H$$

n-(1-) Propanol   Propionic (Propanoic) Acid

About a quarter of the 100 kilotonnes of propionaldehyde now produced is oxidized to propionic acid and over half hydrogenated to *n*-propanol. Outlets for these products are mainly into specialities, such as agricultural chemicals.

In Europe, propionic acid is more important as a grain preservative. The significant quantities co-produced in naphtha oxidation are supplemented by the 'hydrocarboxylation' of ethylene:

$$C_2H_4 + CO + H_2O \xrightarrow{Ni\,catalyst} C_2H_5CO_2H$$

The propionaldehyde route is operated on only a small scale.

Texaco (U.S.A.) are seeking to homologate acetic acid:

$$CH_3CO_2H + CO + 2H_2 \rightarrow CH_3CH_2CO_2H + H_2O$$

11.9 C$_4$ products

11.9.1 *Butenes and butadiene*

The use of naphtha for ethylene production provides about 30–35% w/w C$_4$ hydrocarbons relative to ethylene, with the following approximate composition

40% butadiene (1, 3-)

30% isobutene (2-methylpropene)

20% n-butenes

10% n- and isobutanes

Butadiene is first recovered by extractive distillation, employing furfural or an amide, such as dimethylformamide. Some 1·4 Mt and 0·7 Mt of butadiene is obtained in Western Europe and Japan respectively.

With less naphtha and gas oil cracking in the U.S.A., a part of the 1·4–1·5 Mt demand for butadiene must be met by the catalytic or oxidative dehydrogenation of n-butenes, and some n-butane, and imports.

The major outlets for butadiene in all areas are rubbers, as polybutadiene or co-polymers with styrene or acrylonitrile. Co-polymer resins, chloroprene and adiponitrile (excluding Japan) account for the remainder.

The remaining components are common to many refinery streams, and thus isobutene and n-butenes may be derived from either source.

Isobutene is generally recovered by chemical means, being far more reactive than other components. For example, by contacting a mixture with 65% sulphuric acid, t-butyl alcohol is formed in the acid phase, from which fairly pure isobutene (for butyl rubber) may be regenerated. Alternatively, viscous 'polybutenes' are produced by contacting C$_4$ mixtures with Lewis acid catalysts, while methyl t-butyl ether (MTBE), an expanding gasoline additive, is obtained by reaction with methanol in the presence of an acidic ion-exchange resin:

$$(CH_3)_2C = CH_2 + CH_3OH \rightarrow (CH_3)_3COCH_3$$

Such uses now consume over 1 Mt of isobutene in both the U.S.A. and Western Europe.

By far the largest chemical use for n-butenes is sec-(2-) butanol manufacture. The separation of 1-butene is also becoming quite important, primarily for co-monomer use in LLDPE (section 11.3.4).

11.9.2 *Sec-butanol (2-butanol) and methyl ethyl ketone (2-butanone)*

After isobutene removal, n-butenes in the mixture with butanes may be converted to 2-butanol by the old two-stage sulphuric acid process or by using an acidic ion-exchange resin catalyst. Essentially all 2-butanol so produced is converted to 2-butanone by dehydrogenation (cf. isopropanol to acetone). 2-Butanone is a major solvent in the preparation of coatings. Production figures are about 300 kilotonnes and 250 kilotonnes in the U.S.A. and Western Europe respectively.

11.9.3 Tert-butanol

Most t-butanol is now obtained as a co-product of propylene oxide manufacture (section 11.8.2). For gasoline use, dehydration to isobutene, for conversion to MTBE, is increasingly preferred over direct use of the alcohol.

11.9.4 Maleic anhydride (cis-butenedioic anhydride)

Maleic anhydride (MA) is important in the production of unsaturated polyesters, used mainly for glass-fibre reinforced resin products. The long-established oxidation of benzene over vanadia catalysts at 400–450°C is very wasteful.

$$C_6H_6 + 4\tfrac{1}{2}O_2 \rightarrow \begin{matrix} CH \cdot CO \\ \| \quad\quad\quad O \\ CH \cdot CO \end{matrix} + 2CO_2 + 2H_2O$$

The selectivity is at best 60% of theory, and hence the carbon utilization is no more than 40%. Mitsubishi in Japan (and a small European plant) oxidize a stream rich in n-butenes with a selectivity of about 45%. However, a number of companies in the U.S.A. developed catalysts to give acceptable yields from n-butane, and all U.S. production (180 kilotonnes per annum) has been switched to this feedstock.

11.9.5 Chloroprene (2-chlorobuta-1, 3-diene)

Butadiene is chlorinated in the gas phase to give mixtures of 1, 2-dichlorobut-3-ene and 1, 4-dichlorobut-2-ene, which are interconvertible in the presence of copper catalysts. The former is dehydrochlorinated by contact with aqueous sodium hydroxide, to give chloroprene (2-chlorobuta-1, 3-diene) for polymer production (neoprene). The earlier acetylene-dimer route has been phased out. Production figures are of the order 200 kilotonnes in the U.S.A. and 100 kilotonnes in Western Europe.

11.9.6 Methacrylates (2-methylpropenoates)

Methyl methacrylate (methyl 2-methylpropenoate) is still manufactured by the old acetone cyanhydrin route:

$$(CH_3)_2CO + HCN \rightarrow (CH_3)_2C - CN \xrightarrow[c.H_2SO_4]{CH_3OH} CH_2 = C - CO_2CH_3$$
$$\qquad\qquad\qquad\qquad\quad | \qquad\qquad\qquad\qquad\qquad\qquad\quad |$$
$$\qquad\qquad\qquad\qquad\;\; OH \qquad\qquad\qquad\qquad\qquad CH_3 \;\; (+NH_4HSO_4)$$

The oxidation of isobutene to methacrylic acid has been commercialized in Japan, and several alternative routes have attracted serious attention of late. The homopolymer provides clear sheet products (Perspex, Lucite), and copolymers are used in coatings and paints. U.S. production is approaching 500 kilotonnes per annum.

11.9.7 Butyraldehydes (butanals) and primary butanols

During the 1960s, older routes based on acetaldehyde (section 11.7.2) largely gave way to the hydroformylation of propylene, and displacement is now virtually complete. The use of cobalt catalysts at 145–180°C and 100–400 atm. still predominates, giving n- and iso-butyraldehydes in ratios of 3 or 4 : 1.

$$C_3H_6 + CO + H_2 \rightarrow CH_3CH_2CH_2CHO \text{ or } (CH_3)_2CHCHO$$

In addition to the C_4 alcohols, most companies produce 2-ethylhexanol by the aldolization of n-butyraldehyde and subsequent hydrogenation; in the Aldox process, the aldolization occurs within the propylene hydroformylation reactor.

The co-production of isobutyraldehyde has always proved embarrassing, and the introduction of the rhodium/phospine catalyst system by Union Carbide not only permits milder operating conditions (c. 100°C, 7–20 atm.), but also gives n-/iso-ratios of 8 : 1 or more.

U.S. production of n-butanol is presently over 400 kilotonnes per annum for solvent use and a variety of esters, while 2-ethylhexanol (nearly 300 kilotonnes per annum in the U.S., but higher in Europe) is used mainly for the phthalate ester as a plasticizer for PVC. The n- and iso-butyraldehydes are also subjected to aldol condensation/crossed Cannizzaro reactions with formaldehyde, to give the polyols trimethylol-propane (2, 2-bishydroxy-methyl-1-butanol) and neopentyl glycol (2,2-dimethyl-1,3-propanediol).

11.9.8 C_4 diols and related products

1,4-Butanediol is the main intermediate to tetrahydrofuran and γ-butyro-lactone, but is finding increasing use in the production of polybutylene terephthalate, an engineering plastic. Production in the U.S.A. (over 150 kilotonnes per annum) and Europe (c. 85 kilotonnes per annum) is via 'Reppe' chemistry, based on acetylene:

$$C_2H_2 \xrightarrow[\text{CuCat.}]{CH_2O} HC \equiv CCH_2OH \xrightarrow[\text{CuCat.}]{CH_2O} HOCH_2C \equiv CCH_2OH$$

propargyl alcohol   1, 4-butynediol

$$\downarrow H_2$$

$$HO(CH_2)_4OH \xleftarrow{H_2} HOCH_2CH = CHCH_2OH$$

1, 4-butanediol   1, 4-butenediol

The intermediates have a number of specialist applications, such as corrosion inhibitors.

Routes based on the hydrolysis of 1,4-dichlorobut-2-ene (see chloroprene), the hydrogenation of maleic anhydride (on a smallish scale) and the diacetoxylation of butadiene have all been commercialized in Japan. Hydrogenation of maleic anhydride can provide γ-butyrolactone, 1,4-butanediol and/or tetrahydrofuran, according to conditions, and has been mooted for future commercialization in the U.S.A. ARCO have announced their intention to commercialize the hydroformylation of allyl alcohol, obtainable by isomerization of propylene oxide. Some tetrahydrofuran is also produced by hydrogenation of furan, derived from natural pentoses.

11.10 C_5 + aliphatics

11.10.1 *Isoprene*

Isoprene (2-methylbuta-1, 3-diene) is produced mainly by the dehydrogenation of 'isoamylenes' (2-methylbutenes) or isopentane (2-methylbutane), in processes similar to those used in the U.S.A. for butadiene manufacture.

In France and Japan, isoprene is also produced from isobutene (or *t*-butanol) and formaldehyde in 1 or 2 stage processes (via a 1, 3-dioxan intermediate). The overall reaction is:

$$CH_3$$
$$|$$
$$(CH_3)_2C = CH_2 + CH_2O \rightarrow CH_2 = C - CH = CH_2 + H_2O$$

U.S. and Western European production figures for polyisoprene, the closest synthetic to natural rubber, appear to be quite small at less than 100 kilotonnes per annum in each area. However, there is a much higher capacity, and presumably greater use, for polyisoprene in Eastern Europe and the USSR.

11.10.2 *Plasticizer alcohols*

Plasticizers for polyvinyl chloride are predominantly the C_7–C_{13} primary alkyl diesters of phthalic acid; diesters of aliphatic diacids are also used as plasticizers and in synthetic lubricants. 2-Ethylhexanol, via *n*-butyraldehyde, was the first alcohol used; now, many others, produced by the hydroformylation of C_6–C_{12} olefins, have entered the scene.

Over acid catalysts, propylene and butenes give homo- and co-oligomers within the required molecular weight range. The derived primary alcohols are highly branched. (The prefix iso- is used for commercial names e.g. iso-octanol). However, linear alpha-olefins (terminal) have found greater demand. Their production by the thermal cracking of waxes, at 500°C upwards, is increasingly giving way to the oligomerization of ethylene, with either aluminium alkyls or Ni(O) complexes as catalysts. The latter are the basis for SHOP, the Shell Higher Olefin Process. These processes produce olefins with a broad carbon number range; the lower molecular weight fractions are used for plasticizer alcohol production, and the higher fractions for detergents (next section). Shell can tailor the range of olefins produced, by converting unwanted higher fractions to internal olefins and dismutation with ethylene.

In addition to cobalt and Union Carbide's rhodium-phosphine catalysts for the hydroformylation process, Shell's cobalt/phosphine system has potential advantages, providing higher n-/iso-ratios than cobalt alone while producing mainly the *alcohol* in one step.

$$RCH = CH_2 + CO + 2H_2 \xrightarrow{Co/phosphine} RCH_2CH_2CH_2OH$$

11.10.3 *Detergent intermediates*

The highly branched propylene-butenes oligomers were also used to alkylate benzene and phenol, to provide precursors for sodium alkylbenzene

Petrochemical routes to synthetic detergents

sulphonates and polyethoxylated alkylphenols. However, biodegradability is favoured by linear alkyl groups, and the range of products has now increased, as shown in the scheme.

Suitable C_{12} and higher internal linear olefins can be produced by monochlorination and subsequent thermal dehydrochlorination of n-paraffins; higher alpha-olefins are produced by wax cracking or ethylene oligomerization. Either alkyl chlorides or olefins may be used to alkylate benzene or phenol, or the olefins hydroformylated to give primary alcohols for sulphation or ethoxylation:

$$RCH_2OH + SO_3 \rightarrow RCH_2OSO_3H \xrightarrow{NaOH} RCH_2OSO_3Na$$

$$RCH_2OH + nC_2H_4O \rightarrow RCH_2O(C_2H_4O)_nH$$

The olefins can also be sulphonated with SO_3, while free-radical addition of sodium bisulphite provides sodium alkanesulphonates directly.

$$RCH = CH_2 + NaHSO_3 \rightarrow RCH_2CH_2SO_3Na$$

A more direct (Japanese) route from n-paraffins entails Bashkirov oxidation (liquid phase in the presence of boron oxides at $c.$ 160°C) to secondary alcohols, for ethoxylation.

U.S. production of n-paraffins is over 300 kilotonnes per annum with about 70% accounting for most of the linear alkylbenzene ($c.$ 250 kilotonnes) produced. Alpha-olefin production is about 500 kilotonnes per annum, providing both C_{12} + alcohols ($c.$ 300 kilotonnes—but over half from natural oils) and alkanesulphonates. Only about 100 kilotonnes per annum ethoxylated alkylphenols are produced.

11.11 Aromatics

11.11.1 *Hydrocarbons*

The separation of aromatic hydrocarbons (as a group) from mixtures with close-boiling paraffins and olefins, such as 'catalytic reformate' or 'pyrolysis gasoline', is generally effected by extraction with furfural (2-formylfuran), liquid SO_2 or other solvents. Benzene, toluene and mixed xylenes are then separated by distillation.

The requirement for benzene is about 5·5 Mt per annum in both the U.S.A. and Western Europe. Coal-tar operations make only modest contributions. In Western Europe the main source is pyrolysis gasoline, whereas catalytic reforming (section 11.2.2.3) is the main source in the U.S.A., where the catalytic (clay) or thermal hydrodealkylation of toluene is also more important.

$$PhCH_3 + H_2 \xrightarrow{550—650°C} PhH + CH_4$$

By far the largest outlet for benzene ($c.$ 50%) is styrene (phenylethene); the largest plant has a capacity of 680 kilotonnes per annum! Ethylbenzene is produced by the reaction of benzene with ethylene; a variety of liquid and gas phase processes, with mineral or Lewis acid catalysts, are used. The ethylbenzene is then dehydrogenated to styrene at 600–650°C over iron or other metal catalysts in over 90% selectivity. Co-production with propylene oxide (section 11.8.2) also requires ethylbenzene, but a route based on toluene and ethylene has received some attention:

$$2\,PhCH_3 \xrightarrow[600°]{PbO} PhCH = CHPh \xrightarrow[WO_3]{C_2H_4} 2\,PhCH = CH_2$$

60–70% of the styrene is used for homopolymers, the remainder for co-polymer resins. An approximate energy balance for rigid polystyrene in a modern production unit might be:

Ethylene (1·06 mol, say 2 MJ/mol)	2·12
Benzene (1·05 mol, say 1·5 × fuel value)	4·98
Processing utilities	1·04
Input for Styrene monomer	8·14 MJ/mol (78·3 GJ/tonne)
Styrene monomer (1·01–1·02)	79–80
Processing (v. small)	2–3
Input for Polystyrene	81–83 GJ/tonne

Other major uses of benzene are cumene (over 20%, see phenol), cyclohexane (c. 15%) and nitrobenzene (5–6%). Major outlets for toluene are for solvent use and conversion to dinitrotoluene.

Of the remaining aromatics, only p-xylene and o-xylene have major chemical uses. o-Xylene is generally separated first by distillation; the use of low temperature (-70°C) crystallization for recovery of p-xylene has largely given way to selective adsorption on to molecular sieves (Zeolites; Parex process). With periodic shortages of mixed xylenes, the remaining m-xylene-rich material has often been isomerized with Lewis acid (HBF$_4$ liquid phase, in Japan) or silica-alumina (gas-phase) catalysts; newer Zeolite catalysts can give p-xylene-rich mixtures.

11.11.2 Phenol

Alkylation of benzene with propylene produces cumene (isopropylbenzene). Oxidation of cumene at about 110°C to modest conversions (c. 20%) gives the hydroperoxide, which on treatment with sulphuric acid at 70–80°C produces phenol and acetone in equimolar quantities (90–92% of theory).

$$PhH + C_3H_6 \xrightarrow{Acid} Ph-\underset{\underset{CH_3}{|}}{\overset{\overset{CH_3}{|}}{CH}} \xrightarrow{O_2} Ph-\underset{\underset{CH_3}{|}}{\overset{\overset{CH_3}{|}}{COOH}} \xrightarrow{Acid} PhOH + \underset{\underset{CH_3}{|}}{\overset{\overset{CH_3}{|}}{CO}}$$

Dow has commercialized a 2-stage process from toluene. Benzoic acid is produced by liquid-phase oxidation in the presence of cobalt, and is then subjected to 'oxidative decarboxylation' with a copper catalyst, again in the liquid phase:

$$PhCH_3 \overset{O_2}{\to} PhCO_2H \overset{O_2}{\to} PhOH + CO_2$$

Of the 1·5 Mt of phenol produced in the U.S.A. in 1987, about 40% went into phenol-formaldehyde resins (for plywood adhesives) and some 13% to caprolactam. About 25% was converted into 'bisphenol A', the basis for polycarbonate glazing and engineering resins:

bisphenol A   polycarbonate

In Western Europe, phenol production is 1 Mt per annum, with less into phenol-formaldehyde resins (more urea-formaldehyde is used) and rather more into cyclohexanone for nylons.

11.11.3 Benzyls

Several other companies oxidize toluene to benzoic acid, as above. Except for Dow and Snia Viscosa (nylon-6), outlets are mainly into speciality uses. Benzaldehyde is often recovered as a by-product, while Rhone-Poulenc and others oxidize toluene to give mainly benzaldehyde and benzyl alcohol (phenylmethanol). However, many derivatives, including the major benzyl esters, are produced via the side-chain chlorination of toluene.

$$PhCH_3 \overset{Cl_2}{\to} PhCH_2Cl \overset{Cl_2}{\to} PhCHCl_2 \overset{Cl_2}{\to} PhCCl_3$$

(Butylbenzyl phthalate – a plasticizer)

11.11.4 Nitro-compounds and amines

Traditional processes are still used for the nitration of benzene and toluene, and the hydrogenation of the nitro-derivatives to amines. The production of polyurethanes, for foams or coatings, is a major outlet for aniline (c. 60%) and virtually the sole use for 2,4/2,6-toluenediamines.

Aniline is coupled by reaction with formaldehyde to produce 4,4'-diaminodiphenylmethane:

Reaction of a diamine with phosgene then produces a di-isocyanate:

$$R(NH_2)_2 + 2COCl_2 \rightarrow R(NCO)_2 + 4\,HCl$$

The di-isocyanates finally react with 'polyols', either simple (e.g. glycerol, trimethylolpropane) or, more generally, adducts with propylene oxide, to give the cross-linked addition polymers; for each group:

$$RNCO + HOR' \rightarrow RNH \cdot CO_2R'$$

U.S. production figures for aniline and toluenediamines exceed 400 kilotonnes and 200 kilotonnes per annum respectively.

11.11.5 *Phthalic (benzene-1,2-dicarboxylic) anhydride*

Phthalic anhydride is manufactured largely by the gas-phase oxidation of *o*-xylene at 350–400°C over vanadia catalysts. Selectivities are now about 80% of theory.

In the U.S.A., the fluid-bed oxidation of naphthalene, from both coal tar and petroleum sources, is significant.

About half the phthalic anhydride is used to produce diesters as plasticisers for PVC. The remainder is used mainly for alkyd (unsaturated) and other polyester resins. U.S. production has climbed back to 470 kilotonnes per annum; Western European production is nearly 50% higher.

11.11.6 *Terephthalic (benzene-1, 4-dicarboxylic) acid*

In the U.S.A., the Far East and the U.K. (ICI), most terephthalic acid (1,4-benzenedicarboxylic acid) is produced by the liquid-phase oxidation of *p*-xylene in acetic acid with a cobalt/manganese/bromide catalyst system, originated by Amoco. Several non-bromide variants, requiring the addition of readily oxidizable precursors of acetic acid, were commercialized, but most, if not all, have been shut down. Increasingly, the final product is the high-purity diacid, whereas most older plants converted the crude acid to the dimethyl ester for purification and sale. (Small quantities of isophthalic acid are manufactured in a similar manner.)

The major alternative, still very important in Europe, is the Hercules–Witten process; *p*-xylene is oxidized without a solvent to *p*-toluic (4-methylbenzoic) acid, which is esterified before further oxidation:

$$H_3C-\underset{}{\bigcirc}-CH_3 \xrightarrow{O_2} H_3C-\underset{}{\bigcirc}-CO_2H$$

$$\Big\downarrow CH_3OH$$

$$HO_2C-\underset{}{\bigcirc}-CO_2CH_3 \xleftarrow{O_2} H_3C-\underset{}{\bigcirc}-CO_2CH_3$$

Final esterification again gives dimethyl terephthalate for purification. Finally, two Henkel processes were commercialized in Japan, but the current status is again unclear:

$$\underset{}{\bigcirc}\!\!\!\begin{array}{c}CO_2K\\CO_2K\end{array} \xrightarrow[c.\,400^\circ C]{Zn/Cd} \underset{}{\bigcirc}\!\!\!\begin{array}{c}KO_2C\\CO_2K\end{array}$$

$$2PhCO_2K \xrightarrow{Zn/Cd} C_6H_4(CO_2K)_2 + C_6H_6$$

Combined annual production figures for pure terephthalic acid and dimethyl terephthalate are about 3·7 Mt in the U.S.A., nearly 2 Mt in Western Europe and over 1 Mt in Japan. About 95% is converted to polyethylene terephthalate by reaction with ethylene glycol, with fibres as the major end-use.

11.12 Nylon intermediates

Although nylon intermediates are aliphatic, the main starting materials are aromatic hydrocarbons. In the earliest processes, phenol was hydrogenated (over nickel or palladium) to cyclohexanone/ol mixtures, known as 'KA oil', which gave adipic (hexanedioic) acid on oxidation with nitric acid. Half of this adipic acid was converted to the dinitrile, by vapour phase reaction with ammonia. Finally the dinitrile was hydrogenated to hexamethylenediamine (1, 6-diaminohexane). Nylon-6, 6 is produced by the condensation of essentially equimolar quantities of adipic acid and the diamine.

In subsequent processes, now predominating, benzene is hydrogenated to cyclohexane for aerial oxidation (cobalt catalysed) to 'KA oil' (or Bashkirov oxidation to cyclohexanol). One or two companies appear to have substituted aerial oxidation of KA oil for the nitric acid oxidation process, while Celanese produce their 1, 6-diaminohexane via 1, 6-hexanediol, rather than adiponitrile.

Routes to nylon 6,6 and nylon 6 intermediates

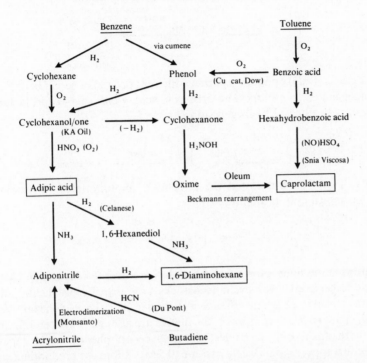

However, the most significant changes in nylon 6,6 production lie in the newer routes to adiponitrile introduced by Monsanto and Du Pont into both the U.S.A. and Western Europe. The former entails the electroreductive dimerization of acrylonitrile, while Du Pont use the addition of HCN to 1,3-butadiene catalysed by Ni(O) complexes:

$$CH_2 = CH \cdot CH = CH_2 \xrightarrow{HCN} CH_2 = CHCH_2CH_2CN \xrightarrow{HCN} NC(CH_2)_4CN$$

This process is by no means as simple as this equation implies, with the formation of several isomers, and hence the need for interconversions at each stage. Both processes take the pressure off benzene, from which the multistage processes gave poor yields of dinitrile (c. 50% of theory). In a somewhat similar manner, BASF have operated a semi-commercial plant to produce adipic acid by the bis-hydrocarboxylation of 1,3-butadiene:

$$CH_2 = CH \cdot CH = CH_2 + 2CO + 2H_2O \xrightarrow{Gp\ VIII\ metal} HO_2C(CH_2)_4CO_2H$$

Cyclohexanone alone is the main starting point for the newer nylon 6; the hydrogenation of phenol can be directed to give this material, but KA oil (or cyclohexanol) must be dehydrogenated. The sulphate or, more recently, phosphate of the oxime, formed by reaction with hydroxylamine (as its salt), undergoes the Beckmann rearrangement in the presence of oleum. (Excess acid is neutralized with ammonia for sale into fertilizers.)

Caprolactam

Snia Viscosa (Italy) operate an entirely different process, wherein benzoic acid is hydrogenated to cyclohexanecarboxylic acid, which gives caprolactam by reaction with nitrosyl hydrogen sulphate:

Finally, Toray have reported the pilot-scale photoinitiated nitrosation of cyclohexane to cyclohexanone oxime:

U.S. production figures for nylon-6,6 and -6 are over 0·7 Mt and 0·5 Mt per annum respectively. Non-cyclic feedstocks now provide practically all the diamine for nylon-6,6 and phenol well under half the cyclohexanone for caprolactam. In Western Europe, the figures are about 0·7 Mt per annum for each, with again only a modest dependence on phenol, while Japanese production appears to be solely nylon-6 (0·5 Mt). Fibre use predominates, but use as an engineering resin may become more important. Small quantities of higher nylon analogues are produced.

11.13 The future

11.13.1 The products

Firstly, we must consider the future of current products of the petrochemical industry. An overriding consideration is the future production level of all material goods, a case for the crystal ball. However, reduced material and energy usages, and a longer life, will certainly be sought in all applications. A more modest and stable, but by no means technologically sterile, period of fabrication activity may be anticipated.

In such circumstances, how do our products stack up against conventional materials? For a particular material to remain in use, it should be producible with a lower energy utilization than competitive materials for the same application. The following table contains energy input data for a number of fabrication and construction materials (in cast form, unless otherwise indicated). They are derived from a variety of sources[16], and the bases for individual figures may vary somewhat; the lowest figures for metals may not include energy used in mining and the preliminary refining of ores. The renewable solar energy input, represented by the fuel value of wood, is not included for cellulosic products.

Even the most energy intensive 'engineering' resins are competitive with light casting alloy components on a volumetric basis, the most relevant in many applications. The tough film polymers, such as LLDPE and polypropylene, are competitive with cellulosic films and even paper sacks, which must be over double the weight for comparable strength. A more striking example is the plastic bottle, which need weigh only 50g for 1-litre contents compared with 500g in glass; the plastic bottle is also subject to far fewer breakages and saves considerable energy in transportation. Longer life and weather resistance (e.g. PVC versus wood for window frames), general resistance to corrosion and the ability to produce light, rigid composites (such as glass-fibres reinforced resin panels for lighter vehicles), all possibly support recent predictions that plastics will continue to displace other materials in some uses.

In the long term some of the predictions based on current energy inputs may prove a little shaky. Power and steam generation for the 'heavy' industries can readily be based on coal, and increased efficiency or alternative technology for

	Density tonne/m^3	Energy Utilization, GJ	
		per tonne	per m^3
Titanium	4·5	c. 550	2 500
Copper	8·9	c. 130	1 150
Aluminium	2·7	250–300	700–800
Magnesium	1.74	270–400	470–700
Zinc	7·1	c. 55	400
Steel (carbon)	7·8	28(45)	220(350)
Polyacetal	1·43	240	340
Polyphenylene oxide			290
Nylons 6, 6 and 6	1.14	235	270
Polyester			240
Polycarbonate	1.2	180	220
Acrylic	1·2	180	220
Polypropylene (film and cast)	0.90	110–180	100–160
ABS resin	1.07	120	130
PVC (rigid)	1·38	65–80	90–110
Polystyrene	1·06	75–105	80–110
HDPE	0·95	85–120	80–115
LDPE (film and cast)	0·92	80–110	75–100
Cellulose film		180	
Paper (sack)		60	
Glass (sheet)	(2·5)	29	(70)
(containers)	(2·5)	19–20	(50)
Portland cement	(2·1)	5–9	(10–20)
Bricks	(2·5)	4–5	(10–12)
Concrete		c. 1	

the generation of electrical energy may have a greater impact. Recycle of selected scrap could also reduce the average energy requirements considerably. In contrast, a need to produce organics from coal would presently increase energy requirements significantly, though cheaper electrical power might favour acetylene routes. Nonetheless, some organics will still prove advantageous.

11.13.2 *Future raw materials and production routes*

As we are all too aware, supplies of petroleum and natural gas will become exhausted in a matter of tens of years at current consumption rates. The investment of enormous sums of money in seeking oil in the most inhospitable areas of the World, such as the North Sea and Alaska, highlights the lengths to which the petroleum industry is driven to extend supplies. Methods for enhanced oil recovery, shale oil distillation etc. may eke out resources a little further.

Total consumption is, of course, an important factor. Despite the current low prices, there must be greater emphasis on conservation; increased efficiency, thermal insulation, lighter vehicles and other improvements should have an increasing and more permanent impact. For power generation, efficiency shows little dependence on the nature of the fuel. Despite higher capital costs, for handling and environmental protection, an extensive switch back to coal has occurred, particularly in Europe. The production of stable dispersions of coal, in oil or even water, could extend its use to smaller furnace applications, a topic being pursued by BP and others. Alternative energy sources will possibly enter the scene more slowly. In contrast, for maximum efficiency, the major organic products must presently be derived from light hydrocarbon feedstocks such as natural gas, LPGs or naphtha. Hence it may be logical to reserve supplies for the more modest requirements of the petrochemical industry over a longer timescale.

In this context, the automobile may prove to be either a foe or a friend. Gasoline production takes the naphtha fraction, the preferred feedstock for a full range of olefinic intermediates, and lead phase-out places increasing demand on aromatics. Changes cannot occur in either the petrochemical or automobile industries overnight; both represent a considerable investment of capital, and therefore energy, and rapid obsolescence of usable equipment in itself represents a waste of energy. However, in recognition of a continuing need for suitable liquid fuels, many companies are working on gasoline substitutes or extenders, which may free sufficient naphtha or themselves be more amenable to chemical conversions.

Hydrogenation and solvent extraction of coal represent direct conversions. However, presently favoured schemes start with the steam reforming of natural gas, where this represents the major resource (as commercialized in

New Zealand[18]) and the gasification of coal (West Germany and the U.S.A.) to produce syngas for the following options:

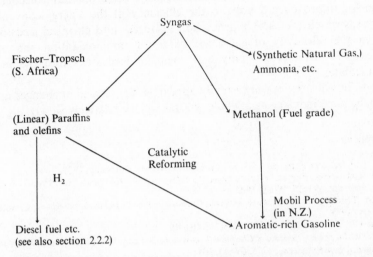

The methanol/syngas combination currently provides an entrée to acetic acid and anhydride, with possibly vinyl acetate, acetaldehyde and ethanol not far behind. Fischer–Tropsch products fit in with plasticizer alcohol and detergent requirements, and may provide a desirable cracker feedstock in the longer term. The Mobil methanol-to-gasoline process is particularly attractive for aromatic feedstocks. However, if naphtha supplies fail, lower olefins move somewhat down the line in such schemes; in view of the importance of polyethylene, considerable work on the conversion of syngas or methanol to lower olefins is in progress worldwide[17]. Ethylene-based routes to other vinylic polymers may therefore survive, despite a predicted comeback for acetylene.

The author is less convinced that fermentation will provide fuel for many years to come, except in special circumstances. Waste material must be used for process energy; unless the collection of such waste forms part of the current activity (like the production of bagasse from sugar cane), the extra operations could well dissipate much of the energy realized. Nevertheless, the production of chemical intermediates by fermentation may become more competitive.

Hence the most dramatic immediate changes in the petrochemical industry will be found in attitudes and research. There are obviously many uncertainties in the longer-term future, and the watchwords will be adaptability and feedstock flexibility, be it by new routes or revitalization of old ones. The continuing story of acetic acid and anhydride, and many other products, illustrates how the bulk organic chemicals industry can still respond to change.

However, except where new processes show an immediate and significant economic benefit, the uncertainty of long term market forecasts and availability of good plants for current routes point to a fairly modest rate of constructional activity in the future. Equally, there remains considerable scope for improvement, both in the efficiency of the 'energy-conversion' processes which translate hydrocarbon resources into chemical feedstocks, and in the selectivity of gas-phase oxidation, ammoxidation and other processes which convert individual chemical feedstocks into the major intermediates.

Thus, new processes will certainly appear on the scene, but changes in the petrochemical industry as a whole will take place relatively slowly.

References

1. (a) *Our Industry Petroleum*, 5th edn., British Petroleum Company, 1977.
   (b) *Kirk-Othmer Encyclopedia of Chemical Technology*, (*Patroleum, Refinery Process Survey*), 3rd edn., Vol. 17, Wiley, 1982.
2. J. M. Tedder *et al.* (eds.), *Basic Organic Chemistry, Part 5 Industrial Products*, Chapter 3, Wiley, 1975.
3. *The Oil and Gas Journal*, 1972, (Sept. 4), 103–10.
4. *Kirk–Othmer Encyclopedia of Chemical Technology*, 3rd edn. Vol. 1 (Acetylene), Wiley, 1978.
5. *Hydrocarbon Processing*, 1975, (Nov.), 104.
6. (a) *Energy in the 80's*, The Institution of Chemical Engineers, Symposium Series No. 48, 1977.
   (b) *Energy Conservation in the Chemical and Process Industries*, The Institution of Chemical Engineers, G. Godwin Ltd., 1979.
   (c) *Process Energy Conservation*, *Chem. Eng. Magazine*, McGraw-Hill, 1982.
7. H. D. Marsch and H. J. Herbort, *Hydrocarbon Processing*, 1982, (June), 101–5.
8. *Kirk–Othmer Encyclopedia of Chemical Technology*, 3rd edn., Vol. 2 (Ammonia), Wiley, 1978.
9. *Kirk–Othmer Encyclopedia of Chemical Technology*, 3rd edn., Vol. 15 (Methanol), Wiley, 1981.
10. *Hydrocarbon Processing*, 1981, (Nov.), 129–32.
11. (a) E. Supp, *Hydrocarbon Processing*, 1981, (March), 71–5.
    (b) *Hydrocarbon Processing*, 1981, (Nov.), 184.
12. J. F. Roth *et al.*, *Chemtech*, 1971, 1, 600–5.
13. (a) J. L. Ehrler and B. Juran, *Hydrocarbon Processing*, 1982, (Feb.), 109–13.
    (b) H. W. Coover Jr. and R. C. Hart, *Chem. Eng. Progress*, 1982, (April), 72–5
14. (a) G. H. Emert and R. Katzen, *Chemtech*, 1980, (Oct.), 610–4.
    (b) P. B. Weisz, *Chemtech*, 1980, (Nov.), 653–4.
15. *C&EN*, 1983, (April 11), 41–2.
16. (a) Ref. 6(a)
    (b) R. M. Ringwald, *C&I*, 1982, (May 10), 281–6
    (c) *Chemtech*, 1983, (Feb.), 128.
    (d) *Chemtech*, 1980, (Sept.), 550–1 and 557
    (e) *Engineering Materials and Design*, 1980, (April).
17. (a) T. Inui and Y. Takegami, *Hydrocarbon Processing*, 1982, (Nov.), 117–20
    (b) C. B. Murchison and D. A. Murdick, *Hydrocarbon Processing*, 1981, (Jan.), 159–64.
18. (a) S. L. Meisel, *Chemtech*, 1988, (Jan.), 32–37.
    (b) C. J. Maiden, *Chemtech*, 1988, (Jan.), 38–41.
    (c) J. Z. Bem, *Chemtech*, 1988, (Jan.), 42–46.

Bibliography

Basic Organic Chemistry, Part 5, Industrial Products (ed. J. M. Teddar, A. Nechvatal and A. H. Jubb), John Wiley and Sons, 1976.

Chemicals from Petroleum, 4th edn., A. L. Waddams, John Murray, 1980.

Faith, Keyes and Clark's Industrial Chemicals, 4th edn., F. A. Lowenheim and M. K. Moran, John Wiley and Sons, 1975.

Industrial Organic Chemicals in Perspective, Part 1, *Raw Materials and Manufacture*, H. A. Witcoff and B. G. Reuben, John Wiley and Sons, 1980.

Kirk–Othmer Encyclopedia of Chemical Technology, 3rd edn. (now complete), Wiley Interscience 1978 onwards. (Publication of the first volumes of a new 4th edition should commence in 1989/90.)

Directory of Chemical Producers, Western Europe, 8th edn., Vol 1 (Index to Companies) and Vol 2 (Products and Regions), SRI International, 1985. (Volume 2 lists companies' nominal production capacities/locations for many bulk chemicals. New editions are fairly frequent.)

Periodical Special Issues and Supplements

Hydrocarbon Processing publishes the *Petrochemical Handbook* as part of the November issue in odd-numbered years. This provides one-page (i.e. very brief) descriptions, with simplified flow-sheets and some information on economics and commercial status, for a considerable number of processes being offered for license.

C&EN (*Chemical & Engineering News*, a U.S. weekly) publishes 'Facts and Figures' as part of a June issue every year (June 20th in 1988). This lists the previous year's production figures for over 50 chemicals, polymers and chemical sectors (and some company information) for the U.S., and for a somewhat smaller range of materials for many other countries.

ECN (*European Chemical news*, a U.K. weekly) periodically issues separately-bound supplements, under the title *Chemscope*, on specific topics, e.g. chemical sectors, the chemical industries in particular European countries and related matters (plant construction, distribution etc.).

Every issue of the U.S. weekly *Chemical Marketing Reporter* contains a 'Chemical Profile' for a specific chemical, giving U.S. companies' capacities/locations and a brief rundown on production, price, outlets and growth potential (for the U.S. only). (There is a 2–3 year gap between updates on any particular chemical.)

Index